# AI NAR

From the Greek god Hephaestus's golden handmaidens to the Terminator, what we now call artificial intelligence (AI) has long been a part of Western culture. As real AI begins to touch on all aspects of our lives, these stories shape our expectations, providing a backdrop of culturally entrenched hopes, fears and possible futures. *AI Narratives* is the first book to explore this history of imagining intelligent machines.

Part I provides a historical overview from ancient Greece to the start of modernity. These chapters explore the revealing prehistory of key concerns of contemporary AI discourse, from the nature of mind and creativity to issues of power and rights, from the tension between fascination and ambivalence to investigations into artificial voices and technophobia. Part II focuses on the twentieth and twenty-first centuries in which a greater density of narratives have emerged alongside rapid developments in AI technology. These chapters reveal not only how AI narratives have consistently been entangled with the emergence of real robotics and AI, but also how they offer a rich source of insight into how we might live with these revolutionary machines.

This book explores the relationship between imaginative narratives and contemporary debates about AI's social, ethical and philosophical consequences, including questions of dehumanization, automation, anthropomorphization, cybernetics, cyberpunk, immortality, slavery, and governance. The contributions from leading humanities and social science scholars show that narratives about AI offer a crucial site for exploring contemporary debates about these powerful new technologies.

# AI NARRATIVES

## A History of Imaginative Thinking about Intelligent Machines

Edited by

*Stephen Cave, Kanta Dihal, Sarah Dillon*

OXFORD
UNIVERSITY PRESS

Great Clarendon Street, Oxford, OX2 6DP,
United Kingdom

Oxford University Press is a department of the University of Oxford.
It furthers the University's objective of excellence in research, scholarship,
and education by publishing worldwide. Oxford is a registered trade mark of
Oxford University Press in the UK and in certain other countries



First published 2020
First published in paperback 2024



Published in the United States of America by Oxford University Press
198 Madison Avenue, New York, NY 10016, United States of America

British Library Cataloguing in Publication Data
Data available

Library of Congress Cataloging in Publication Data
Data available

ISBN 978–0–19–884666–6 (Hbk.)
ISBN 978–0–19–891470–9 (Pbk.)

Printed and bound by
CPI Group (UK) Ltd, Croydon, CR0 4YY

# Contents

## Part II Modern and Contemporary

# Notes on Contributors

**Olivia Belton** is a postdoctoral research associate at the Leverhulme Centre for the Future of Intelligence at the University of Cambridge. She recently completed a PhD on posthuman women in science fiction television at the University of East Anglia. She is currently researching public perceptions of and media representations of autonomous flight.

**Stephen Cave** is Executive Director of the Leverhulme Centre for the Future of Intelligence, Senior Research Associate in the Faculty of Philosophy, and Fellow of Hughes Hall, all at the University of Cambridge. After earning a PhD in philosophy from Cambridge, he joined the British Foreign Office, where he spent ten years as a policy adviser and diplomat, before returning to academia. His research interests currently focus on the nature, portrayal, and governance of AI.

**Kate Devlin** is Senior Lecturer in Social and Cultural Artificial Intelligence at King's College London. Her research in Human–Computer Interaction and Artificial Intelligence investigates how people interact with and react to technologies, both past and future. She is the author of *Turned On: Science, Sex and Robots* (Bloomsbury, 2018), which examines the ethical and social implications of technology and intimacy.

**Kanta Dihal** is a postdoctoral researcher at the Leverhulme Centre for the Future of Intelligence, University of Cambridge. She is the Principal Investigator on the Global AI Narratives project, and the Project Development Lead on Decolonizing AI. In her research, she explores how fictional and nonfictional stories shape the development and public understanding of artificial intelligence. Kanta's work intersects the fields of science communication, literature and science, and science fiction. She is currently working on two monographs: *Stories in Superposition* (2021), based on her DPhil thesis, and *AI: A Mythology*, with Stephen Cave.

**Michael Dillon** is Emeritus Professor of Politics at Lancaster University. Inspired by Continental Philosophy and the intersection of modern politics, science, and religion, his work has focused on Security, on how Security has become a generative principle of formation for modern politics, and on the biopoliticization of Security in the digital age. His last book was entitled *Biopolitics of Security: A Political Analytic of Finitude* (2015). His current book project is entitled *Making Infinity Count*. Michael Dillon coedits the *Journal for Cultural Research* (https://www.tandfonline.com/rcuv20, Routledge), and is coeditor of a monograph series on *Political Theologies* for Bloomsbury (https://www.bloomsbury.com/uk/series/political-theologies/).

**Sarah Dillon** is a feminist scholar of contemporary literature, film and philosophy in the Faculty of English, University of Cambridge. She is author of *The Palimpsest: Literature, Criticism, Theory* (2007); *Deconstruction, Feminism, Film* (2018); and co-author (with Claire Craig) of *Storylistening: Narrative Evidence and Public Reasoning* (2021). She has edited *David Mitchell: Critical Essays* (2011); and co-edited *Maggie Gee: Critical Essays* (2015) and *Imagining Derrida* (2017), a special issue of *Derrida Today*. She is the general editor of the Gylphi Contemporary Writers: Critical Essays book series, and also broadcasts regularly on BBC Radio 3 and BBC Radio 4.

**Ben Halliburton** is a doctoral candidate at Saint Louis University. He is currently working on his dissertation, 'Utriusque illorum illustrium regum, pari gradu consanguineus': *The Marquisate of Montferrat in the Age of Crusade,* c.*1135–1225*, under the supervision of Thomas Madden.

**Minsoo Kang** is an associate professor of history at the University of Missouri—St. Louis. He is the author of *Sublime Dreams of Living Machines: The Automaton in the European Imagination* (2011, Harvard

University Press) and *Invincible and Righteous Outlaw: The Korean Hero Hong Gildong in Literature, History, and Culture* (2018, University of Hawaii Press). He is also the translator of the Penguin Classic edition of the classic Korean novel *The Story of Hong Gildong*, and the author of the short story collection *Of Tales and Enigmas* (2006, Prime Books).

**Kevin LaGrandeur** is a professor of English at the New York Institute of Technology, and a Fellow of the Institute for Ethics and Emerging Technology. He specializes in literature and science, digital culture, and Artificial Intelligence and ethics. His writing has appeared in both professional venues, such as *Computers and the Humanities*, and *Science Fiction Studies*, and in the popular press, such as the *USA Today* newspaper. His books include *Artificial Slaves* (Routledge, 2013), which won a 2014 Science Fiction and Technoculture Studies Prize, and *Surviving the Machine Age* (Palgrave Macmillan, 2017).

**Genevieve Liveley** is a Turing Fellow and Reader in Classics at the University of Bristol, where her particular research interests lie in narratives and narrative theories (both ancient and modern). She has published widely in books, articles, and essays on narratology, on chaos theory, cyborgs, AI, and how ancient myth might help us to better anticipate the future.

**Paul March-Russell** is Lecturer in Comparative Literature at the University of Kent, Canterbury. He is the editor of *Foundation: The International Review of Science Fiction* and commissioning editor of the series, SF Storyworlds (Gylphi Press). His most recent book is *Modernism and Science Fiction* (Palgrave, 2015), whilst other relevant publications have appeared in *The Cambridge History of the English Short Story* (2016), *The Cambridge History of Science Fiction* (2019), *The Edinburgh Companion to the Short Story in English* (2018), and *Popular Modernism and Its Legacies* (Bloomsbury, 2018). He is currently working on contemporary British women's short fiction and animal/ecocritical theory.

**Graham Matthews** is an assistant professor in English at Nanyang Technological University, Singapore. His most recent book is *Will Self and Contemporary British Society* and his work on twentieth-century literature has appeared in journals and edited collections such as *Modern Fiction Studies, Textual Practice, Journal of Modern Literature, Critique, English Studies, Literature & Medicine*, and *The Cambridge Companion to British Postmodern Fiction.*

**Anna McFarlane** is a British Academy Postdoctoral Fellow at Glasgow University with a project entitled 'Products of Conception: Science Fiction and Pregnancy, 1968–2015'. She has worked on the Wellcome Trust-funded Science Fiction and the Medical Humanities project and holds a PhD from the University of St Andrews in William Gibson's science fiction novels. She is the coeditor of *Adam Roberts: Critical Essays* (Gylphi, 2016) and *The Routledge Companion to Cyberpunk Culture* (2019).

**Julie Park** is a scholar of seventeenth- and eighteenth-century material and visual culture working at the intersections of literary studies, information studies and textual materialism. She is Assistant Curator/Faculty Fellow at the Bobst Library of New York University, author of *The Self and It* (Stanford University Press, 2010), and co-editor of *Organic Supplements* (University of Virginia Press, 2020). Her current projects are *My Dark Room*, a study of the camera obscura as a paradigm for interiority in eighteenth-century England's built environments and *Writing's Maker*, an examination of self-inscription technologies and their materials in the long eighteenth century.

**Gabriel Recchia** is a research associate at the Winton Centre for Risk and Evidence Communication at the University of Cambridge, where he is currently studying how best to communicate information about risks, benefits, statistics, and scientific evidence. Previously, he was at the Centre for Research in the Arts, Social Sciences and Humanities, where he developed techniques for the analysis of large corpora of historical texts. He also conducts research on statistical models of language and their applications in the cognitive and social sciences.

**Beth Singler** is the Junior Research Fellow in Artificial Intelligence at Homerton College, Cambridge. In her anthropological research she explores popular conceptions of AI and its social, ethical, and religious implications. She has been published on AI apocalypticism, AI and religion, transhumanism, and digital ethnography. As a part of her public engagement work, she has produced a series of short documentaries on AI, and the first, *Pain in the Machine*, won the 2017 AHRC Best Research Film of the Year award. Beth is also an Associate Research Fellow at the Leverhulme Centre for the Future of Intelligence.

**Will Slocombe** is Senior Lecturer in the Department of English at the University of Liverpool, the Director of the MA pathway in Science Fiction Studies, and one of the directors of the Olaf Stapledon Centre for Speculative Futures. He teaches modern and contemporary literature and literary theory, with a focus on science fiction. His main current research is concerned with representations of AI, and his second monograph, *Emergent Patterns: Artificial Intelligence and the Structural Imagination*, is forthcoming in 2020.

**Sam Thomas** completed her PhD in Classics and Ancient History at the University of Bristol, exploring the narrative ethics of Lucan's epic Civil War.

**E. R. Truitt** is an associate professor of Comparative Medieval History at Bryn Mawr College (Pennsylvania, USA) and the author of *Medieval Robots: Mechanism, Magic, Nature, and Art* (University of Pennsylvania Press, 2015), as well as the author of numerous scholarly articles on the history of automata and clock making, pharmacobotany, and *materia medica*, astral science, and courtly technology in the medieval period. She has written for *Aeon*, *The TLS*, and *History Today*, and she consulted on the Science Museum's exhibit "You, Robot."

**Megan Ward** is an assistant professor of English at Oregon State University and the author of *Seeming Human: Artificial Intelligence and Victorian Realist Character* (The Ohio State University Press, 2018). In addition, she codirects *Livingstone Online* (www.livingstoneonline.org), a digital archive of the Victorian explorer David Livingstone.

# Acknowledgements

This book originates in the AI Narratives project, a joint initiative of the Leverhulme Centre for the Future of Intelligence at the University of Cambridge and the Royal Society. In 2017 and 2018, this project held four workshops to explore how AI is portrayed, what impact that has, and what we can learn from how other emerging technologies have been communicated. Each workshop convened a highly interdisciplinary group, including not only academics from a wide range of fields, but also representatives of industry, government, news media, and the arts. The participants were united by a sense of the urgent importance of engaging critically with the narratives surrounding AI at this crucial historical moment when the technology itself is advancing so rapidly. The energy and insight these events generated inspired us to prepare this collection. We would like to thank the Cambridge and Royal Society teams who prepared these workshops, and all those who participated so enthusiastically. In particular, we would like to thank Claire Craig, then Chief Science Policy Officer at the Royal Society, for her crucial role in supporting the AI Narratives project.

We would also like to thank the many people who helped in the preparation of this volume. First, the work of our research assistant, Clementine Collett, was invaluable, particularly in the final weeks before the submission deadline. We are immensely grateful to those colleagues who reviewed our (the editors') contributions to this collection: Olivia Belton, Rachel Adams, Will Slocombe, and Michael Shapiro. We would also like to thank the team at Oxford University Press, who have so smoothly guided this project to publication, in particular Ania Wronski, Francesca McMahon, and Sonke Adlung. Finally, we acknowledge the generous support of the Leverhulme Trust, through their grant to the Leverhulme Centre for the Future of Intelligence, and the support of the University of Cambridge Faculty of English's Research Support Fund.

# Introduction

## *Imagining AI*

*Stephen Cave, Kanta Dihal, and Sarah Dillon*

> 'I mean, those parables or whatever they are,
> maybe they mean a lot to you but, uh . . .'
>
> (Sladek 1980, p.52)

## 0.1 AI Narratives

In 1985, David Pringle included John Sladek's *Roderick*—the story of a robot who wanders across a near-future America—in his list of the one hundred best science fiction novels. Pringle declares the novel, and its sequel *Roderick at Random* (1983), 'a treatise on the whole theme of mechanical men, homunculi, automatons and machine intelligence—the ultimate robot novel' (1985). The novel opens at a brilliantly complex moment of crisis built around a situation no doubt familiar to many artificial intelligence (AI) researchers—the question of funding. A small lab in a little-heard-of university in a near future where everything has been automated—from the grading of university papers to police detection—has been receiving financial support from NASA for an AI research project. Only it turns out that the whole funding setup has been a scam designed to line the pockets of a NASA employee who commits suicide on being exposed. Nevertheless, the academics have made breathtakingly exciting progress with the research and have created Roderick, 'a learning system'

(Sladek 1980, p.24) described by one character as 'this artificial intelligence' (p.23). 'He's alive,' insists one of the researchers to his colleague, 'Roderick's alive. I know he's nothing, not even a body, just content-addressable memory. I could erase him in a minute—but he's alive. He's as real as I am [...]. He's realer. I'm just one of his thoughts' (p.48). The research team is now in the unenviable position of having to appeal to the University's Emergency Finance Committee in order to continue with the research, but not all the committee members are persuaded by the value of their work.

Rogers, a sociologist on the committee, visits the robotics lab to see what it is all about, but is disappointed by what he finds: 'a lot of computers and screens and things, I could see those anywhere, and what are they supposed to mean to a layman? I expected—I don't know—' (Sladek 1980, p.24). He arrives with preconceptions about what he would find, preconceptions easily deduced by the AI researcher, Fong:

> 'You wanted a steel man with eyes lighting up? "Yes Master?", that kind of robot? Listen, Roderick's not like that. He's not, he doesn't even have a body, not yet, he's just, he's a learning system [...] A learning system isn't a thing, maybe we shouldn't even call him a robot, he's more of a, he's like a *mind*. I guess you could call him an artificial mind.' (Sladek 1980, pp.24–25)

Rogers scoffs at the esoteric nature of Fong's claim, and at the absence of the embodied AI he is expecting to see: 'am I supposed to tell the committee I came to see the machine and all you could show me was the ghost?' (Sladek 1980, p.25). But it turns out Rogers is less interested in the artificial mind than he is in the mind of the researcher, quizzing Fong about 'this Frankenstein goal' (p.25) and whether he has ever considered 'the social impact of your work' (p.26).

*Roderick* is laden with references to the AI narratives that have preceded it, from Kurt Vonnegut's dystopia of automation, *Player Piano* (1952), to the myth of Francis Bacon's brazen head, to

Albertus Magnus' automaton, to 'the prophet Jeremiah and his son, making the first *golem*' (Sladek 1980, p.52). Dr Jane Hannah, an anthropologist on the Emergency Finance Committee, evidences a detailed knowledge of AI narratives across the globe. In fact, she turns to these traditions in order to try to understand the desires behind the aspiration to create intelligent machines, and to decide whether she should approve funding the continued development of one at her university:

> '... maybe the Blackfeet boy, Kut-o-yis, cooked to life in a cooking pot, but isn't that the point? Aren't they always fodder for our desires? Take Pumiyathon for instance, going to bed with his ivory creation [...] take Hephaestus then, those golden girls he made who could talk, help him forge, who knows what else... Or Daedalus, not just the statues that guarded the labyrinth, but the dolls he made for the daughters of Cocalus, you see? Love, work, conversation, guard duty, baby, plaything, of course they used them to replace people, isn't that the point? [...] And in Boeotia, the little Daedala, the procession where they carried an oaken bride to the river, much like the *argeioi* in Rome, the puppets the Vestal Virgins threw into the Tiber to purge the demons; disease, probably, just as the Ewe made clay figures to draw off the spirit of the smallpox, so did the Baganda, they buried the figures under roads and the first [...] person who passed by picked up the sickness. In Borneo they drew sickness into wooden images, so did the Dyaks [...] Of course the Chinese mostly made toys, a jade automaton in the Fourth Century but much earlier even the first Han Emperor had a little mechanical orchestra [...] but the Japanese, Prince Kaya was it? Yes, made a wooden figure that held a big bowl, it helped the people water their rice paddies during the drought. Certainly more practical than the Chinese, or even the Pythagoreans, with their steam-driven wooden pigeon, hardly counts even if they did mean it to carry souls up to – but no, we have to make do with the rest, and of course the golem stories, and how clay men fashioned by the Archangel – [...] There were the Teraphim of course, but no one knows their function. But the real question is, what do we want this robot *for*? Is it to be a bronze Talos, grinning as he clasps people in his red-hot

> metal embrace? Or an ivory Galatea with limbs so cunningly jointed – ' (Sladek 1980, pp.60–61)

Sladek's novel understands the tradition it lies within, that is, a transhistorical, transcultural imaginative history of intelligent machines. These imaginings occur in a diverse range of narrative forms, in myths, legends, apocryphal stories, rumours, fiction, and nonfiction (particularly of the more speculative kind). They have existed centuries prior to the origin of the modern scientific field, which might most simply be located in 1956 at the Dartmouth Summer Research Project on Artificial Intelligence, the term 'AI' having been coined the year prior (McCarthy et al 1955). The term is now used to refer to a heterogeneous network of technologies—including machine learning, natural language processing, expert systems, deep learning, computer vision, and robotics—which have in common the automation of functions of the human brain.[1] However, imaginings of intelligent machines have employed a range of other terms, including 'automaton' (antiquity), 'android' (1728), 'robot' (1921), and 'cyborg' (1960).[2] The chapters in this book engage with AI in its broadest sense, one that encompasses all of the aforementioned terms: that is, any machine that is imagined as intelligent.

The exploration of AI narratives by Dr Hannah in *Roderick* covers imaginings from the literature, mythology, and folklore of Native America, Ancient Greece, Classical Rome, Uganda, Ghana, Borneo, China, Japan, Judaism, and Christianity. The chapters in this book focus on narratives that form part of the Anglophone Western tradition. These include works written in English and works in other languages that have had a strong influence on this narrative tradition. The chapters therefore cover a historical period beginning with the automata of the *Iliad*—the oldest narrative of intelligent machines in this tradition, written around 800 BCE—to the present. The book presents this history of imaginative thinking about intelligent machines in two parts. The chapters in Part I cover the long history of imaginings of

AI from antiquity to modernity. Each chapter in this part focuses on a specific historical period: antiquity, the Middle Ages, the Renaissance, the eighteenth century, the nineteenth century, and the modernist period. Together, they explore the prehistory of key concerns of contemporary AI discourse, including: the nature of mind; the imbrication of AI and power; the duality of our fascination with and yet ambivalence about AI; the rights and remuneration of workers (both human and artificial); the relation between artificial voice and intelligence; creativity; and technophobia. Part II takes up the historical account in the modern period, Karel Čapek's 1921 play *R.U.R.* (from which derives the term 'robot') serving as the hinge between the two parts. The chapters in Part II focus on the twentieth and twenty-first centuries, in which a greater density of narratives emerges alongside rapid developments in AI technology. The chapters in this part are organised thematically, with each chapter driven primarily by a focus on imaginative explorations of specific effects and consequences of AI technologies. These include: the dehumanizing effects of humanizing machines; the consequences of automation and mechanization for society; the cultural assumptions embedded in the anthropomorphization of machines; the importance of understanding AI as a distributed phenomenon; the human drivers behind the desire for technologically enabled immortality; the interaction of AI with the sovereign-governance game that defines modern rule; the relationship between imaginative representations of female robots and AIs, and the perception of real-world sex robot technology; and the fear of losing control of AI technologies. This book covers many of the touchstone narratives that might be most familiar to readers, including Isaac Asimov's robot stories and *The Terminator*, but it also aims to draw readers' attention to less well-known texts, such as *Roderick*, that make an important contribution to the rich imaginative history of intelligent machines.

After her exploration of AI narratives, Dr Hannah comes to a conclusion regarding how she will vote at the Emergency Finance Committee: 'As you see,' she concludes, 'I've been turning the problem over, consulting the old stories.... And I've decided to vote against this robot' (Sladek 1980, pp.60–61). For Dr Hannah, AI narratives offer complex explorations of the social, ethical, political, and philosophical consequences of AI, explorations that inform her decision-making about whether or not to fund contemporary scientific research. Many of the chapters in this book support this position, exploring how AI narratives have addressed, and offer sophisticated thinking about, some of the legitimate concerns that AI technologies now raise. But AI narratives are not universally viewed in this way. A counter-narrative is found in *Roderick* in the form of Dr Hannah's colleague, Dr Helen Boag. Her position cuts through the ancient myths and the expectations they create, ones she perceives to be often misaligned with the technology itself: 'really isn't the computer more or less an overgrown adding machine? A tool, in other words, useful of course but only in the hands of human beings. I feel the role of the computer in our age has been somewhat exaggerated, don't you?' (Sladek 1980, p.73). Other chapters in this book engage with the challenges posed by, primarily, dominant AI narratives. These engage with a wider landscape in which prevalent AI narratives are mistrusted or criticised for example for their extremism—utopian or dystopian—or for their misrepresentation of current technology, for instance in their tendency to focus on anthropomorphic representations. In the next section of this introduction, we want to survey that wider landscape of contemporary views about prevalent AI narratives, their functions and effects, as well as the impacts they have, in order to map one of the terrains into which we hope this book will intervene.

## 0.2 The Impact of Narratives

Sheila Jasanoff's concept of 'sociotechnical imaginaries' provides a dominant paradigm for understanding the relationship between technology and the social order. Jasanoff defines these as the 'collectively held, institutionally stabilized, and publicly performed visions of desirable futures, animated by shared understandings of forms of social life and social order attainable through, and supportive of, advances in science and technology' (2015, p.4). Absent from this theorisation, however, is an explicit account of the important role narratives play, both fictional and nonfictional, as fundamental animators of sociotechnical imaginaries. This neglect is at odds with Jasanoff's use of science fiction narratives, for instance, to introduce her theory, and her acknowledgement that science fiction narratives already provide an established site of investigation of sociotechnical imaginaries, long prior in fact to the invention of the term: science fiction already 'situate[s] technologies within [...] integrated material, moral, and social landscapes [...] in such abundance' (p.3). The work of this book contributes to establishing the importance of narratives as constituent parts of any sociotechnical imaginary. Attention to narratives also highlights relationships between society, technology, and the imaginary that are not included in Jasanoff's definition, for instance, visions that are not collectively held, ones that destabilise institutions, and subaltern narratives. Jasanoff acknowledges that sociotechnical imaginaries encompass both 'positive and negative imaginings' (p.3), but only in service of the dominant vision. Foregrounding narratives therefore plays a role in challenging the 'aspirational and normative' (p.5) dimension of Jasanoff's concept, inviting consideration of a much wider range of visions.

Narratives of intelligent machines matter because they form the backdrop against which AI systems are being developed, and against which these developments are interpreted and assessed. Those who are engaged with AI either as researchers or regulators are therefore rightfully concerned, for instance, about the fact that the dominant contemporary imaginings of AI, primarily those of Hollywood cinema and popular news coverage, are often out of kilter with the present state of the technology. The UK House of Lords Select Committee on Artificial Intelligence opens the second chapter of their 2018 report 'AI in the UK: ready, willing and able?' with a sharp critique of prevalent AI narratives:

> The representation of artificial intelligence in popular culture is lightyears away from the often more complex and mundane reality. Based on representations in popular culture and the media, the non-specialist would be forgiven for picturing AI as a humanoid robot (with or without murderous intentions), or at the very least a highly intelligent, disembodied voice able to assist seamlessly with a range of tasks. (p.22)

The Select Committee report continues by observing that 'many AI researchers were concerned that the public were being presented with overly negative or outlandish depictions of AI, and that this could trigger a public backlash which could make their work more difficult' (p.24).

This book contributes to an emerging body of work that goes beyond hype and horror by exploring the more complex ways in which narratives of AI could have significant impact. For instance, the narratives with which AI researchers themselves engage can influence their 'career choice, research focus, community formation, social and ethical thinking, and science communication' (Dillon & Schaffer-Goddard, forthcoming). Within these categories, Dillon and Schaffer-Goddard identify that further investigation is needed into the way in which AI narratives might influence who goes into the field. Scholars such as Alison Adam have been considering for over two decades how masculinity is inscribed into the way AI is conceived (1998) and how this might interplay with a culture in the computing world that is hostile to women. Future research might consider, for instance, whether the consistent portrayal of fictional AI developers as men, from Hephaestus to *Metropolis*'s (1927) Rotwang to Robert Ford of the *Westworld* TV series (2016–present), makes women feel that this role is not for them. This question of who is in the room, or who is in the lab, impacts which systems are developed, how, and for whom.

The dominance of anthropomorphic portrayals of AI also exacerbates the tense relationship between the technology and issues of equality and diversity. Anthropomorphic machines, from fictional androids like the 'hosts' of *Westworld* to the human voices of virtual personal assistants, are mirrors to our societies, perpetuating existing biases and exclusions. They reflect and so reinforce prevalent cultural narratives of different groups' allotted role and worth, delineating further who counts as fully human (Rhee 2018).[3] Extending work on AI narratives to social justice issues around race and ethnicity is crucial. Further research here might consider, for instance, whether stock images of AI as Caucasian male humanoid robots distort views of who AI is for, and whose jobs will be impacted by this technology, obscuring the potential effect on disadvantaged communities.

In order to understand what kinds of technologies are being developed, those outside the professional AI field rely on narratives that mediate between the technology world and the public sphere. Narratives can therefore strongly influence public acceptance and uptake of AI systems, and a significant amount of science popularisation has this goal explicitly in mind (Gregory 2003). The Select Committee report notes that the role of AI narratives as an intermediary between research and the public was a concern raised by AI researchers, who 'told [the Committee] that the public have an unduly negative view of AI and its implications, which in their view had largely been created by Hollywood depictions and sensationalist, inaccurate media reporting' (Select Committee on Artificial Intelligence 2018, p.44). This view is supported by Dillon and Schaffer-Goddard's findings, who argue that 'such stories have a strong influence on the researchers' science communication activity—researchers often need to argue against the pictures they paint, but can also use this as an incentive and a springboard to paint more positive, or at least realistic, pictures of AI-influenced futures'. They found that AI researchers expressed 'a strong desire for more sophisticated stories about AI, which would be of benefit to the research community and public discourse, as well as to literary and cinematic quality and production' (Dillon & Schaffer-Goddard, forthcoming). Many of the chapters in this book identify such stories, encouraging attention to them, in contrast to the narratives that currently dominate.

Prevalent narratives are of course not just those of the popular media and the press. Large corporations such as Microsoft and Google invest significant resources in developing ethics principles and other narratives aimed at fostering public acceptance of AI. Whether this technology is adopted has implications far beyond these companies' bottom lines. Use of AI in healthcare for instance, will impact the advancement of the medical field and individual well-being and mortality rates. At the same time, these technologies pose significant concerns regarding privacy, social justice, workers' rights, democracy, and more. In this context, AI narratives have played, and will continue to play, a crucial role in determining the future of AI implementation.

By influencing the perceptions of policymakers, and by steering public concerns, narratives also affect the regulation of AI systems. For example, there is ongoing debate about the ways narratives influence public policy on highly advanced or superintelligent AI (Johnson & Verdicchio 2017). On the one hand, there are those who argue that current real-world risks are being obscured by Terminator-style stories. The Select Committee report notes that prevalent AI narratives 'were concentrating attention on threats which are still remote, such as the possibility of "superintelligent" artificial general intelligence, while distracting attention away from more immediate risks and problems' (Select Committee on Artificial Intelligence 2018, p.23). In contrast, there is the potential for those who oppose restraints on the development of superintelligent AI to propagate narratives sceptical of its capacities, so that policymakers see no need for regulation (Baum 2018).

Many authors and scholars consciously use narratives to explore the possibilities for a future with intelligent machines and disrupt existing tropes. Imaginative thinking about AI can probe both dystopian and utopian scenarios, showing flaws in overly unidirectional thinking, and anticipating consequences before they affect the lives of millions of people. Asimov, for instance, developed his robot stories in response to a plethora of science fiction works in the 1920s and 30s that presented, in his view, unhelpful depictions of robots as either extremely menacing or extremely pathetic characters (1995 [1982]). Currently, initiatives around the world are producing a growing body of work—especially science fiction—that is commissioned with the specific mandate to explore the impact of particular technologies, or to imagine how current technologies can be developed to create a utopian future (Amos & Page 2014; Coldicutt & Brown 2018; Finn & Cramer 2014). At the same time, the field of futures studies uses narratives to explore the consequences of specific decisions in policymaking and scientific development (Avin 2019; B. D. Johnson 2011). Their scenarios are intended to be realistic forecasts directing the focus of AI ethics research.

The way AI is portrayed is therefore a social, ethical, and political issue. Through shaping the technical field, the acceptance of the resulting technology, and its regulation, and through encoding normative sociopolitical assumptions, these portrayals have far-reaching implications. It is therefore essential that prevalent narratives be critically examined, and that they be contested by the privileging of more sophisticated and complex narratives of AI, both fictional and nonfictional. These more complex stories can be and are being newly invented, but this book draws attention to the extensive history of imaginative thinking about intelligent machines upon which they might build, and by which contemporary thinking about AI should be informed. This book looks to past imaginings—of the future and of alternate realities—in order to inform present thinking about AI.

## 0.3 Chapter Guide

Philosopher Daniel Dennett offers one illustration of the impact of AI narratives on the scientific field. In 1997, Dennett wrote an eight-page fan letter to Richard Powers after reading *Galatea 2.2*

(Powers 1995). Explaining his actions later, Dennett observes of Powers's novel:

> His representation of AI is wonderful. It is remarkable how this very interested bystander has managed to cantilever his understanding of the field out over the abyss of confusion and even throw some pioneering light on topics I thought I understood before. (2008, pp.151–52)

For Dennett, Powers's novel serves to illuminate the field: 'What particularly excited me,' he continues, 'was how Powers had managed to find brilliant ways of conveying hard-to-comprehend details of the field, details people in the field were themselves having trouble getting clear about' (Dennett 2008, p.152). Dennett understands Powers' novel as a contribution to scientific research and knowledge. 'The novel,' he says, 'is an excellent genre for pushing the scientific imagination into new places' (p.160). This is because it challenges 'personal styles of thinking, and a style is (roughly) a kit of *partially disabling* thinking-habits' (p.160). Dennett is keen to caveat this acknowledgement of the power of novels, however, declaring that 'you have to be a powerful thinker to pull off the trick. That's why most science fiction doesn't repay the attention of scientists' (p.161). He even goes so far as to suggest that we need a new name 'for the rare novels like *Galatea 2.2* that manage to make a contribution to the scientific imagination' (p.161). No new name is needed, however, and *Galatea 2.2* is not so rare a contribution as Dennett seems to think. The chapters in this book explore a wide range of AI narratives and the many roles they play not only in extending the scientific imagination, but the ethical, political, and social imagination as well.

Part I opens with a chapter on the very earliest references to intelligent machines. Through close analysis of Homer's *Iliad* and *Odyssey*, Genevieve Liveley and Sam Thomas trace the various gradations of weak to strong machine 'intelligence' that these ancient poems describe, and explore the ancient mind models that they assume. They conclude that the Homeric mind model—which sees both humans and machines possessing programmable thoughts (*phrenes*) and minds (*noos*)—helps to explain why Homer is part of our own cultural AI programming. They propose that Homer's ancient tales of intelligent machines have established a 'programme' for us to follow in the retelling and rescripting of our present and future AI narratives. Part I then moves from antiquity to the long medieval period. First, E. R. Truitt explores the imagination of objects we might recognize as AI: from artificial servants to elaborate optical devices, from augmented human perception to sentient machines. She argues that throughout this era, AI appears in imaginative contexts to gain, consolidate, and exercise power. It does so in ways remarkably recognizable from a twenty-first-century standpoint, from maintaining class and gender hierarchies, to aiding military or political dominance, to gaining knowledge of the future that could be used to a person's advantage. In the next chapter, Minsoo Kang and Ben Halliburton focus on this question of foreknowledge by examining the major appearances of the story of the speaking head of philosopher Albertus Magnus (c. 1193–1280) in medieval and early modern writings. The head is a wondrous object able to converse and even reason, confirming it as a kind of medieval AI, animated for the purpose of divination. Kang and Halliburton demonstrate that, in a way that is reflective of contemporary anxiety and concerns about AI, the different versions of the story move between questioning whether Albertus was dealing in illicit knowledge in making the object, secularizing it as a purely mechanical device, or demonizing it as a work involving diabolical beings. The changing narratives around the head exhibit both a fascination with, as well as an anxious ambivalence towards, the object, revealing that our current attitudes towards AI originate from long-standing and primordial feelings about artificial simulacra of the human.

In the next chapter, Kevin LaGrandeur engages with a different talking brass head, this one built by a natural philosopher in Robert Greene's Elizabethan comedy *Friar Bacon and Friar Bungay*

(1594). Engaging with Greene's play as well as Renaissance stories of the golem of Prague and of Paracelsus's homunculus, LaGrandeur demonstrates the fears and hopes embedded in Renaissance culture's reactions to human invention. In particular, his chapter exposes the entanglement of AI narratives with discourses of slavery, showing how intelligent objects of that period are almost uniformly proxies for indentured servants. The tales with which LaGrandeur engages signal ambivalence about our innate technological abilities—an ambivalence that long predates today's concerns. The promise represented by these artificial servants, of vastly increased power over natural human limits, are countervailed by fears about being overwhelmed by our own ingenuity. Julie Park shifts attention from the artificial head to the artificial speech that emanates from it, investigating what the eighteenth-century history of artificial voices tell us about the relationships between machines, voice, the human, and fiction. Given the rise in our daily lives of voice-operated 'intelligent assistants', this investigation is especially pertinent. By examining the eighteenth-century case of a speaking doll and the cultural values and desires that its representation in a 1784 pamphlet entitled *The Speaking Figure, and the Automaton Chess-Player, Exposed and Detected* reveals, Park's chapter provides a historical framework for probing how the experiences and possibilities of artificial voice shed light on our deep investments in the notion of voice as the ultimate sign of being 'real' as humans.

In designing his foundational test of AI, Alan Turing refers to Ada Lovelace's Victorian pronouncement that a machine cannot be intelligent because it only does what it is programmed to do. This idea continues to shape the field of computational creativity as the 'Lovelace objection'. In her chapter, Megan Ward argues that this term is a misnomer and sets out to show that Ada Lovelace actually proposes a much more nuanced understanding of human-machine collaboration. By situating Lovelace's work within broader Victorian debates about originality in literary realism, especially in relation to Charles Dickens and Anthony

Trollope, Ward demonstrates how Lovelace's ideas participate in a broader Victorian debate that redefines originality to include technologically enhanced mimesis. Ward proposes that understanding computational creativity not in opposition to the Lovelace objection, then, but as the development from Victorian originality to contemporary creativity, may open a way forwards to a concept of creativity inclusive of mechanicity. Paul March-Russell's chapter concludes Part II's historical sweep from antiquity to the present with a focus on the technophobia of modernist literature towards the question of machine intelligence. Ranging across texts by Edmund Husserl, Albert Robida, Emile Zola, Ambrose Bierce, Samuel Butler, H. G. Wells, E. M. Forster, and Karel Čapek, March-Russell demonstrates that, whilst modernism may have been enthusiastic towards other forms of technological innovation, the possibility of a machine that could think—and, above all, talk—like a human stirred age-old responses about the boundaries between human and nonhuman life-forms. March-Russell concludes that, despite Čapek's popularization of the term 'robot', it is only with the rise of science fiction that literary authors begin to supersede the technophobia of their modernist predecessors.

Čapek's play provides the hinge from Part I, covering antiquity to modernity, to Part II, which focuses on modern and contemporary AI narratives, predominantly, although not exclusively, science fictional. In her chapter, Kanta Dihal engages with Karel Čapek's *R.U.R.* (1921), alongside Ridley Scott's *Blade Runner* (1982) and Jo Walton's Thessaly trilogy (2014–2016), in order to contextualize the imagining of the robot uprising in fiction against the long history of slave revolts. Doing so, she shows that in these three works, the revolt is depicted as a justified assertion of personhood by the intelligent machines. These stories show how intelligent machines are just as likely to be denied personhood as historically oppressed groups have been: those in power are unwilling to grant personhood if this were to threaten their personal comfort. In the next chapter, Will Slocombe also draws together texts from the early twentieth century to the present day. Slocombe engages with four different imagined 'Machines' in E. M. Forster's 'The Machine Stops' (1909); Paul W. Fairman *I, The Machine* (1968); Asimov's 'The Evitable Conflict' (1950); and the popular television series *Person of Interest* (2011–2016). He examines the ways in which representations of AI during the twentieth century—particularly in nonandroid form (what might be termed a 'distributed system')—dovetail with pre-existing perceptions of society operating as a 'machine', and become imbricated in broader social discourses about loss of autonomy and individuality.

Graham Matthews' chapter focuses on the mid-twentieth century, a crucial period for the technological development of AI, the moment at which, it might be said, AI moved from the realm of myth to become a real possibility. Matthews contends that midcentury AI narratives must be situated in relation to concomitant technological developments in automation and cybernetics. These elicited widespread concern among government institutions, businesses, and the public about the projected transition from industry to services, the threat of mass unemployment, technologically driven sociopolitical change, and the evolving relationship between humans, religion, and machines. The rhetoric of the Fourth Industrial Revolution closely echoes these midcentury debates. Matthews analyses the varied representation of AI in midcentury novels such as Michael Frayn's *The Tin Men* (1965); Len Deighton's *Billion Dollar Brain* (1966); and Arthur C. Clarke's *2001: A Space Odyssey* (1968). He argues that these novels problematize AI narrative tropes by resisting anthropomorphic tendencies and implausible utopian and dystopian scenarios; instead, they address the societal ramifications—both positive and negative—for humans faced with technological breakthroughs in AI.

Beth Singler's chapter moves the historical focus into the second half of the twentieth century and introduces a series of new methodological approaches in the book's later chapters. Singler employs historical and cognitive anthropological approaches to examine the cultural influences on our stories about AI, focusing in particular on anthropomorphization, and attending to the cultural assumptions about the human child, and the AI 'child', present in our AI narratives. She works her exploration through engagement with a range of texts, including the novel *When HARLIE Was One* (1972); the films *D.A.R.Y.L.* (1985), *AI: Artificial Intelligence* (2001), *Star Trek: Insurrection* (1998), and *Tron: Legacy*, (2010); the television series *Star Trek: Next Generation* (1987 to 1994); and a speculative nonfiction account of the future relationship of humanity and AI, Hans Moravec's *Mind Children* (1988). Anna McFarlane approaches AI narratives through the lens of genre, focusing on cyberpunk science fiction and the possibilities opened, both in fictional narratives and wider cultural narratives, by the genre's sustained interrogation of AI as a phenomenon that is dispersed throughout networks. She argues that the narrative innovations and tropes of early cyberpunk writers such as William Gibson and Samuel R. Delany, and the continued mutation of these ideas in 'post-cyberpunk', such as in the work of Cory Doctorow, respond to exponentially changing technology, and shape contemporary understandings of AI's cutting edge, for instance algorithmic decision-making. Stephen Cave explores representations of immortality in science fiction texts primarily from the cyberpunk tradition—works by William Gibson, Greg Egan, Pat Cadigan, Robert Sawyer, Rudy Rucker, and Cory Doctorow. These authors offer subtle, frequently sceptical portrayals of the psychological, philosophical and technological challenges of using technology to free oneself from the constraints of the body. Cave shows that these works offer a particularly important site of critique and response to techno-utopian narratives by influential contemporary technologists—in particular Hans Moravec and Ray Kurzweil.

In their chapter, Sarah Dillon and Michael Dillon bring political theory into dialogue with literary criticism in order to explore the interaction between AI and the ancient conflict between sovereignty and governance, in which sovereignty issues the warrant to rule and governance operationalizes it. They focus on three novels in which games, governance, and AI weave themselves through the text's fabric: Iain M. Banks's *The Player of Games* (1988) and *Excession* (1996), and Ann Leckie's *Ancillary Justice* (2013). These novels play out the sovereign-governance game with both artificial and human actors. In doing so, Dillon and Dillon argue that the novels question what might be politically novel about AI, but in the end, reveal that whilst AI impacts the pieces on the board, reducing some and advancing others, it does not materially change the logic of the game. They conclude that these texts therefore raise questions but do not provide answers with regard to what might be required for AI technologies to change the algorithms of modern rule. Moving from the politics of sovereignty and governance to the politics of gender, Kate Devlin and Olivia Belton's chapter explores the relationship between fictional representations of female robots and AIs and the perception of real-world sex robot technology. They offer a critical analysis of science fiction media representations of gendered AI, focusing on Ava from *Ex Machina* (2014), Samantha from *Her* (2013), and Joi from *Blade Runner 2049* (2017). They show how these AIs' gender identities are often reinforced in stereotypical ways, and how both embodied and disembodied AIs remain highly sexualized. Their chapter demonstrates how discourses around real-life sex robots are deeply informed by prevalent fictional narratives, and advocates for a more gender-equal approach to the creation of both fictional and factual robots, in order to combat sexist stereotypes.

The final chapter of the collection introduces a digital humanities methodology to approaching AI narratives, that is, one that combines the tools of traditional literary analysis with computational techniques for identifying key themes and trends in very large quantities of text. Gabriel Recchia presents a computationally assisted analysis of the English-language portion of the Open Subtitles Corpus, a dataset of over 100,000 film subtitles ranging from the era of silent film to the present. By applying techniques used to understand large corpora within the digital humanities, Recchia presents a qualitative and quantitative overview of several salient themes and trends in the way AI is portrayed and discussed in twentieth- and twenty-first-century film. Recchia's analysis confirms many of the dominant themes and concerns discussed in the previous chapters of the book, whilst also opening the way for new ways of approaching and analysing AI narratives in the future.

## 0.4 Conclusion

Through our personal engagement with contemporary debates on the impact of AI, the three of us have witnessed many Roderick moments in the last few years. Misplaced expectations and mutual incomprehension between stakeholders have been as much a part of the present moment's AI revolution as extravagant utopian and dystopian visions. In a recent survey on perceptions of this technology, when asked how the respondent would explain AI to a friend, several responses were simply expressions of anxiety, such as 'creepy' or 'scary robots' (Cave, Coughlan, & Dihal 2019). Despite these concerns, it is likely that AI technologies will be highly consequential for the shape of society in the near and long term. If their effects are to be positive rather than negative, it will be essential to reconcile the multiple discourses of different publics, policymakers, and technologists, and lay bare the assumptions and preconceptions on which they rest. We hope this book, in beginning to unpick the fascinating and complex history of AI narratives, will contribute to that goal.

## Notes

1. This definition of 'AI' is informed by Marcus Tomalin's introductory talk at the workshop 'The Future of Artificial Intelligence: Language, Gender, Technology', 17 May 2019, University of Cambridge.
2. These various terms and their related imaginaries have been explored in a rich body of scholarly work with which this collection is in dialogue, and to which it aims to contribute. Several of the contributors to this collection have written key works in the field, including Minsoo Kang (2011), Kevin LaGrandeur (2013), E. R. Truitt (2015), and Megan Ward (2018).

3. Clementine Collett and Sarah Dillon's report 'AI and Gender: Four Proposals for Future Research' (2019) outlines the challenges AI technologies, including humanoid robotics and virtual personal assistants, present to gender equality; identifies current research and initiatives; and proposes four areas for future research.

## Reference List


Adam, A. (1998) *Artificial knowing: gender and the thinking machine*. London, Routledge.

Amos, M. & R. Page (eds.) (2014) *Beta Life: short stories from an a-life future*. Manchester, Comma Press.

Asimov, I. (1995 [1982]) Introduction. In: *The complete robot*. London, HarperCollins. pp.9–12.

Avin, S. (2019) Exploring artificial intelligence futures. *Journal of AI Humanities*. https://doi.org/10.17863/cam.35812.

Baum, S. (2018) Superintelligence skepticism as a political tool. *Information*. 9(9), 209. https://doi.org/10.3390/info9090209.

Cave, S., K. Coughlan, & K. Dihal (2019) 'Scary robots': examining public responses to AI. *Proceedings of the 2019 AAAI/ACM Conference on AI, Ethics, and Society*. https://doi.org/10.17863/CAM.35741.

Coldicutt, R. & S. Brown (eds.) (2018) *Women invent the future*. Available from: https://doteveryone.org.uk/project/women-invent-the-future/ [Accessed 18 September 2019].

Collett, C. & S. Dillon (2019) *AI and gender: four proposals for future research*. Cambridge, Leverhulme Centre for the Future of Intelligence.

Dennett, D. (2008) Astride the two cultures: a letter to Richard Powers, updated. In: S. J. Burn & P. Dempsey (eds.) *Intersections: essays on Richard Powers*. Champaign, IL, Dalkey Archive Press. pp.151–61.

Dillon, S. & J. Schaffer-Goddard (forthcoming). What AI researchers read: the influence of stories on artificial intelligence research.

Finn, E. & K. Cramer (eds.) (2014) *Hieroglyph: stories and visions for a better future*. New York, William Morrow.

Gregory, J. (2003) Understanding 'science and the public': research and regulation. *Journal of Commercial Biotechnology*. 10(2), 131–39.

Jasanoff, S. (2015) Future imperfect: science, technology, and the imaginations of modernity. In: S. Jasanoff & S.-H. Kim (eds.), *Dreamscapes of modernity: sociotechnical imaginaries and the fabrication of power*. Chicago: University of Chicago Press. pp.1–33.

Johnson, B. D. (2011) Science fiction prototyping: designing the future with science fiction. *Synthesis Lectures on Computer Science*. 3(1), 1–190. https://doi.org/10.2200/S00336ED1V01Y201102CSL003.

Johnson, D. G. & M. Verdicchio (2017). Reframing AI discourse. *Minds and Machines*. 27(4), 575–90. https://doi.org/10.1007/s11023-017-9417-6.

Kang, M. (2011) *Sublime dreams of living machines: the automaton in the European imagination*. Cambridge, MA, Harvard University Press.

LaGrandeur, K. (2013) *Androids and intelligent networks in early modern literature and culture: artificial slaves*. New York, Routledge.

McCarthy, J., M. L. Minsky, N. Rochester, & C. E. Shannon (1955) *A proposal for the Dartmouth Summer Research Project on Artificial Intelligence*. 31 August. Available from: http://raysolomonoff.com/dartmouth/boxa/dart564props.pdf [Accessed 18 September 2019].

Powers, R. (1995) *Galatea 2.2*. New York, Picador.

Pringle, D. (1985) *Science fiction: the 100 best novels* (Kindle). London, Science Fiction Gateway.

Rhee, J. (2018) *The robotic imaginary: the human and the price of dehumanized labor*. Minneapolis, University of Minnesota Press.

Select Committee on Artificial Intelligence (2018) *AI in the UK: ready, willing, and able?* (No. HL 100 2017–19). Available from: House of Lords website: https://publications.parliament.uk/pa/ld201719/ldselect/ldai/100/100.pdf [Accessed 18 September 2019].

Sladek, J. (1968) *The reproductive system*. London, Granada.

Sladek, J. (1980) *Roderick, or, the education of a young machine*. New York, Carroll & Graff.

Sladek, J. (1983) *Roderick at random*. London, Granada.

Truitt, E. R. (2015) *Medieval robots: mechanism, magic, nature, and art*. Philadelphia, University of Pennsylvania Press.

Ward, M. (2018) *Seeming human: Victorian realist character and artificial intelligence*. Columbus, Ohio State University Press.

# PART I
# ANTIQUITY TO MODERNITY

# 1

# Homer's Intelligent Machines

## *AI in Antiquity*

*Genevieve Liveley and Sam Thomas*

## 1.1 Introduction

Intelligent machines have been a staple of narrative fiction for the past 3,000 years and, in the classical Greek and Roman myth kitty, we find a significant corpus of stories featuring devices which exhibit varying degrees of artificial intelligence (see Mayor 2018; Bur 2016; Liveley 2005; Liveley 2019; Rogers & Stephens 2012; Rogers & Stephens 2015). Representing the earliest phase of this ancient literary tradition, the poet Homer describes relatively simple mechanical devices such as self-pumping bellows (*Iliad* 18.468) and self-opening gates (*Iliad* 5.748–52 and 8.392–6) that appear able to anticipate the desires of their users and perform basic repetitive tasks spontaneously and with a moderate degree of autonomy. These 'almost intelligent widgets' (Pfleeger 2015, p.8) exhibit what Gasser and Almeida characterize as 'weak (or narrow)' rather than 'strong (or general) AI' (2017, p.59).[1] However, Homer also describes slightly more complex contraptions that exhibit correspondingly stronger and more developed levels of (quasi) intelligent automation: multipurpose tripods—serving as tables, altars, and stands—that are represented as *automatos* or 'self-acting' (*Iliad* 18.373–19.379) as they move back and forth between the homes of the gods.[2] Yet more sophisticated automata are deemed to possess something in addition to this power simply to anticipate their users' needs and to move 'of themselves' or 'without visible cause'—as Homer typically describes the operations of these devices. In Homeric epic, we also find a higher order of 'intelligent' machines whose narrative representation suggests that something more significant and cognitively complex than rudimentary automation is being imagined. Homer apparently refers to a pair of silver and gold watchdogs which guard the palace of Alcinoos (whose name actually means 'Strong mind') not with sharp teeth and claws but with their own supposedly 'intelligent minds' (*Odyssey* 7.91–7.94). The ships that eventually take Odysseus home to Ithaca not only move as 'fast as a thought' (*Odyssey* 7.36) but 'navigate by thought' (*Odyssey* 8.556). And the slaves made of gold who serve as personal assistants to the god Hephaestus (*Iliad* 18.418–18.422) exhibit a still higher order of intelligence: they not only possess human form but have the power of movement, of speech, and of thought too. These machines have voices, physical strength, and—uniquely among Homer's wondrous machines—intelligent *minds*.

Through close literary analysis of these Homeric devices, tracing the various gradations of weak to strong machine 'intelligence' that the epic poems describe and the mind models that these gradations assume, this chapter considers what these ancient narratives might tell us about the ancient history of AI. Beginning with a re-examination of Homer's weak AI, his simple automata and autonomous vehicles, in order to provide context and so help us better to appreciate the more sophisticated models of artificial mind and machine cognition attributed to Homer's stronger, embodied AI, this chapter asks: What kinds of priorities and paradigms do we find in AI stories from Homeric epic and (how) do these still resonate in contemporary discourse on AI? In particular, what (if any) distinctions does Homer draw between artificial and human minds and intelligences? And what (if any) is the legacy of Homer's intelligent machines and the ancient narrative history of AI?

## 1.2 Homer's Automata

Although not *sensu stricto* intelligent machines, Homer's eighth-century-BCE descriptions of self-opening gates (*Iliad* 5.748–5.752 and 8.392–8.396), self-pumping bellows (*Iliad* 18.468–18.473), and self-propelling tripods (*Iliad* 18.373–18.379) provide an important background measurement against which to take our reading of Homer's more sophisticated AIs.

The Homeric gods do not keep slaves for manual work (Garlan 1988, p.32) but could hardly be expected to open and close their own gates, so Homer grants them a pair of automatic gates to their heavenly citadel which 'of themselves groan on their hinges' (*automatai de pulai mukon*) as they spontaneously open for the goddess Juno and her chariot at the very same moment as she touches her horses with a whip (*Iliad* 5.748–5.749; the same formula is repeated at 8.393). Similarly, the metalworking god Hephaestus does not keep slaves to work his furnace, but he has an automated machine which controls the variable intensity of the heat supply to his crucibles, wherein he can thereby simultaneously smelt bronze, tin, silver, and gold (*Iliad* 18.468–18.473). Although these bellows are not explicitly characterized by Homer as *automatai*, the fact that there is a total of twenty in operation here, servicing four separate smelting processes, indicates that the two-handed, club-footed god is not pumping them all himself. In such a context, and at such a semi-industrial scale, these devices are evidently working at some level—like the gates of heaven—autonomously. What is more, we are told that the bellows work not in response to Hephaestus' manual pumping but at his command—he *orders* them to work (*Iliad* 18.469)—and that they vary their outputs according to his wishes/instructions, to suit what Hephaestus desires (*Iliad* 18.473). We are not told *how* the bellows might 'know' what Hephaestus wants or needs, or how his commands are communicated, received, and processed. However, the key verb used to describe the object of Hephaestus' commands (*ergazesthai*) is revealing here: it is a term typically used by Homer and his contemporaries to refer to the manual labour of slaves. Hera apparently opens the automated gates of heaven with the crack of her horse whip, and Hephaestus engages with this automated machine as if it were his slave.

The bellows, like the gates of heaven, are represented as tools that replace human slaves and, as such, are assumed to function both mechanically and cognitively on a basic level akin to that of their human counterparts. In this context—reflecting an ancient culture in which slavery was widespread—Homer and his audiences would readily understand that the bellows are able to 'know' what their user desires them to do in the same way that a slave is able to 'know' what its master wants it to do. The user master gives an order and the machine slave obeys. The human (or, in this case, the divine) user master demonstrates his capacity for higher-order cognitive function in his powers of judgement, of technical expertise, of decision-making, and the like; the machine slave demonstrates its lesser cognitive capacity in doing what it is told. User master and machine slave think and work separately, on different cognitive planes, yet synergetically, towards the same goals.

This same synergetic—and hierarchical relationship—is also suggested in Homer's description of the thing that user master and machine slave are together employed in producing here. For Homer's Hephaestus uses one of these basic-model machine slaves to aid him in the task of manufacturing automata of even greater technical ingenuity and (quasi) intelligence. For Hephaestus uses his self-pumping and self-regulating bellows to make a set of self-moving tripods (*Iliad* 18.372–18.381):

> He moved to and fro about his bellows in eager haste; for he was manufacturing tripods (*tripodas*), twenty in all, to stand around the wall of his well-built hall. He had set golden wheels (*kukla*) on to the base of each one so that of themselves (*automatoi*) they could enter the assembly of the gods for him/at his bidding (*hoi*) and return again to his house, a wonder to see (*thauma idesthai*). They were almost fully finished, but the clever/cunning (*daidalea*)

> handles/ears (*ouata*) were not yet fixed upon them. He was making these now, and was cutting the rivets (*kopte de desmous*), working away with intelligent understanding (*iduiesi prapidessi*), . . .

The connection between the bellows and the artefacts they produce is reinforced here by the reference to their same generous number: there are twenty bellows (*Iliad* 18.468–18.473) powering the furnace that Hephaestus is using to manufacture these twenty tripods (*Iliad* 18.374). And, just as the bellows represent both ingenious product and process, these tripods are explicitly represented by Homer as the products of Hephaestus' mechanical prowess—objects which demonstrate and even share in his technological ingenuity, here characterized as his 'clever' or 'cunning skill' (*daidalea*).[3]

Homer gives us a relatively detailed picture of how the tripods are fashioned: golden wheels (*kukla*) are fixed to the base of each tripod, and their elaborate handles or 'ears' (*ouata*) are attached with metal rivets. Yet Homer tells us relatively little about their operation. Again, like the automated bellows which work at Hephaestus' command (*Iliad* 18.469) and 'in whatever way' required (*Iliad* 18.473), the tripods are also supposed to move 'at [Hephaestus'] bidding (*hoi*)' (*Iliad* 18.376). However, Homer does not spell out for us what mind model might enable these devices (with their 'clever' ears) to know or anticipate what Hephaestus wants or needs. Crucially, nor does he tell us what the tripods are supposed to do. The practical functionality of the automatic doors and automated bellows was clear. What is less clear from Homer's narrative is why Hephaestus' self-moving tripods, these *automatoi*, are enhanced not only with golden wheels (in and of itself a marker of relative technological sophistication) but with the power to move both of their own accord and at their master's bidding. Why are these further enhancements desirable in this context?

Some of this uncertainty arises from the ambiguity of the artefact. Tripods are regularly featured in Homeric epic and appear to have had a range of purposes, reflecting their status as high-value objects in archaic Greek society (see Rutherford 2018, p.173; Papalexandrou 2005). In the *Iliad* tripods are often awarded as prizes (*Iliad* 11.700, 12.259, 12.264), and Priam includes 'two bright tripods' among the ransom he pays Achilles (*Iliad* 24.248–24.264); in the *Odyssey*, Menelaus' Egyptian treasure haul includes a selection of tripods he has been given as guest gifts (*Odyssey* 4.129–4.131), and the Phaeacians ceremoniously present their guest Odysseus with gifts of both tripods and cauldrons to set on top of them (*Odyssey* 13.13). Indeed, this appears to have been a core function of the tripod: to serve as a stand or support for bowls used for cooking, burning incense or sacrifice, and other ritual activity. Rutherford also suggests an additional purpose—to impress: 'The combination of tripod-plus-cauldron provided an excellent medium in which to show off wealth in metal; the more elaborate the workmanship, the more the object would impress' (Rutherford 2019, p.173). Hephaestus' tripods (all twenty of them) with their golden wheels are certainly capable of fulfilling this function, particularly as it remains unclear what practical purpose they could provide in service of the Homeric gods, who—after all—have no need of tripods or cauldrons for cooking, burning incense, or any of the other purposes to which Homeric mortals would have put them.

However, there is an alternative rationale suggested by Homer's account of these wheeled machines not merely moving but moving 'of their own accord': the AI upgrade may be designed (or, at least, used) neither to enhance functionality nor to impress but *to cheat*. The other contexts in which we encounter Homeric tripods highlight their status as a kind of proto-currency: they are given and received as reward, as guest gift, as ransom, and as an offering to a god—typically in return for something (Seaford 2004). Their core function is to serve as token and as object of exchange—to move between people and their houses (or temples). Hephaestus' innovation, therefore, is not simply to add wheels to his tripods. In fact, the archaeological record confirms that wheeled tripods already existed in the archaic Greek world: fragments dating to the ninth or eighth centuries BCE have been identified in Ithaca

(Benton 1934–1935, pp.88–89, 99), and the British Museum dates an example from Cyprus back to the twelfth century BCE (Rutherford 2019, p.174). In putting wheels onto his tripods, then, Homer's Hephaestus is demonstrating an already familiar, traditional technology. What *is* remarkable about his engineering feat is not the wheels, therefore, but the fact that these tripods can 'of themselves enter the assembly of the gods for him/at his bidding and *again return to his house*' (*Iliad* 18.377–18.378). We find this return of tripods to their original owner and donor nowhere else in Homer. And we should be surprised by it here. The fact that Hephaestus' autonomous tripods can be given to others but then make their own back again suggests that their simple AI enhancements enable Hephaestus to commit a kind of proto–cyber crime against his fellow gods. The epithet that Homer grants Hephaestus and his work here—*daidalea* (*Iliad* 18.379)—is well placed: it suggests not only technological ingenuity, but a kind of trickery is at work. A technologically minded trickery, moreover, that is synergetically shared between Hephaestus and the devices he designs.

In Homer's account of the autonomous tripods, emphasis is placed primarily upon the 'intelligent understanding' (*iduiesi prapidessi*) of the maker rather than upon that of the made (*Iliad* 18.381).[4] Any artificial 'intelligence' belonging to the tripods is only implied. Homer tells us nothing about how these or any of his other basic *automatoi* 'know' what to do or 'know' what their user requires of them. At most, he ascribes to these machine tools a simple mind model akin to that of the slave who obeys the spoken orders of a master—and we notice that Hephaestus' tripods with their 'clever ears' are particularly well designed in this regard; these machine slaves have clever ears all the better to hear the commands of their clever user master as he employs his intelligent understanding (*iduiesi prapidessi*) to direct their operations.

In Homer's *Odyssey*, however, we encounter a case in which the 'intelligent understanding' of Hephaestus is not limited to the creator but is potentially transferred to one of his creations. The animated guard-dog statues of silver and gold that Hephaestus makes to stand beside the gates to Alcinoos' palace on Phaeacia are—like the tripods—said to be crafted by him 'with intelligent understanding' (*iduiesi prapidessi*) (*Odyssey* 7.91–7.94; cf. *Iliad* 18.381). The Greek text makes it difficult to determine whether the 'intelligent understanding' here is to be attributed to the metal dogs or to their maker: *prapides* (understanding) appears in the plural, but, like Homer's use of the equivalent mind word *phrenes* (the plural form of which is used typically to designate 'thought' or 'thoughts'), this could apply equally to Hephaestus or to the dogs. In fact, exactly the same formulation, 'intelligent understanding' (*iduiesi prapidessi*), appears elsewhere in Homeric epic simply to describe Hephaestus as he works (*Iliad* 1.608 and 20.12). Nevertheless, translators and commentators of Homer have been willing—even eager—to elide the intelligence of the maker (and the making) with that of the made in the case of the Phaeacian dogs, debating whether or not their representation imagines some kind of ancient AI.[5] However, the Homeric narrative offers too little detail, and the Greek is too ambiguous, to authorize any definitive resolution to this question. Yet, if we look elsewhere in the Homeric poems we do find descriptions of intelligent artefacts and the direct attribution of thought—and, therefore, of 'artificial intelligence'—to some of Homer's other mechanical devices, opening up the possibility that Hephaestus' gold and silver watchdogs were also imagined to possess 'intelligent understanding' of some kind.

## 1.3 Homer's Autonomous Vehicles

Indeed, we find the unambiguous attribution of intelligent thought to a machine—or, more explicitly, to a vehicle—in another key narrative involving Homer's Phaeacians, who are said to possess ships that display clear signs of intelligence (*Odyssey* 8.555–8.564). Anticipating the satnavs and driverless vehicles of our own century, these Phaeacian ships need neither pilot nor steering oar—they know exactly where in the world their passengers want to go and how to avoid all obstacles along the way, and they are indefatigable and invulnerable. That is, until they are turned to stone by an angry Poseidon (*Odyssey* 13.163)—an apt and tragic peripeteia for these thinking machines which used to move as 'fast as a thought' (*Odyssey* 7.36).

We are first introduced to these imaginary autonomous vehicles in Homer's *Odyssey* with an apt simile that soon proves to be much more than metaphoric (*Odyssey* 7.36): '[the Phaeacian's] ships are swift as a bird on the wing or as a thought (*noema*)'. Like the human mind (*noos*) which is also perceived to fly as swiftly as a bird in ancient Greek models, these ships unambiguously possess a 'natural' kind of (artificial) intelligence.[6] Preparing to send his guest safely home in one of these remarkable ships, the Phaeacian King Alcinoos asks Odysseus for his home address so that he can (effectively) programme the ships' 'satnav' (*Odyssey* 8: 555–64):

> Tell me your country, your people, and your city, so that my ships can take you there, plotting the course with/in their minds (*tituskomenai phresi*). For we Phaeacians have no pilots (*kuberneteres*), nor steering oars (*pedali*) such as they have on other ships; the ships themselves (*autai*) know (*isasi*) the minds (*noemata*) and thoughts/intentions (*phrenas*) of men, and they know (*isasi*) every city and every rich land. They cross the expanse of the sea with great speed (*takhisth*), hidden in mist and cloud, and they never have to fear any damage or shipwreck.

In place of pilot and steering oar, these ships have *phrenes*—here representing the intellectual programming by which the ships 'know' (*isasi*) the minds (*noemata*) and intentions/thoughts (*phrenas*) of their human passengers. Indeed, the programming of these ships is represented as equivalent to—and as actively substituting for—the human minds both of pilots (who will be absent for the voyage) and passengers (who, in Odysseus' case, will be asleep for the voyage). These ships actively, purposively, and independently plot (*tituskomai*) their course across the sea; they *know* where they need to go and they *know* the best way to get there safely.[7] These intelligent ships are programmed by the knowledge contained in human minds, but because these ships have *phrenes*, they can use this programming to plan ahead and negotiate practical problems; they can think independently (*autai isasi*).

We have encountered *automatos* machines in Homer before, but, as Tassios observes, this is 'the first time that the techno-mythical thought of the Greeks dares to figure out *human* ... achievements' (Tassios 2008, p.6) in this domain; all the other Homeric examples of automata are either designed by or employed in the service of gods. In this light, Kalligeropoulos and Visileiadou (2008) can further claim that: 'a new technological vision appears here: the constructed thought, the artificial intelligence, the ability of programming, the development of a technology capable of controlling the route of a ship' (Tassios 2008, p.79). The similarities between the artificial intelligence attributed to Homer's Phaeacian ships and modern technological visions of the same (particularly in the arena of driverless vehicles) are certainly striking. However, what should be no less striking, though easily overlooked, are the subtle similarities between the model of cognition and of intelligence that Homer assigns here to both humans and their machines.

In this light, we notice that of the various mind words used in this passage, Homer ascribes *phrenes* (thought) to the ships but not *noos* (mind), whereas humans are deemed to possess both *phrenes* and *noos*/*noemata*. Is this significant? *Phren* or *phrenes* in Homer (where it is typically found in the plural form) can signify both the thought processor ('the thing that thinks') and the output of that processor ('the thing that is thought'). It can designate both the container for thoughts and ideas and those thoughts themselves. But it is particularly associated with acts of *practical* reasoning and thinking; the kind of cognition that readily translates into useful, purposeful, action (see Sullivan 1988, pp.7–8; Bremmer 1983, p.61; Snell 1982, pp.59–65). Such practical reasoning and knowledge is, of course, precisely the kind of intelligence that an autonomous vehicle requires to successfully and safely achieve its goals. The cognitive demands placed upon its processing are significantly higher than the cognitive load of, say, a pair of automatic gates or set of bellows—even of a self-moving wheeled tripod. Yet, while certainly more sophisticated in its operations than these basic machines, the cognitive powers required of an autonomous vehicle are not here deemed straightforwardly representative of higher-order (human) thinking. A distinction is maintained between the cognitive capacity of the ships and the humans who programme them, and, once again, we recognize a hierarchy of cognitive function and ability at work in Homer's 'intelligent' machines: the thoughts (*phrenes*) of the Phaeacian ships are programmed by the knowledge and intelligence found in human minds (*noos*/*noemata*). The ships accordingly think like the humans who programme them, but—just like the machine-slave devices in Hephaestus' workshop—are dependent upon the higher-order cognition of their human masters to command and control them. The ships' 'independent' thinking and 'independent' movement is limited by their human programmers and users.

Highlighting this feature of the ships' AI programming, Homer includes a number of significant but easily overlooked details in his description of Alcinoos' commands for the deployment of the ship that is to take Odysseus home. When Alcinoos orders a ship to be made ready for Odysseus, he specifies that it should be one 'which has never yet made a voyage' (*Odyssey* 7.34–7.36). The ship that will carry Odysseus home will be untried, sailing out on its maiden voyage. That is, the ship will have no pre-programming or learned experience of its own; its *phrenes* will be clean and clear. Alcinoos further orders (*Odyssey* 7.40) that the ship is to be manned by fifty-two of the very best of young Phaeacian manhood (*Odyssey* 7.36)—youths who do *exactly* what Alcinoos commands (*Odyssey* 7.49). 'Strong minded' Alcinoos controls everything in this operation, and his instructions are carried out to the letter.

In this light, the human mind model that Homer uses to explain how the ships 'know' what they do is particularly apt. In Homer's representation of human intelligence, *phrenes* are sometimes represented in an active role, and they may occasionally 'pilot' the human *thumos* or spirit (as they pilot the Phaeacian ships), but the primary characteristic of *phrenes* is passive 'responsiveness'; *phrenes* are acted upon, rather than acting or initiating; gods place good (or bad) plans and ideas into human *phrenes* (see Padel 1992, pp.21–22); god-like humans place the plans and directions conceived in their own minds (*noemata*/*noos*) into the *phrenes* of these Phaeacian ships. As literal vessels which are 'swift as a bird on the wing or as thought (*noema*)' (*Odyssey* 7.36; cf. 8.562), the Phaeacian ships make excellent models for the particular kind of AI that we see here—that is, a lower-order mode of human intelligence (the ability to process and follow directions and commands) transferred to a machine.

What is more, the ships possess *phrenes* but not *noos*, and whereas *noos* is never attributed to any animal in Homer (only to gods and humans), *phrenes* are also found in some animals (*Iliad* 4.245 and *Iliad* 17.57). Again, this suggests that something other (and possibly less sophisticated) than ordinary human intelligence is imagined to be at work in these ships: animals and slaves in this model have instinctive, practical intelligence; they know how to avoid danger and achieve simple goals. Endowed with *phrenes* but not *noos*, therefore, the ships are deemed to possess a basic kind of artificial intelligence—a kind of machine memory—that is not only simple but programmable; humans place itineraries into their *phrenes* to which they respond and react (as gods place ideas into human *phrenes*); humans initiate the machine thinking process and benefit from the practical reasoning and outputs that result.

In this context we remember that in Homer, as in Greek literature and philosophy thereafter, 'Memory is knowledge that can seem to come from outside, yet it is also part of "mind". Information can be "written on" [*phrenes*]. You remember by inscribing "inside your *phrenes*," "on the tablets of your *phrenes*"' (Padel 1992, p.76). In the *Odyssey*, the epic poet Phemius (avatar of the storytelling Odysseus and of Homer himself) claims that: 'I am self-taught (*autodidaktos*), but the god has put into my mind (*phresin*) all kinds of songs' (*Odyssey* 22.347).[8] And in the *Iliad*, in Homer's invocation to the Muses (daughters of Memory) before reciting the Catalogue of Ships (*Iliad* 2.484–2.493), he asks that they 'remind' him of this—literally to 'put it into his mind' (*mnesaiath'*). The *phrenes* of the Phaeacian ships therefore appear to function in the same way as human *phrenes.* However, Homer's representation of these thinking machines does not simply equate natural with artificial intelligence, or assume that human and machine thinking are straightforwardly the same. The ships may have *phrenes* (thoughts) but they do not possess *noos* (minds), whereas humans have both *phrenes* and *noos/noemata*. There appears to be some important distinction being drawn in Homer between the cognitive potential of an object and a person. Is this a 'natural' intuition or instinct on Homer's part—that gods possess superior intelligence to humans, who possess superior intelligence to animals (and enslaved humans), who possess equivalent and/or greater intelligence to some machines? And is the relatively superior cognitive potential of humans somehow enabled (or enhanced) by their physical embodiment *qua* human? To address these questions, we must turn to examine the highest order of AI found in Homer's poetry: the embodied AI encountered in Homer's account of humanoid thinking machines.

## 1.4 Homer's Embodied AI

In the *Odyssey*, as we saw, Homer's intelligent machines are endowed with *phrenes* only, but in the *Iliad* we encounter machines granted both *phrenes* and *noos* (or *nous* as it appears in later Greek writing). Hephaestus' anthropomorphic automata possess both mind and the power of intelligent thought—or, as Homer puts it, they have 'intelligence (*phrenes*) in their minds (*noos*)' (*Iliad* 18.420). As Hephaestus sets aside the automated bellows and other tools used to manufacture his *automatos* tripods, a new category of intelligent machine appears in order to help him (*Iliad* 18.412–18.422):

> moving swiftly to support their master (*anakti*), there came handmaids (*amphipoloi*) made of gold, looking just like (*eioikuiai*) living (*zoesi*) girls (*neenisin*). They have in them intelligence (*phresin*) in their minds (*noos*), and human speech (*aude*) and strength (*sthenos*), and they know (*isasin*) how to work (*erga*) thanks to the immortal gods. They hurried out now (*epoipnuon*) to support (*hupaitha*) their lord.

This description is remarkable for many reasons, but not least of all because *noos* in Homer is only assigned to gods and humans; unlike *phrenes*, which animals may possess, *noos* is never attributed to an animal. Bremmer suggests that this is because *noos* conveys something 'more intellectual' than *phrenes* (Bremmer 1983, pp.56–57, cf. p.127 on animal *phrenes*). Certainly, *noos* carries more authority than *phrenes*, and 'the *noos* of Zeus is always more powerful than that of men' (*Iliad* 16.688). Linguistic analysis of the Homeric epics confirms that *noos*, like *phrenes*, can be conceived both as the container in which thought takes place and as thought itself (see Snell 1982, pp.9–22; Padel 1992, pp.32–33; Sullivan 1989). Yet, as Homer's description indicates, *noos* is conceived here specifically as the container or processor in which *phrenes* are located. Textual and linguistic analysis of these terms also highlights lexical associations between *noos* and 'visual activity', prompting critics and commentators to suggest that *noos* for Homer 'is analogous to the eye' and that 'to know' means 'to see' (Snell 1982, p.18). Sullivan also sees *noos* as a form of 'inner vision' (1988, pp.130–31), and for Padel, *noos* is 'an essentially perceiving force: "intention," "sense." [*Noos*] sees and hears . . . It is intellect and intelligence . . . or it is an act of intellect, a "thought," a "plan" ' (Padel 1992, p.32).[9] As McDonald notices, *noos* in Homer is 'associated with what we now think of as higher-order cognitive functions . . . which we might perhaps want to render as "intellect".' (McDonald 2003, p.16).

There are two kinds of (strong) artificial intelligence in Homer, therefore: one associated with *phrenes* and another, higher-order intellect, associated with *noos*. Like the Phaeacian ships, Hephaestus' anthropomorphic machines 'know' (*isasin*; cf. *Odyssey* 8.560–8.561) how to do the work required of them. In this case, the *phrenes* of these machines (like those of the Phaeacian ships) are programmed by an external agent of higher (here divine) intelligence: the *amphipoloi* 'know how to work thanks to the immortal gods' (*Iliad* 18.421).[10]

But this knowledge is also represented here in association with the higher-order cognitive function of *noos*—perhaps because, as the embodiment of that function in human form suggests, they are better equipped in terms of their hardware for seeing, knowing, and higher-order thinking than are inanimate, nonanthropomorphic objects such as a ships, tripods, bellows, or gates. As embodied AI, the *amphipoloi* have eyes and ears, as well as *phrenes* and *noos*—although, for Homer and his ancient audiences, thought and mind would not have been associated with the human head or brain. Rather, the descriptions of body parts found in Homeric epic and other archaic Greek poetry suggests that the *phrenes* would have been located somewhere in the human chest—in the region of the lungs and diaphragm.[11] Therefore, these embodied, humanoid, AI have human-like *phrenes* (thought) because they have human-like form: their physical design supports their cognitive capacity. These embodied, humanoid, AI have human-like *noos* (mind) too: *noos*, we recall 'is analogous to the eye' for Homer (Snell 1982, p.18); and the human eye, it seems, is the body part most associated with intellect and intelligence in Homer's model of mind. Made to look 'just like real girls', the *amphipoloi* concomitantly have the body parts needed for thinking 'just like real girls'.[12]

Yet, despite the potential for higher-order cognitive function that Homer allows Hephaestus' embodied AI, the status of these golden machines *qua* objects—albeit as intelligent, talking, moving objects—is repeatedly emphasized in this narrative. As Hephaestus puts away his other tools (*hopla*), these personal assistant AIs come out—one set of tools effectively, functionally, replacing another. Indeed, the language of labour and slavery infuses the representation of these humanoid machine tools just as it did the other devices in the god's workshop: Hephaestus is described as their master (*anakti*)—a term frequently used in Homer to denote the relation of master to slave;[13] these embodied AIs themselves are his personal assistants/attendants, his handmaidens (*amphipoloi*)—a term used in Homer only to denote a female slave role (see *Odyssey* 1.331, 6.199); they may have *phrenes* and *noos*, and even the power of speech (*aude*), but their core function is work (*erga*)—another term typically used by Homer to denote the labour of slaves, hence the emphasis upon the physical strength (*sthenos*) in addition to the intellectual powers of the embodied AI *amphipoloi* here.[14]

Thus, two and a half thousand years before Čapek (1923) introduced the word 'robot' to the world stage in *RUR*, we find Homer representing his own autonomous artefacts and artificial humans as slaves: they are 'robots' *avant la lettre*. Indeed, it may be significant that Plato and Aristotle already drew similar and suggestive correlations between Homer's intelligent machines and human slaves. In the *Meno* (97d–98a), Plato's Socrates explicitly likens the automata created by Hephaestus and Daedalus in ancient myth (characterized here as 'Daedalian statues'—*Daidalou agalmasin*) to slaves: both, he claims, will run away unless they are chained up. What is more, he introduces this analogy to explain a crucial distinction between two kinds or modes of intelligence: *episteme* (understanding/knowledge) and *doxa* (opinion/belief). Plato's definitions of these two concepts are notoriously complex, but what matters for us here is the hierarchical distinction being drawn between higher and lower modes/domains of intellectual power. The slave, for Plato's Socrates, is aligned with the automaton, and both are aligned with the lower-order mode of knowledge that Socrates assigns to *doxa*. Indeed, as Devecka explains, for both

Socrates and Plato—and subsequently for Aristotle too—*doxa* signifies 'a state of knowledge not dissimilar to the "participation in *logos* without understanding" that characterizes a natural slave' (2013, p.63).

We saw evidence indicating a similar hierarchy of intelligence in Homer. In *Odyssey*, the distinction that Homer drew between the *phrenes* that the Phaeacian ships were deemed to possess and the *noos*/*noemata* possessed by the humans who owned and used them differentiated between two different orders of intellectual capacity. Indeed, Aristotle appears to have this specific Homeric example of AI in mind when he draws an explicit correlation between Homer's fictional automata and real-world slaves in the *Politics* (1253b). Focusing upon the possession of life/soul (*psucha*) rather than of mind (*noos*/*nous*) in this case, Aristotle suggests that:

> some tools are lifeless/soulless and others living/have souls and so, for a ship's pilot, the steering oar is a lifeless/ soulless tool and the look-out man is a living tool/a tool with a soul . . . If every tool (*organon*) could do its own work (*ergon*), as ordered or by anticipation (*proaisthanomenon*), like the statues of Daedalus or the tripods made by Hephaestus, of which the poet says that 'of themselves (*automatoi*) they could enter the assembly of the gods', then shuttles could weave and plectrums pluck by themselves (*autai*). In this case, managers would not need assistants/subordinates and masters would not need slaves.

Already in antiquity, then, automata and intelligent machines were conceived as artificial slaves, occupying a special social status no less than ontological category: above that of an inanimate tool—an oar, say; but below that of a human, Alcinoos, for example, who is lower again in both status and cognitive power than a god such as Zeus, whose *noos*, we recall, 'is always more powerful than that of men' (*Iliad* 16.688). Accordingly, most of Homer's automata appear capable of performing only a limited range of functions autonomously, and a pattern of simple repeated motion itself repeats throughout these descriptions. In Homer's fiction, gates open and close; bellows rise and fall; tripods move back and forth between Hephaestus' home and those of the other gods; and ships travel back and forth across the sea. In Aristotle's counterfactual, shuttles weave, and plectrums pluck, back and forth across loom and lyre, respectively. The weak AI, the 'intelligence' (or quasi-intelligence) attributed to such devices is ostensibly restricted too: as in Aristotle's definition of the 'natural slave' as one who lacks understanding, these 'artificial slaves' simply do their own work (*ergon*), as ordered (by their user master) or by anticipation (*proaisthanomenon*).[15] Even the Phaeacian ships, which were represented as exhibiting a higher cognitive capacity than Hephaestus' machines, were endowed with *phrenes* rather than *noos*, with the equivalent of *doxa* rather than *episteme*: they do what they are programmed to do. And, although Hephaestus' *amphipoloi* were attributed both *phrenes* and *noos*, as humanoid 'tools', they too fit Aristotle's definition of 'a living tool/a tool with a soul'—or, in this case, a mind (*Politics* 1253b). What is more, as artificial 'girls', the embodied gender of the *amphipoloi* automatically qualifies them (for both Homer and Aristotle) for the equivalent status of 'natural slave'; as women, as slaves, and as machines, they do what they are told to do.[16]

## 1.5 Homer's AI Legacy

We twenty-first-century humans—like Homer's intelligent machines—also appear to do what our (literary) programming tells us to do. In the domain of modern AI narratives, we too seem capable of recognizing only a limited range of AI functions, and a pattern of simple repetition repeats through both real-world and story-world discourses on AI.

The AI narratives that appear in Homeric epic ostensibly exhibit several of the same patterns and tropes of narrative representation and thinking about AI that we find repeated in many nineteenth-, twentieth-, and twenty-first-century representations. In particular: 'a focus on embodiment; a tendency towards utopian or dystopian extremes; and a lack of diversity in creators, protagonists, and types of AI' (Cave et al 2018, p.4). Homer's fictional *amphipoloi* appear to be the ancestors of the subservient female AI assistants encountered in so many modern films and TV series—such as Anita/Mia (one of the 'synths' from *Humans* [Nettheim, D. 2015–2018], Samantha (from *Her* [Jonze et al 2014]), or Joi (from *Blade Runner 2049* [Villeneuve, D. et al 2018])—no less than those found in so many actual modern homes—such as Siri, Alexa, or Cortana. The embodiment of Hephaestus' *amphipoloi* as 'living girls' also anticipates an enduring sci-fi trope that plays out in numerous modern AI fictions (from the utopian story world of *Lars and the Real Girl* [Kimmel, A. et al 2011] to the dystopian fantasy of *Ex Machina* [Garland 2015]). Hephaestus' wondrous (*thaumasta*) technological creations are all encountered in distinctly wondrous *utopian* story worlds—in the service of the immortal, Olympus-dwelling, gods. The Phaeacian ships represent a rare example of human engagement with AI in Homer's imaginary, but this is tempered by the fact that the Phaeacians are also represented in Homer as inhabiting a kind of island utopia—these humans live *like* the gods (*Odyssey* 7.205). In Homer we also find intelligent embodied slave machines created and programmed 'with intelligent understanding' (*iduiesi prapidessi*) by individual male creators (Hephaestus and Alcinoos), and used in the service of other, predominantly male protagonists (Odysseus, Hephaestus—and other gods). We encounter a familiar range of AI prototypes exhibiting increasing levels of 'intelligent' capabilities too: ranging from the mundane household gadget (the automated gates), the industrial tool (Hephaestus' automatic bellows), the exclusive/expensive AI device designed to impress (Hephaestus' automated tripods), the AI device with potential for criminal exploitation (Hephaestus' automated tripods again), the security system (the Phaeacian dogs), the self-steering autonomous vehicle (the Phaeacian ships), and the anthropomorphic AI assistant (Hephaestus' *amphipoloi*). The parallels between the AI of our own century and Homer's intelligent machines seem *thaumasta*—wondrous—indeed.

But perhaps these parallels are not entirely wondrous, and have a prosaic (even mechanical) explanation—one aptly supported by the Homeric mind model of both humans and machines possessing programmable *phrenes* and minds in which our/their higher-order cognitive functions are coded by what we/they see and hear. According to one brief history of artificial intelligence (Buchanan 2005, p.53):

> *Ever since Homer wrote of mechanical 'tripods' waiting on the gods at dinner, imagined mechanical assistants have been a part of our culture.*

Homer is part of our cultural AI programming. Indeed, the theory of AI 'schemata' or 'scripts' promotes the idea that the basic operations of both machine and human learning depend upon correlating new data and experiences with data and experiences already stored in a 'memory' bank. Both humans and machines are understood to draw upon an 'experiential repertoire' (Herman 1997, p.1047) or 'patrimonial hoard' of previous human experience (Barthes 1974, p.204) when it comes to making sense of anything unfamiliar, making sense of the new by comparing it with or contrasting it to the old, by testing its relation to a stored data set of 'knowledge frames' or 'knowledge scripts'.[17]

Such narratologically inflected insights into machine intelligence and machine learning in turn help us to understand how and why Homer's ancient tales of intelligent machines appear to have established this 'programme' for us to follow in the retelling and rescripting of our present and future AI narratives. Homer's tales of intelligent machines are stored as part of the AI data set in the human 'memory bank' that constitutes the classical myth kitty. Homer's intelligent machines are programmed into our cultural *episteme* (understanding/knowledge) and *doxa* (opinion/belief) about AI. Homer's AI narratives, it seems, have been placed into modern human *phrenes*, into the *noos* of the Western cultural imaginary.

# Notes

1. Gasser and Almeida (2017, p.59), discussing the use of the term 'AI' as 'an umbrella term to refer to a certain degree of autonomy' exhibited in various AI applications, define these two AI types thus: 'Weak AI describes the current generation of applications that are focused on a relatively narrow task...Strong AI, in contrast, refers to machines with genuine intelligence and self-awareness in the sense that the machine has the ability to apply intelligence to any problem.'
2. All translations in this chapter are our own, and all Greek is transliterated. Definitions and translations of key terms in Greek are taken from the 2011 edition of *The Online Liddell-Scott-Jones Greek-English Lexicon*—hereafter *LSJ*.
3. This is the first instance of this word in the Greek lexicon and the etymological root for Hephaestus' mortal equivalent 'Daedalus'.
4. The *LSJ* defines *prapides* as equivalent in Homer to *phrenes*: 'the seat of mental powers and affections...understanding, mind'; *iduia* is related to the verb *oida*—to know. See *LSJ* s.v.v.
5. Faraone (1987, pp.269–71) points to ancient Assyrian apotropaic statues—figures 'programmed' with magical inscriptions to ward-off intruders—as possible 'real-world' counterparts to Hephaestus' creations in Homer (see also Garvie 1994, pp.181–82; Liveley 2005, p.279; Mayor 2018, pp.143–44).
6. On the swiftness of thought see Padel 1992, p.129; Bremmer 1983; p.127; and *Iliad* 15.79–15.83, where Hera flies from Mt Ida to Olympus as fast as the mind (*noos*) of a nostalgic man remembering his travels.
7. Notice the repetition of the verb 'to know' (*isasi*), which for Garvie (2008, p.345) 'emphasises the mental capacity of the ships'.
8. Notice the juxtaposition here (as in the descriptions of Homer's intelligent—and quasi-intelligent—machines) between autonomy and programming.
9. Compare this with Appollonius' account of the robot Talos, who is destroyed when Medea remotely takes control of his mind through his eyes: *Argonautica* (4.1635–4.1688); see also Buxton 2013; Liveley 2005; Mayer 2018, pp.7–32.
10. Exactly the same formulation is used by Homer's contemporary Hesiod to describe the manufacture and 'programming' of the first human woman, Pandora (Hesiod *Works* 61–64). On Pandora as cyborg see Hammond (2018).

11. The *LSJ* associates *phrenes* with the 'midriff' and 'heart', in addition to the mind. See *LSJ* s.v.
12. At least, they have the artificial body parts needed to convince *us* that they are capable of artificial intelligence.
13. The *LSJ* defines *anakti* as equivalent in Homer to 'master'. See *LSJ* s.v.
14. The *LSJ* defines *epoipnuon* as describing specifically the busy 'bustling' work of subordinates and slaves. See *LSJ* s.v.
15. On the history of such 'artificial slaves', see LaGrandeur (1999); LaGrandeur (2011); and LaGrandeur (2013).
16. Indeed, Aristotle uses the 'natural' inferiority of women to 'prove' the reality of the 'natural' slave (*Politics* 1254b 12-16). The bibliography on Aristotle's thoughts on slavery and on women is vast, but, in particular, see DuBois (2003); Heath (2008); and Parker (2012).
17. See Herman (2002) and Sandford and Emmott (2012).

## Reference List

Barthes, R. (1974) *S/Z*. New York, Hill and Wang.

Benton, S. (1934–1935) Excavations in Ithaca III. *Annual of the British School at Athens*. 35, 45–73.

Bremmer, J. N. (1983) *The early Greek concept of the soul*. Princeton, NJ, Princeton University Press.

Buchanan, B. G. (2005) A (very) brief history of artificial intelligence. *AI Magazine*. 26(4), 53–60.

Bur, T. (2016) *Mechanical miracles: automata in ancient Greek religion*. Sydney, Australia, University of Sydney. Available from: https://ses.library.usyd.edu.au/bitstream/2123/15398/1/bur_tcd_thesis.pdf [Accessed 29 January 2019].

Buxton, R. G. A. (2013) *Myth and tragedies in their ancient Greek contexts*. Oxford, Oxford University Press.

Čapek, K. (1923) *RUR (Rossum's universal robots): a fantastic melodrama*. Garden City, NY, Doubleday, Page.

Cave, S. et al (2018) *AI narratives: portrayals and perceptions of artificial intelligence and why they matter. Royal Society*. Available from: https://royalsociety.org/topics-policy/projects/ai-narratives/ [Accessed 29 January 2019].

Devecka, M. (2013) Did the Greeks believe in their robots? *Cambridge Classical Journal*. 59, 52–69.

DuBois, P. (2003) *Slaves and other objects*. Chicago, University of Chicago Press.

Faraone, C. A. (1987) Hephaestus the magician and near eastern parallels for Alcinous' watchdogs. *Greek, Roman and Byzantine Studies*. 28, 257–80.

Garlan, Y. (1988) *Slavery in ancient Greece*. Ithaca, NY, Cornell University Press.

Garland, A., et al (2015) *Ex Machina*. Universal Pictures.

Garvie, A. F. (1994) *Odyssey. Books VI–VIII*. Cambridge, Cambridge University Press.

Gasser, U. & V. A. F. Almeida (2017) A layered model for AI governance. *IEEE Internet Computing*. 21 (6) (November), 58–62.

Hammond, E. (2018) Alex Garland's *Ex Machina* or the modern Epimetheus. In: J. Weiner, B. E. Stevens, & B. M. Rogers (eds.) *Frankenstein and its classics: the modern Prometheus from antiquity to science fiction*. London, Bloomsbury. pp.190–205.

Heath, M. (2008) Aristotle on natural slavery. *Phronesis*. 53, 243–70.

Herman, D. (1997) Scripts, sequences, and stories: elements of a post-classical narratology. *Proceedings of the Modern Language Association*. 112(5), 1046–59.

Herman, D. (2002) *Story logic: problems and possibilities of narrative*. Lincoln, University of Nebraska Press.

Jonze, S. et al (2014) *Her*. Warner Bros Entertainment.

Kalligeropoulos, D. & S. Visileiadou (2008) The Homeric automata and their implementation. In: S. A. Paipetis (ed.) *Science and technology in Homeric epics*. Dordrecht, Springer Netherlands. pp.77–84.

Kimmel, A. et al (2011) *Lars and the Real Girl*. 20th Century Fox Home Entertainment.

LaGrandeur, K. (1999) The talking brass head as a symbol of dangerous knowledge in Friar Bacon and in Alphonsus, King of Aragon, *English Studies*. 80(5), pp.408–22.

LaGrandeur, K. (2011) The persistent peril of the artificial slave, *Science Fiction Studies*. 38(2), pp. 232–52.

LaGrandeur, K. (2013) *Androids and intelligent networks in early modern literature and culture: artificial slaves*. New York: Routledge.

Liveley, G. (2005) Science fictions and cyber myths: or, do cyborgs dream of Dolly the sheep? In: V. Zajko & M. Leonard (eds.) *Laughing with Medusa: classical myth and feminist thought*. Oxford: Oxford University Press.,pp.275–94.

Liveley, G. (2019, forthcoming) Beyond the beautiful evil? The ancient/future history of sex robots. In: G. M. Chesi & F. Spiegel (eds.) *Classical literature and post-humanism*. London, Bloomsbury.

Liddell, H. G., R. Scott, H. S. Jones, & R. Mckenzie (2011) *The online Liddell-Scott-Jones Greek-English lexicon*. Irvine, University of California, Irvine. Available from: http://www.tlg.uci.edu/lsj [Accessed 29th January 2019].

Macdonald, P. S. (2003) *History of the concept of mind*. Aldershot, Ashgate.

Mayor, A. (2018) *Gods and robots: the ancient quest for artificial life*. Princeton, NJ, Princeton University Press.

Nettheim, D. (2015–2018) *Humans*. Kudos Film and Television.

Padel, R. (1992) *In and out of the mind: Greek images of the tragic self*. Princeton, N.J., Princeton University Press.

Papalexandrou, N. (2005) *The visual poetics of power: warriors, youths and tripods in early Greece*. Lanham, Lexington Books.

Parker, H. N. (2012) Aristotle's unanswered questions: women and slaves in *Politics* 1252a-1260b. *Eugesta: European Network on Gender Studies in Antiquity*, 2. Available from: https://eugesta-revue.univ-lille3.fr/pdf/2012/Parker-2_2012.pdf [Accessed 29 January 2019].

Pfleeger, S. (2015) Spider-Man, hubris, and the future of security and privacy. *IEEE Security & Privacy*. 13, 3–10.

Rogers, B. & B. Stevens (2012) Classical receptions in science fiction. *Classical Receptions Journal*. 4(1), 127–47.

Rogers, B. & B. Stevens (2015) *Classical traditions in science fiction*. Oxford, Oxford University Press.

Rutherford, R. B. (2019) *Iliad. Book XVIII*. Cambridge, Cambridge University Press.

Sanford, A. & C. Emmott (2012) *Mind, brain and narrative*. Cambridge, Cambridge University Press.

Seaford, R. (2004) *Money and the early Greek mind: Homer, philosophy, tragedy*. Cambridge, Cambridge University Press.

Snell, B. (1982) *The discovery of the mind in Greek philosophy and literature*. New York, Dover.

Sullivan, S. D. (1988) *Psychological activity in Homer: a study of frēn*. Ottawa, Canada, Carleton University Press.

Sullivan, S. D. (1989) The psychic term *νόος* in Homer and the Homeric Hymns. *Studi Italiani di Filologia Classica*. 7, 152–95.

Tassios, T. P. (2008) Mycenean technology. In: S. A. Paipetis (ed.) *Science and technology in Homeric epics*. Dordrecht, Springer Netherlands. pp.3–34.

Villeneuve, D., et al (2018) *Blade Runner 2049*. Alcon Entertainment et al.

# 2

# Demons and Devices

## *Artificial and Augmented Intelligence before AI*

*E. R. Truitt*

## 2.1 Introduction

The imaginative landscape of the Latin Middle Ages teems with examples of artificial intelligence. Some appear as silent metal servants, discerning and adept in highly socialized environments. Still others take shape as elaborate, far-sighted guardians. Yet examples of artificial intelligence in medieval Latin culture are not limited to the kinds of artefacts that neatly mirror twenty-first century examples of AI (usually defined as machines that demonstrate human-like cognitive abilities, such as learning and problem-solving).[1] Additional examples include technology to augment human ability, such as extraordinary optical devices, and include the use of astral science, demons, and natural philosophy (the text-based branch of knowledge relating to the natural world and the laws that governed it) to discern the future, apprehend distant events, and uncover hidden information. These examples, despite differences in causation and operation, exhibit close links within medieval texts: across genres, including romance, legend, history, and philosophical treatise, magic mirrors appear alongside prophetic metal statues, and the same men who cast horoscopes and divined their meaning were also credited with making sentient automata.

The striking cultural coherence of what initially appears to be a disparate group of objects and methods reveals the differences between how terms like 'artificial' and 'intelligence' were understood in the medieval Latin West and how they are understood today. The contemporary definition of 'AI' fails to capture the full range of artificial intelligence in the medieval period, which encompasses a spectrum from augmented human intelligence to sentient machines. Instead, applying a current understanding of AI imposes a partial perspective on the past and flattens the nuances and complexities of a period of time encompassing a millennium and extending over north-western Eurasia and the Mediterranean region.[2] At the same time, there is no word or even set of words used in the medieval period to describe this category of objects. But because the examples coalesce in the textual record, it makes sense to consider them as part of the same broad category. Although it is a departure from actors' categories to call them 'artificial intelligence', this anachronistic nomenclature signals continuity with contemporary AI, especially in the areas of surveillance and the exercise of power and maintenance of control, and constitutes this heterogeneous group of examples into a cohesive historical subject. 'Artificial' and its cognates in the medieval period appear embedded in a semantic field that includes something that is made (artificial) by a person (an artisan), as well as something that is fraudulent or inherently deceptive (artifice), and refer to a branch of knowledge, such as the liberal arts.[3] Although there was no single definition of 'intelligence' in the medieval period, knowledge gained through sense experience was generally seen as vital to learning and cognition: medieval philosophers grappled with Aristotelian texts and commentaries that framed sensory knowledge as necessary to rationality, despite being a method of knowledge-acquisition shared with animals (Crisciani 2014; Marenbon 2007; Pasnau 1997). Therefore, instruments—both real and fantastical—that could augment human capability in vision or hearing, for example, are at one end of the spectrum of artificial intelligence in the medieval period. At the other end of the spectrum we find sentient machines: artificial objects that demonstrate understanding when interacting with humans and their environment. This notion of understanding finds echoes elsewhere, as 'intelligence' in this period can also refer to an agreement between two people. Furthermore, 'intelligence' could also refer to demons or other noncorporeal spirits, and these disembodied intelligences were believed to have hidden knowledge or information about the future, and could be persuaded, compelled, or tricked into providing it (Kieckhefer 1997 [1989], pp. 151–75). Thus, objects that appear as arbiters of truth and falsehood, or that can reveal future events—including the examples of prophetic statues and talking heads discussed by Kevin LaGrandeur, Minsoo Kang and Ben Halliburton in this volume—must also be considered on the spectrum of artificial intelligence in this period.

One definition of 'intelligence' that does not appear in the medieval period, in Latin or in vernacular languages, is the sense of acquiring information through espionage or other covert means. The link between intelligence and information, already apparent in the medieval sense that an intelligence (a demon) might convey secret information to a person, grew more pronounced in the sixteenth century as 'intelligence' took on the meaning of information or news, and especially information gathered through espionage to gain an advantage in politics, warfare, or trade. However, this definition of 'intelligence' inheres in several of the examples of medieval artificial intelligence. In addition to creating oracular artificial intelligence or using astral prediction to get information about the future, many of the examples related to augmented intelligence—such as using a magic mirror, or a practice called far-casting to view events in distant places—involve gaining information covertly in order to gain political or personal advantage.

These examples and others I consider appear in several different kinds of sources, including historical chronicle, poetry, philosophical and scientific treatise, as well as nonnarrative textual sources, like horoscopes and charms, and visual sources. The wide array of sources from the Latin Middle Ages in which artificial intelligence appears is evidence of the breadth of interest in the subject. Many of the examples include aspects that one might now find fantastical or impossible. However, medieval writers configured the categories of 'factual' and 'fictional' differently from how they are configured today. The variety of different possible routes to artificial intelligence, from summoning demons to carefully observing the stars to making new kinds of machines, reveals a more generous understanding of causality: all these were equally plausible because some phenomena could have multiple possible causes. Just because natural phenomena were often regular, predictable, and explicable did not mean that was always the case. According to medieval Latin natural philosophy, singular events or unusual natural phenomena, like comets or strange births, fell outside the ordinary, regular course of nature. Such phenomena were considered *praeter naturalis* or 'beyond nature'. The causes of preternatural events could be natural: many philosophers in this period conceived of Nature as a powerful force that, while ultimately subordinate to God, had enormous capability for variation and surprise (Park 1999). One clear example of the preternatural from this period is magnetism, an inexplicable yet undeniable force. Preternatural phenomena might also have demons as their cause, as it was understood that demons had superhuman abilities and intelligence that could create stunning effects. Astral science, including prediction through interpreting the positions and movement of celestial bodies, dealt with natural causes; in this case, the matter of the cosmos and laws that governed it (Daston 1991, pp. 96–98). Ultimately, however, preternatural and natural phenomena were understood to have, at their foundation, the same primary cause: God. Therefore, preternatural and natural events might differ in their likelihood, but not in their inherent plausibility (Truitt 2015, p. 6). This means that the ability to spy on people using an optical device or far-casting (visualizing distant events via a reflective surface) were taken to be equally plausible.

Although there are different possible paths to artificial intelligence in this period, it appears in a fairly limited number of contexts. First, it could be used to uncover deception or corroborate the truth. Second, artificial intelligence in the form of armed sentries appears in a number of different texts in this period. More broadly, it also often appears in contexts of espionage and surveillance. Finally, artificial intelligence promised a route to foreknowledge. These contexts all have in common its use to attain and exert power, found across the spectrum of examples. When artificial intelligence appears in texts from the Latin Christian West, it is often used to maintain class and gender relations, to gain and assert military or political dominance, or to acquire knowledge that confers an advantage.

## 2.2 Vigilant Sentries and Arbiters of Truth

Artificial intelligence appears in a variety of guises and multiple kinds of texts to maintain existing class and gender hierarchies. In a thirteenth-century continuation of the late-twelfth-century (unfinished) epic poem *Perceval*, by Chrétien de Troyes, at a large tournament hosted by King Arthur, two marvellous artificial intelligences guard the entrance to a knight's pavilion. One servant was made of silver, and one of gold, and one held a spear, and the other held a harp. Their job was to screen potential visitors to the knight Alardin's tent. If anyone not of noble lineage were to attempt entry, the figure with the spear would bar the way. And if a young woman who was not a virgin tried to enter the tent, the figure with the harp would sound a discordant note, and its strings would break (*Perceval* 1977, vol. 3, p. 149). Both of these figures are described as 'enchanted beings', and their job is to ascertain the hidden truth and to uphold notions of purity, either of blood or of virtue (Truitt 2005, p. 173). In a poem written for a courtly audience, they testify to the centrality of the aristocracy, and to the ideas that distinctions between social ranks are meaningful and that women's chastity is important. They also testify to anxieties of the aristocracy: the boundary between nobleman and churl required policing, lest some jumped-up peasant take on airs and authority above his station. An enchanted figure that could not be tricked or bribed could be trusted to police the boundaries of rank. The second artificial intelligence conveys, via musical note, to everyone in earshot whether or not the approaching woman is a virgin. In this example, the defining feature of a woman's identity is her virginity. Furthermore, that information does not belong to her alone but to the wider community.

An artificial intelligence that could provide certainty about a woman's sexual history found different expression in the form of the *bocca della verità*, the 'mouth of truth'. This marvel, attributed to Virgil, was able to determine if a woman was truly chaste. The poet Virgil, during the Latin Middle Ages, had a reputation as a learned man and a sorcerer, capable of creating all manner of marvels. This reputation had little to do with his poetry, and actually eclipsed his fame as a poet in this period (Spargo 1934; Wood 1983; Ziolkowski & Putnam 2008, p. 829). According to the legend, which appears in several different texts and languages from the fourteenth century onwards, Virgil installed a device with a mouth (a statue in some versions, a face set into a wall in others) into which a woman accused of adultery or fornication would place her hand.[4] If she was guilty of either the device would bite off her fingers, 'so that [she] must live from then on in disgrace' (Ziolkowski & Putnam 2008, p. 872; Spargo 1934, pp. 208–27). Both this example and the artificial harper in *Perceval* present moving statues made by magic that are capable of perception and discernment and of discerning a truth that is nonobvious and hidden from the human observer. Moreover, the public nature of these virginity and chastity tests underscores that monitoring female sexuality was of enormous importance to elites in this period.

Ideal servants, like Alardin's sentinels, first appear in the imaginative record of the Latin Christian West in the late eleventh century, and often as armed guards, indicating an enduring link between artificial intelligence and policing. Although actual moving statues, automata, were not manufactured in the Latin Christian West until the turn of the fourteenth century, they were known of there much earlier, initially through diplomatic and economic contact with the empires of Central Asia and North Africa (Byzantium and the ʻAbbasid Caliphate, especially) where moving statues (wine-servants, musicians, and animals) were built and installed in courtly and public spaces. These objects took hold in the Latin Christian imagination and appear regularly in texts written in Latin and romance languages from the late eleventh century onward. In a number of these texts, automata appear as sentries, guarding bridges, castles, or tombs, similar to the figure of Talos from Greek myth. In part, this draws attention to their status as liminal figures that are lifelike, but not alive (Truitt 2005). Yet it also speaks to the desire for infallible, untiring, loyal armed retainers to maintain safety and order. Though their function is deadly serious, they are dazzling objects, like the golden archer in the Old French version of the Aeneid, *Le Roman d'Éneas* (ca. 1160), that guards the tomb of Camille, the warrior queen of the Amazons (*Éneas* 1891). In some of these examples, these artificial guardians are vanquished by the hero. For example, in the Old French verse history of Alexander the Great, *Le Roman d'Alexandre* (ca. 1180), Alexander, thanks to the help of one of his Persian advisors and guides, defeats two enchanted golden sentries—one tall and thin, the other short and fat, and each armed with a large hammer—that guard a bridge into a magic forest (*Alexandre* 1994, pp. 506–08). And in the Old French prose version of the story of Lancelot, *Lancelot do lac* (ca. 1220), Lancelot's quest to find the Grail leads him to a castle guarded by copper knights that have been created via demonic magic. Once Lancelot defeats the last one, the enchantment on Dolereuse Garde ('Fortress of Misery') is broken, and the castle changes to Joyeuse

Garde ('Joyful Keep') (*Lancelot* 1980, vol. 1, p. 283; Truitt 2015, p. 59). In these examples, and similar ones, the artificial figures can be defeated by the hero *because* he is the hero—one indication of his prowess as a warrior is his ability to subdue the mechanical guardians; the implication is that any other challenger would be defeated. And in all of these instances, the artificial knights, archers, or pikemen are seen as better than their human counterparts. Unlike human sentries, the artificial ones will never get tired or ill, lazy or lecherous, and, like the sentries guarding Alardin's tent, they cannot be tricked or bribed.

The repeated motif of the armed robot guard suggests concern about security and about loyalty, trust, and control. This concern extended beyond keeping a castle or mausoleum or bridge secure to keeping tabs on an entire empire. 'The Salvation of Rome' (*Salvatio Romae*) was an elaborate surveillance and alarm system that could perceive, assess, and react to events taking place far away. This device, attributed, like the *bocca della verità*, to Virgil, appears in several different texts from the late twelfth century onward. The most extensive description of this marvel appears in the natural philosophical treatise *On the Natures of Things* (*De rerum natura*, ca. 1190) by the English scholar and courtier Alexander Neckam (1157–1217); according to him, Virgil had built in Rome a palace which contained a number of wooden statues, each representing a different province of the empire, and each holding a bell in its hand.[5] 'Whenever any province dared to foment a plot against the majesty of the Roman Empire, the image of the intemperate traitor began to bang on the little bell. A soldier of bronze, seated on a bronze horse, on the topmost gable of the aforementioned palace, turned itself in the very direction where it might look toward that province' (Ziolkowski & Putnam 2008, p. 856). Neckam's description appears in a chapter about the potential of the liberal arts and the places where those arts flourished, thereby tying Virgil's invention to his mastery of the liberal arts rather than forbidden branches of study, such as demonic magic. Furthermore, the function of the *Salvatio Romae*, to safeguard the empire from revolt, recalls Virgil's epic poem about the foundation of the Roman Empire, the *Aeneid*. Although Virgil may have been known more for his reputation as a magician and learned man in the Latin Christian West than for writing the *Aeneid*, he remained linked to the Roman Empire and was seen as having some kind of unofficial yet critically important oversight of it. Moreover, Neckam was writing during a period when the ruling dynasty of England, the Plantagenets, controlled an empire that included England, Ireland, and most of what is now France. Henry II (1133–1189; r. 1154–1189) faced numerous rebellions—from castellans in his domains, from his Archbishop of Canterbury, Thomas Becket, and from his wife and sons. Henry's son and successor, Richard I (1157–1199; r. 1189–1199), likewise spent a significant amount of time responding to challenges to his authority and rule. The invention of the *Salvatio Romae* mirrors the practical concerns of maintaining control over a large empire and the need to have reliable, timely information about threats to its stability.

## 2.3 Espionage and Surveillance

Artificial intelligence appears in multiple guises in textual and visual sources to provide surveillance and foster espionage, indicating a long-standing desire to use artificial means to gather information to gain advantage. The *Salvatio Romae* is one example, but there are many others. Catroptomancy (using a reflective surface, such as a bowl of water or a polished metal mirror, as a viewfinder) was one way to see events distant in time and place. All of the mantic arts (the arts of divination)—chiromancy (palm reading), geomancy (interpreting earth signs), pyromancy (fire signs), and scapulomancy (interpreting the marks on the shoulder blades of animals, often sheep or goats)—promised foreknowledge, but catroptomancy also promised knowledge of the distant now. In multiple versions of the history of Alexander the Great (widely translated and circulated in this period),

Nectanebus, mage and ruler of Egypt, was the true father of Alexander, not Phillip of Macedonia. According to multiple texts, Nectanebus had used magic to ascertain events in Persia and Macedonia, and to find the right time to rape and impregnate Olympias, Phillip's wife, thereby becoming the founder of a great dynasty (*Alexandre* 1994, p. 82; Thomas of Kent 2003, p. 8, Kelly 2002, p. 43, Stoneman 2008, p. 15). In most versions of this account, Nectanebus uses a bowl of water and a wand to scry. Yet in the illustration from an early fourteenth-century French prose history, the artist has depicted him with these (by his right hand) along with what looks like a spyglass held up to his eye, despite the fact that they are unattested to in the historical record until the early seventeenth century (Figure 2.1).

By depicting an optical device, the artist shows the reader that the purpose of the artificial intelligence is to help Nectanebus see what is happening in a different place.[6] Far-casting (using magic to see distant events) remained linked to the strange and fantastical magical practices of people who lived far to the east of London or Reims at least into the fifteenth century. Two texts from the fourteenth century that circulated widely in the centuries following, *The Travels of Sir John Mandeville* (ca. 1357; This text is also sometimes known as *Mandeville's Travels*), and the *Canterbury Tales* (ca. 1387–1397), both mention far-casting in connection with the Mongols. The narrator of *The Travels of Sir John Mandeville* asserts that in far-casting [*fercastynge*], the Mongols 'surpass all men under Heaven' (*Mandeville's Travels* 1967, p. 157; Truitt 2015, pp. 36–38), while in one of the *Canterbury Tales*, 'The Squire's Tale', a mysterious delegate from the ruler of India brings a magic mirror as a birthday gift to the Mongol ruler Cambyuskan, also known as Chinggis (Genghis) Khan. (Chaucer 1987; pp. 171–72; Lightsey 2001).

Though far-casting remained linked to distant lands, especially those to the east, it was not exclusive to that part of the world, and it appears often enough to suggest a persistent fascination with the possibility of overcoming the limitations of human

**Figure 2.1** Nectanebus, on top of his palace in Egypt, uses a wand, bowl, and spyglass to determine events taking place in Macedonia. MS 78 C 1, fol. 5v, Kupferstichkabinett, Staatliche Museen; bpk Bildagentur/Dietmar Katz/Art Resource, NY

vision, especially to gain knowledge covertly. An early twelfth-century Anglo-Norman compendium of medical recipes includes one to discover what a friend is doing; the recipe includes a cipher and the use of a talisman, and must be undertaken at a particular time to be effective.[7] This recipe was included with others for medications, as well as a uroscopy treatise and pseudo-Galenic texts, because prognostication was (and remains) a vitally important element of medical practice.

Just over a century later, the engagement of Latin Christian scholars with Arabic work on optics and astral science (in Latin translation) led some to argue that optics made it possible—at least theoretically—to see things both small and far away, and gave rise to fantastical, almost wistful accounts of optical devices that are at one end of the spectrum of artificial intelligence in this period. Scholastic philosopher and Franciscan Roger Bacon (ca. 1214–1292) concluded that it was possible to make amazing optical devices that could reveal things hidden from view either by distance or size, using only the laws of nature and its inherent capabilities, rather than demons or trickery. 'For glasses can be so shaped that things placed at the greatest distances appear most near and conversely; so that we could read the most minute letters at an incredible distance, count things however small' (Bacon 1859, p. 534). Other lenses or mirrors could be shaped and placed 'so that hidden things appear evident' (Bacon 1859, p. 535). Bacon came to these conclusions after he spent decades studying Latin translations of Arabic works on optical devices and principles, read widely in the ancient tradition, and sought out natural knowledge that could be gained only through experience. He articulated these speculative devices (along with many others) in his works for courtly patrons, Pope Clement IV and (likely) Henry III of England, and emphasized the utility of these devices for the defence of Christendom and England alongside their utility to the cause of science. Gathering information, whether about enemy encampments or the movement of the planets, required artificial devices that could augment human intelligence. Additionally, relying on a person to act as scout or spy carried with it risks of exposing valuable personnel to capture and counter-measures; these risks disappear with the use of a spyglass or other optical device.

Bacon was known in later centuries, in part, for his writing on optics and on artificial marvels, and gained his own reputation as a sorcerer capable of creating and using artificial intelligence. In addition to the oracular head for which he was known in the early modern period (LaGrandeur 2013, and this volume), he was also credited with making two magical mirrors. These appear in a history of the Franciscan Order, written by Peter of Trau, one of Bacon's brethren, about a hundred years after his death. According to Peter's account, one mirror was used to light candles, and the other showed 'what people were doing in any part of the world'. This mirror proved so distracting to his students at Oxford—they were driven to abandon their studies after seeing their loved ones in difficulty back home—that the university authorities insisted it be destroyed (Molland 1974).[8]

Ascertaining if a friend or family member is in distress or otherwise needs assistance is a benevolent end to which surveillance might be turned. Artificial intelligence for surveillance and to exert social control and enforce normative standards of behaviour, dress, and even thought appears in one particularly stunning example in a mid-twelfth-century Old French verse account of the history of the Trojan War, *The Romance of Troy* (*Le Roman de Troie* ca. 1165). Courtly romances, like this one, were written for elite secular audiences and often took some kind of historical matter as their subject (Spiegel 1993, pp. 61–62; Kelly 1974; Petit 2004). This text was written for French-speaking courtly elites, members of a warrior-aristocracy, and the refined Trojan court is presented as an ideal mirror to the courtly society that was its audience (Benson 2007, p. 1337; Blumenfeld-Kosinski 1997). In *The Romance of Troy*, the centre of the Trojan court, 'the Alabaster Chamber', had in it golden automata, installed to ensure that the moral, aesthetic, and behavioural codes of courtly conduct were adhered to by the inhabitants (Sullivan 1985; Truitt 2005, pp. 173–75).[9] The golden figures (two young men and two young women, along with an eagle and satyr that are ancillary to our discussion here) were created by three learned men ('sages dotors') who were also poets, and well-versed in necromancy ('nigromance') (Sainte-Maure 1904–1912, vol. 2, p. 377). These three men are not artisans or metal-workers; their labour is of a different cast. 'Dotors' refers to a text-based education in the liberal arts, natural philosophy, or medicine; these were all courses of study that would confer knowledge in astral science, languages, and properties of natural things. Poets, like enchanters, cast a spell on their audience through the power of the human voice and language. And 'nigromance' is, in this period, a term that denotes text-based magic, whether demonic, celestial, or natural magic (Kieckhefer 1997, pp. 8–9; Weill-Parot 2002; Zambelli 1992). Even before the audience knows anything about what these figures do, it knows that these objects are the result of intellectual craft, rather than artisanal labour (Truitt 2015, pp. 55–61). One of the young female figures holds up a magic mirror, in which inhabitants of the Chamber could see the way their dress and appearance conformed to courtly standards. Another, a juggler, tumbler, and conjurer, engages in what is now called 'attention capture': its function is to keep people in the Chamber so distracted and entertained that they could not leave before it would be polite or appropriate to do so. 'It is difficult for anyone to leave the Chamber while the image is conjuring tricks, as it stands on the pillar' (Sainte-Maure 1904–1912, vol. 2, p. 382). One of the two male figures plays music to put people in the appropriately courtly frame of mind and to provide cover for private conversations. '[None] could hear the music and remain in low spirits or feel pain. People in the Alabaster Chamber are not gripped by foolish ideas, evil thoughts, or ridiculous desires. The music benefits the listeners, for they can talk aloud, and none can overhear' (Sainte-Maure 1904–1912, vol. 2, p. 384). The fourth figure watches the people in the Chamber and conveys to each, via secret hand signals, how one's behaviour is or is not appropriately refined.

> It would watch over the Chamber and . . . demonstrate to each person what he should do and the necessary action: it would make these things known without other people perceiving it . . . None could be in the Chamber any longer than he ought to be; the figure would clearly show when it was time to leave, and when it was too soon, and when it was too late; it would often take note of this. It prevented on the part of all in the Chamber, and

> all who entered or left it, behavior that was boorish, uncouth, or rash. It was not possible to be foolish or rude or irresponsible because the figure skillfully guarded all against uncourtly behavior.
> (Sainte-Maure 1904–1912, vol. 2, pp. 388–390)

The emphasis in this section on surveillance and discipline of courtly elites that encompassed appearance, deportment, and regulation of emotion and thought was intended to resonate with the courtly audience. The narrator of the text asserts that the people in the Alabaster Chamber were more confident and less anxious about their deportment and appearance thanks to the golden figures: 'People were hardly ever mocked or accused of behaving boorishly' (Sainte-Maure 1904–1912, vol. 2, p. 392; Truitt 2005, p. 173). The court was the centre of government, finance, and military planning; it was a site of epistemic and cultural production; and it was also a domestic space, where the ruler and his family lived. Skilful courtiers needed to abide by the social norms and be adept at navigating the many tacit and explicit rules that governed courtly life. The automata in the Alabaster Chamber dramatically illustrate the presence of behavioural norms in courtly culture, and the importance of those norms as a form of social control. Moreover, these automata combine the function of surveillance with the enforcement of social norms, along with attention capture and entertainment. Imagine a virtual assistant that could play music or audio books, and stream your favourite films to watch whenever you want, but it would also force you to change your behaviour or clothing before going out in public—perhaps by refusing to unlock the door to your home—and would report your thought crimes, ethical lapses, uncharitable impulses, and poor wardrobe choices to the government authorities.[10]

These artificial intelligences are superior to their human counterparts. Just as artificial intelligence that might provide hidden or out-of-reach information was seen as preferable to using human scouts or spies, and the *bocca della veritá* was a far more reliable way of getting the truth than taking a woman at her word, the four golden automata are better than the humans that might perform their tasks. The acrobat never makes a mistake; the musician never plays a wrong note or forgets the tune; and the figures that watch the inhabitants for lapses in dress, speech, or manner cannot be tricked or evaded. Similar to the *Salvatio Romae* or the figures outside Alardin's tent, these golden AIs are the ideal servants. They serve the rulers of Troy and surveil the court unerringly.

The figures in the Alabaster Chamber are found at the centre of the Trojan court, and many of the other examples noted here also appear in a courtly context. Bacon's speculative inventions appear in treatises written for courtly patrons; the artificial intelligences that guard tombs, like in *Éneas*, do so within courtly contexts, in texts written for courtly audiences. In part, this reflects the status of the court—secular and ecclesiastical—as an important institution in this period. It also reflects that in the Latin Middle Ages the court fostered the creation of scientific knowledge. The courtly sciences were those that offered the possibility of intervention in the natural order to improve human life and to exercise power, including astral science, engineering, and the mantic arts. A formal text-based education in the liberal arts, natural philosophy, or medicine was not necessary to excel in some aspect of the courtly sciences: for example, the people in charge of horse and dog breeding, or the artillery masters, would not need such an education. But, as was the case with Roger Bacon, sometimes those who received a thorough grounding in the university curriculum of natural philosophy, with its emphasis on theory over practice, combined their text-based erudition with practical expertise in the service of political and religious authority. Regardless of one's educational background, the courtly sciences privilege utility. The display of scientific or technological virtuosity was also part of courtly life in this period, and knowledge was often staged as spectacle (Williams 2008). But the court was also a site associated with artificial intelligence in this period because, like now, artificial intelligence was seen as a way to maintain, exercise, or consolidate authority, and, in the Latin Christian West, the court was a locus of power.

## 2.4 Prediction and Foreknowledge

There is one subset of medieval artificial intelligence that, although associated with powerful figures, was for private use. I refer, of course, to the oracular head or prophetic statue. These legends (of creating either a talking head or a talking statue) all refer to men with titanic reputations for erudition, especially in natural philosophy: Gerbert of Aurillac (ca. 950–1003), a tenth-century polymath who was counsellor to kings and emperors before becoming Pope Sylvester II (999–1003); Robert Grosseteste (ca. 1175–1253), a Franciscan and eventual Bishop of Lincoln who was renowned for his work on optical theory and astral science; Albertus Magnus (1200–1280), the Dominican friar and university lecturer whose commentaries on Aristotelian natural philosophy helped to bring Greek and Arabic philosophy in line with Christian dogma (see also Kang & Halliburton in this volume); and Roger Bacon, introduced above. In some versions, the philosopher summons a demon to inhabit the prophetic statue, while in others he studies the positions of the stars (Truitt 2015; Truitt 2012; LaGrandeur 2013; Paravincini-Bagliani 2013; Oldoni 1983). In all of these legends, the querent is ultimately thwarted in his goal of acquiring foreknowledge: he misinterprets the prophecy (Gerbert), sleeps through the moment of revelation (Grosseteste, Bacon), or sees the fruits of his labours destroyed by the ignorant (Albertus). These legends are further united by the fact that in every example, the querent is interested in gaining knowledge of the future, usually in the service of accruing or exercising power. Self-knowledge is not the goal, nor is understanding of another individual's subjectivity. These talking heads are externalizations of the desire for foreknowledge; this same desire is likewise apparent, at the other end of the spectrum of artificial intelligence, in astral prediction or the mantic arts.

## 2.5 Conclusion

In all of these examples drawn from across the Latin Middle Ages, artificial intelligence embodies tantalizing, sometimes terrifying, possibility. Artificial intelligence was used to maintain social boundaries and uphold social norms, acquire or consolidate power over others, and attain knowledge otherwise unavailable to human understanding. Exploring medieval artificial intelligence offers a historical perspective, revealing several themes that endure from the Latin Middle Ages into our own era. The close association between artificial intelligence, power, and control is one such area of continuity, as are the desires for perfectly loyal, unerring servants, and for reliable information about the future. Aspects of constancy offer space to reflect on the persistence of the past, enduring ideas about power, and present circumstances. The desire to use artificial intelligence to spy, to police boundaries and populations, and to act as slaves in the service of power appears equally in the medieval period and in the now, so what does that tell us about now? If artificial intelligence is so strongly linked to control and authority over such a long period, what would it take to imagine AI that works to liberate the many, rather than consolidate the power of the few? Yet the areas of discontinuity are equally evocative. The wide spectrum of artificial intelligence represented in sources from the Latin Christian West, from augmented intelligence to sentient machines, arises from ideas about intelligence, technology, and the causality of natural phenomena that differ significantly from those held by many people today. These differences, as well as the wide, strange array of examples, offer the possibility of additional contemporary narratives about AI. A spectrum of AI, rather than a fairly well-defined category, might yield new insights or new possibilities for its use. Including augmented intelligence as part of AI could challenge our humanity, and humanize AI at the same time, if they are constituted as part of one spectrum. The medieval examples here are only a fraction of those available in other languages and cultural traditions from the same period, yet they provide a compelling account of persistence and transformation in imaginative narratives of artificial intelligence between the medieval period and now.

## Notes

1. I distinguish between contemporary AI and medieval examples of artificial intelligence by referring to the latter as 'artificial intelligence' and the former as 'AI'. The term 'AI' itself has not been stable since its introduction in the 1950s, and it tends to be used to describe technology or capability at the leading edge of possibility (such as autonomous vehicles) and does not generally encompass examples of machine learning (such as OCR) that have since become commonplace.
2. 'The Middle Ages' is an historiographic concept that was first invented in the sixteenth century (though Petrarch first spoke of living in a 'middle age' in the early fourteenth century), and it has been reinvented in every century since then. One well-established (though arbitrary) periodization of 'medieval' is from the deposition of the last Roman emperor in the western half of the empire (476) until the deposition of the last Roman emperor in the eastern half of the empire (1453/1461). This essay takes up the areas in which Latin was the main liturgical and political language, between roughly 750–1450. However, this focus should not be taken to convey that material from elsewhere in the world in this same period either does not exist, or would not be equally rich in terms of what it might yield to our historical understanding of artificial intelligence (for example, Saif 2017; Saif 2015; Magdalino & Mavroudi 2006).
3. For historical definitions of the terms 'artificial' and intelligence' (see, for example, *OED* 2010; Godefroy 1982; Ménage 1973; Du Cange 1937; *Anglo-Norman Dictionary* 2008–2012; Cortelazzo & Zolli 1988, all s.v.v. 'ars' or 'art' and 'intelligence' or 'intelligentia').
4. One example can be found in Rome, next to the church of Santa Maria in Cosmedin.
5. The *Salvatio Romae* appeared in several other texts over the following centuries, including a Latin encyclopedia, *The Book of Etymologies* (*Liber derivationum* ca. 1210), an influential collection of legends in Latin, *The Deeds of the Romans* (*Gesta Romanorum* ca. 1300), and a work of history

written in French, *The Mirror of History* (*Myreur des histors* ca. 1350) (Cilento 1983; Spargo 1934, pp.117–35).

6. Eyeglasses were first invented in the Latin Christian West in the thirteenth century, whereas the telescope does not appear until the start of the seventeenth century. However, there is no technological reason why a single-lens telescope could not have been invented before 1600, as there is ample evidence that celestial observers used viewing tubes (without lenses) to look at the stars at least as early as the tenth century (see Worthen 2006; Poulle 1985; Dalché 2013).
7. Sloane MS 475, London, British Library, fol. 111v–112r.
8. The account appears in a chronicle by Peter of Trau, (1384–1385), MS Oxford, Bodl. Canon. Misc. 525, fol. 202v–203v.
9. 'Romance/*Roman*' here refers to the language of the text (written not in Latin, but in vernacular, or romance), as well as to certain generic distinctions that were just coalescing in the mid-twelfth century but that have little to do with contemporary understanding of 'romance' as a genre (see Kelly 1992; Blumenfeld-Kosinski 1980).
10. Perhaps one day.

## Reference List

Alexander de Paris (1994) *Le Roman d'Alexandre.* Edited and translated by Laurence Harf-Lancner. Paris, Librarie Générale Française.

*Anglo-Norman Dictionary* (2008–2012) AND[2] Online edition. Available from: http://www.anglo-norman.net/D/intelligence [Accessed 13 May 2019].

Bacon, R. (1859) *Epistola de secretis operibus artis et naturae et de nullitate magiae.* In: J. S. Brewer (ed.) *Fr. Rogeri Bacon Opera Quaedam Hactenus Inedita.* London. pp. 523–51.

Benson, C. D. (2007) The 'Matter of Troy' and its transmission in Medieval Europe. In: H. Kittle, J. House, & B. Schulze (eds.) *Traduction: Encyclopédie internationale de la recherché sur la traduction.* Berlin, DeGruyter. pp. 1337–40.

Blumenfeld-Kosinski, R. (1997) *Reading myth: classical mythology and its interpretation in medieval French literature.* Palo Alto, CA, Stanford University Press.

Blumenfeld-Kosinski, R. (1980) Old French narrative genres towards a definition of the *Roman Antique. Romance Philology.* 34, 143–59.

Chaucer, G. (1987) *The Canterbury tales.* In: Larry Benson (ed.) *Riverside Chaucer*, 3rd edition. Boston, Houghton Mifflin.

Cilento, N. (1983) Sulla tradizione della Salvatio Romae. In: A. M. Romanini (ed.) *Roma anno 1300: atti della IV Settimana di studi di Storia dell'arte medieval dell'Università di Roma*. Rome, pp. 695–703.

Cortelazzo, M. & P. Zolli (eds.) (1988) *Dizionario etimologico della lingua Italiana*. French and European Publications.

Crisciani, C. (2014) Universal and particular in the *Communia naturalium*: between 'Extreme Realism' and 'Experientia'. In: P. Bernardini & A. Rodolfi (eds.) *Roger Bacon's Communia Naturalium: a 13th century philosopher's workshop*. Florence, SISMEL/Edizioni del Galluzzo. pp. 57–82.

Dalché, P. G. (2013) Le 'Tuyau' de Gerbert, ou la légende savant del'astronomie: origines, thèmes, échos contemporains (avec un appendice critique). *Micrologus*. 21, 243–76.

Daston, L. (1991) Marvelous facts and miraculous evidence in early modern Europe. *Critical Inquiry*. 18, 93–124.

Du Cange, C. (ed.) (1937–1938) *Glossarum mediae et infimae latinitatis*. Paris, Librarie des Sciences et des Arts.

*Éneas* (1891) Edited by Jean-Jacques Salverda de Grave. Paris, Champion.

Godefroy, F. (ed.) (1982 [1880–1885]). *Dictionnaire de l'ancienne langue française et tous ses dialects du IXe au Xve siècle*. Geneva, Slatkine.

Kelly, D. (2002) Alexander's *Clergie*. In: D. Maddox & S. Sturm-Maddox (eds.) *The medieval French Alexander*. Albany, State University of New York Press. pp. 39–56.

Kelly, D. (1974) *Matière* and *genera dicendi* in Medieval Romance. *Yale French Studies*. 51, 147–59.

Kelly, D. (1992) *The art of medieval French romance*. Madison, University of Wisconsin Press.

Kieckhefer, R. (1997 [1989]) *Magic in the Middle Ages*. Cambridge, Cambridge University Press.

LaGrandeur, K. (2013) *Artificial slaves: androids and intelligent networks in early modern literature and culture*. New York, Routledge.

*Lancelot do lac: the non-Cyclic Old French prose romance* (1980) Edited by Elspeth Kennedy. 2 vols. Oxford, Clarendon.

Lightsey, S. (2001) Chaucer's secular marvels and the medieval economy of wonder. *Studies in the Age of Chaucer*. 23, 289–316.

Magdalino, P. & M. Mavroudi (eds.) (2006) *The occult sciences in Byzantium*. Geneva, Pomme d'or.

*Mandeville's travels* (1967) Edited by M. C. Seymour. Oxford, Clarendon.

Marenbon, J. (2007) *Medieval philosophy: an historical and philosophical introduction*. London, Routledge.

Ménage, G. (ed.) (1973 [1650]). *Dictionnaire étymologique de la langue française.* Geneva, Switzerland, Slatkine.

Molland, A. G. (1974) Roger Bacon as magician. *Traditio.* 30, 445–60.

Oldoni, M. (1983) 'A fantasia dicitur fantasma.' Gerberto e la sua storia. II. *Studi Medievali.* 23, 167–245.

*Oxford English Dictionary (OED)* (2010) 3rd edition. Oxford: Oxford University Press.

Paravincini-Bagliani, A. (2013) La légende médiévale d'Albert le Grand (1270–1435): prèmieres recherches. *Micrologus.* 21, 295–368.

Park, K. (1999) Natural particulars: epistemology, practice, and the literature of healing springs. In: A. Grafton & N. G. Siraisi (eds.) *Natural particulars: nature and the disciplines in Renaissance Europe*. Cambridge, MA, MIT Press. pp. 347–67.

Pasnau, R. (1997) *Theories of cognition in the later Middle Ages.* Cambridge, University of Cambridge Press.

*Perceval le Gallois, ou le conte du Graal* (1866–1871) Edited by Charles Potvin, 6 vols. in 3. Paris, Champion.

Petit, A. 2004. *L'anachronisme dans les romans antiques du XIIème siècle: le Roman de Thèbes, Le Roman d'Éneas; Le Roman de Troie, et Le Roman d'Alexandre.* Paris, Champion.

Poulle, E. (1985) L'astronomie de Gerbert. In: M. Tosi. (ed.) *Gerberto: scienza, storie, e mito: atti del Gerberti Symposium, Bobbio 25–27 Iuglio, 1983, Archivium Bobiense II*. Piacenza, Italy, Archivi Storici Bobiense. pp. 597–617.

Saif, L. (2015) *The Arabic influences on early modern occult philosophy.* London and New York, Palgrave.

Saif, L. (2017) From Ġāyat al-ḥakīm to Šams al-maʿārif wa laṭāʾif al-ʿawārif: ways of knowing and paths of power. *Arabica.* 64, 297–345.

de Sainte-Maure, Benoît. (1904-1912) *Roman de Troie.* Edited by Léopold Constans. 6 vols. Paris: F. Didot.

Spargo, J. W. (1934) *Virgil the necromancer: studies in Virgilian legends.* Harvard Studies in Comparative Literature 10. Cambridge, MA, Harvard University Press.

Spiegel, G. (1993) *Romancing the past: the rise of vernacular prose historiography in Thirteenth Century France.* Berkeley, University of California Press.

Stoneman, R. (2008) *Alexander the Great: a life in legend.* New Haven, CT, Yale University Press.

Sullivan, P. (1985) Medieval automata: the 'Chambre de Beautés' in Benoît de Sainte-Maure's *Roman de Troie. Romance Studies.* 6, 1–20.

Thomas of Kent (2003) *Roman de Toute Chevalerie (RdTC).* Paris, Champion.

Truitt, E. R. (2005) 'Trei poëte, sages dotors, qui mout sorent di nigromance': knowledge and automata in twelfth century French Literature. *Configurations*. 12, 167–93.

Truitt, E. R. (2012) Celestial divination and Arabic science in twelfth-century England: the history of Gerbert of Aurillac's talking head. *Journal of the History of Ideas*. 73, 201–22.

Truitt, E. R. (2015) *Medieval robots: mechanism, magic, nature, and art*. Philadelphia, University of Pennsylvania Press.

Weill-Parot, N. (2002) *Les "images astronomiques" au moyen âge et à la renaissance*. Paris, Champion.

Williams, S. J. (2008) Public stage and private space: the Court as a venue for the discussion, display, and demonstration of science and technology in the later Middle Ages. *Micrologus*. 16, 459–86.

Wood, J. (1983) Virgil and Taliesin: the concept of the magician in medieval folklore. *Folklore*. 94, 91–104.

Worthen, S. (2006) *The memory of medieval inventions, 1200–1600: windmills, spectacles, sandglasses, and mechanical clocks*, PhD dissertation. Toronto, University of Toronto.

Zambelli, P. (1992) *The* Speculum Astronomiae *and its enigma: astrology, theology, and science in Albertus Magnus and his contemporaries*. Dordrecht, Kluwer.

Ziolkowski, J. M. & M. J. Putnam (eds.) (2008) *The Virgilian tradition: the first fifteen hundred years*. New Haven, CT, Yale University Press.

# 3

# The Android of Albertus Magnus

## *A Legend of Artificial Being*

*Minsoo Kang and Ben Halliburton*

## 3.1 Introduction

In descriptions of the current capacity and the future possibilities of AI, the awe and wonder that are elicited are often accompanied by anxiety and terror. That is even more common in science fiction literature and film where AI is constantly depicted as a dangerous entity that willfully commits nefarious acts ranging from the violation of people's privacy and the manipulation of politics and commerce, to the takeover of the world and the destruction of humanity. That theme can be analysed in terms of current concerns about new technology, but it can also be understood historically if we adopt a broad definition of AI as any artificial construct that possesses certain capacities of the human mind.

In this chapter, we will examine a premodern instance of it in the form of a magical device that was reputed to have been created by the medieval philosopher Albertus Magnus (ca. 1200–1280). An artificial construct in the shape of a human body or a head that is animated for the purpose of divination, it is described as capable of speech and the use of reason. The stories involving Albertus have been retold many times, from the first version in a 1373 work entitled *Rosario della vita* to the early modern era. The fantastic object has also been associated with such intellectual figures as Gerbert of Aurillac (Pope Sylvester II, 945–1003), Robert

Grosseteste (ca. 1175–1253), and Roger Bacon (ca. 1214–1292). Recently, a number of scholars have revisited the narratives in studies of automata, robotics, AI, and posthumanity (Truitt 2015, pp.69–95; LaGrandeur 2013, pp.78–108; Kang 2011, 68–79; Borlick 2011, pp.129–44; Kavey 2007, pp.32–58; Lightsey 2007, p.35; Sawday 2007, pp.193–95; Marr 2004, pp.205–22). We will demonstrate that earliest narratives struggled with the question of whether Albertus was dealing in illicit knowledge in making the object, their authors ultimately defending his creation as a product of legitimate intellectual magic. When the story was retold from the sixteenth century onwards, new interpretations of the wonder alternatively secularized it as a purely mechanical device or demonized it as a work involving diabolical beings. We will provide explanations for such changing views of the legend in the historical and intellectual contexts of the late Middle Ages and the early modern period. The fascination with and anxious ambivalence towards the object also reveals our current attitude towards AI as originating from a long-standing and primordial feelings towards the artificial simulacrum of humanity.

## 3.2 The Oracular Head

The nature and legitimacy of magic was a major issue of contention among medieval theologians (Bailey 2017; Copenhaver 2015; Boudet 2006; Eamon 1994; Flint 1991; Kieckhefer 1989). Within the established Christian worldview, should all forms of magic be condemned as fraudulent or demonic/diabolical in nature? Or are there certain types of it that are legitimate because they come from angels or even God himself in the form of miracles? What about learned magic that does not deal with supernatural beings at all but involves the manipulation of hidden (occult) forces in nature?

The debate became particularly lively in the thirteenth century with the upsurge of interest in intellectual magic, especially in alchemy and astrology. The anxiety over the legitimacy of such knowledge, as well as admiration for its great practitioners, is evident in stories of the oracular devices made by Gerbert, Albertus, Grosseteste, and Bacon, who were all major scholars of learned magic. They were in possession of devices used for alchemical experiments and astronomical observations, which probably caused concern among those who were unfamiliar with them (Truitt 2015, pp.82–95; Kang 2011, pp.74–75).

Arthur Dickson, who has pioneered the study of the legendary objects, pointed to their sources in ancient stories of talking statues as well as Jewish-Arabic legends of a severed head that was animated for oracular purposes (Dickson 1929, pp.191–93, 201–05). Another source that is often referred to in commentaries on the oracular head/statue is the Greco-Egyptian religious work *Asclepius* (second century CE), which describes the process of endowing statues with spirits. The divine prophet Hermes Trismegistus describes how the ancients used astrology to entice supernatural beings into artificial images to speak prophecy, cure the ill, and bring fortune to the good while punishing the evil (*Hermetica* 1992, p.81). This passage became well known in the Middle Ages as Augustine of Hippo's condemnation of the so-called god-making practice of pagans became one of the most widely quoted passages in writings condemning idolatry (St. Augustine 1984, pp.331–32).

The oracular image represented both the wonder that could be achieved through the use of arcane knowledge and the danger of a practice associated with pagan idolatry. As a result, the ancient object appears in connection with intellectuals who were scholars of magic, reflecting the prevalent uncertainty about the legitimacy of their work. The stories feature contradictory interpretations of the marvel, some seeing intimations of the diabolical, and others justifying it on the basis of the wisdom of their creators. For example, in William of Malmesbury's account of Gerbert's magic, in *The Chronicles of the Kings of England* (1125), the head that prophesizes his death is made through the use of astrology, reminiscent of god making in the *Asclepius* (1847, p.181).

In an earlier narrative of his deeds, however, Gerbert receives the prophecy from a demon he summons (Truitt 2015, p.72). Both accounts portray Gerbert as a perfidious figure who deals with sorcerers and the Devil. Consequently, even though the speaking head is described as a product of learned magic, it is endowed with the air of the illicit. In contrast, the earliest stories involving Albertus Magnus defend his creation as legitimate intellectual work that is misunderstood by the ignorant. Yet the insistent nature of the justification, as well the story element of the device being destroyed, addresses the anxiety aroused by the wonder.

## 3.3 Medieval Origins

The first Albertus story comes from *Rosaio della vita*, a moral treatise that appeared in 1373 that has been attributed on an uncertain basis to the Florentine merchant Matteo Corsini (1322–1402). It is a vernacular work written for a general readership that illustrates various classical and Christian virtues through narrative examples. In a chapter on the subject of *sapientia* (wisdom), a marvelous tale is presented as an illustration of the consequence of the lack of respect for wisdom on the part of the ignorant.

> We find that Albertus Magnus, of the Preaching Friars, had such a great mind that he was able to make a metal statue modeled after the course of the planets, and endowed with such a capacity for reason that it spoke: and it was not from a diabolical art or necromancy – great intellects do not delight in such things because it is something that makes one lose his soul and body; such arts are forbidden by the faith of Christ. One day a monk went to find Albertus in his cell. As Albertus was not there, the statue replied. The monk, thinking that it was an idol of evil invention, broke it. When Albertus returned, he was very angry, telling the monk that it had taken him thirty years to make this piece and "that I did not gain this knowledge in the Order of the Friars." The monk replied, "I have done wrong; please forgive me. What, can't you make another one?" Albertus responded that it would be thirty thousand more years before another could be

> made, as that planet had made its course and it would not return before that time. (Corsini 1373 [1845], pp.15–16)

The device is described not only as an artificial construct but one that is capable of reason and conversation, so a kind of medieval AI. And despite the resemblance of its construction to 'god making' in the *Asclepius*, the narrator insists that Albertus did not engage in 'diabolical art'. In fact, the monk who destroys the wonder is castigated for his ignorant misinterpretation of the object as 'an idol of evil invention'. The narrator is aware, however, that any such work involving arcane knowledge could arouse superstitious horror, leading to the destruction of an invaluable achievement of great knowledge. The story, therefore, simultaneously points to fascination with learned magic and concern about its legitimacy.

While this story is the first involving Albertus Magnus, it was not the most influential version. Arthur Dickson points to a fifteenth-century commentary on the biblical book of Numbers by Alonso Tostado, though he admits that he was not able to consult the latter work directly (Dickson 1929, p.210, n.132). Because of the rarity of the text, modern scholars have not analysed the relevant passage in detail.[1] Tostado, born Alonso Fernández de Madrigal (ca. 1400–1455) in Castile, was one of the most prominent theologians of Spain in his time, known for the prolific nature of his work.

In his commentary on the Numbers, Tostado discusses the passage in which Moses constructs a 'brazen serpent' (1528, ch.21, 4–9) that heals the snakebites of those who gaze upon it. In his meditation on the wonder, Tostado ponders the possibility of endowing inanimate objects with magical power. In a rather startling leap of imagination, he starts with what he considers to be the most magical of effects, the residence of a rational soul within an artificial device. As he poses the question, 'Can it be that some rational virtue, responding to questions, might be able to be caused naturally and astrologically?', which reminds him of two legendary objects, 'the metal head of Spain and another of Albertus Magnus' (Toastati 1528, 15a–15b).

> By the grace of God, it is evident that rational virtue in metal heads – which, constructed by ancient astronomers, respond to all questions – can be caused by astrological observation. In fact, the German Albertus Magnus is said to have constructed a metal head thus. That man was the teacher of the blessed Thomas Aquinas a long time ago and, once when the blessed Thomas entered the chamber of Albertus Magnus, this student of his smashed that head, which was appearing and responding to everything. Moreover, there was in the Numantine territory – namely the city of Zamora, in the place that is called Tábara – a similar metal head, which had heretofore been set up specifically to reveal Jews when they were in the same place. From that place, if at any time anyone from the Jews were in that place – even if known by no one – the head itself would shout that a Jew was there and would repeat it often, never ceasing with its shouting until which time the Jew departed from that place. This head was shattered by the ignorance of the townspeople when they considered it to say a falsehood, while it was shouting the truth. Moreover, it cannot be doubted how such heads could be made astrologically. . . .
>
> (Toastati 1528, 15a–15b)

The 'ancient astronomers' is a reference to the 'god makers' in the *Asclepius*, though Tostado describes their creations as 'metal heads' rather than statues. It is surprising that despite the association of the object to ancient idolatry, Tostado does not immediately reject its legitimacy. There is also the astounding identification of the destroyer of the wonder as none other than Thomas Aquinas, who studied under Albertus starting in 1248, first at the University of Paris and then in Cologne. The greater influence of this version of the story over the original one is testified to by the fact that most future retellings feature a speaking head rather than a full statue and Thomas as its destroyer. No reason is explicitly given for Thomas's action, perhaps to spare the 'blessed doctor' of the accusation of having acted rashly out of superstitious fear.

The major difference between this version and the first one from *Rosaio della vita* is that Tostado does not tell the story as an allegory about the destruction of a great work by ignorance, but rather as a way of pursuing the question of the possibility of making a speaking object. He considers whether it was animated through the agency of demons, angels, or God himself, or by implanting a human soul into it. He admits to the possibility that the wonder was made through 'astrological observation' before denying the involvement of supernatural beings.

> ...it not only exceeds the power of demons to cause rational virtue in a metal head, but similarly it is impossible even for that rational virtue to be caused as a condition or property of a metal head. Indeed, if it is some kind of rational virtue, it is necessary for it to be a demon, a good angel, a rational soul, or God. However, by no means can it be said that God is the thing that responded in the metal head, for God would not serve at the nod of evil men by introducing the most serious errors and responding in such a head. Therefore, it is impossible that God is the thing that speaks in the head, for it is impossible that God speaks, since nothing could act on him through motion.... Likewise, an angel cannot be that which speaks here... since no angel could be compelled in natural or magical operations to be in some place and respond.
>
> Likewise, a human soul cannot be the thing that responds in that head, for now or then, it would either have to have been created only for that purpose, so that it could respond in that head, or it has to be one of the souls created first – namely the soul of some human being. The first way cannot be said, for God does not create the souls of humans for the fulfillment of the desires of the foolish and the wicked. (Tostati 1528, 15a–15b)

It is beyond the powers of demons to inhabit a metal head to make it speak; angels and God cannot be compelled to do so; and it is impossible to implant a human soul into an artificial object, which leaves only one possibility: 'it cannot be doubted how such heads could be made astrologically'. Tostado was fascinated enough with the legend that he wrote about it on two other occasions in similar terms (Tostati 1596, 110v; Tostati 1621, 205–6).

So, in both *Rosaio della vita* and Tostado, the feasibility of constructing the object is affirmed but only through the use of astrology. The first text denies that Albertus was engaged in 'diabolical art' when making the wonder, while the second dismisses the very possibility that it could be achieved through the agency of supernatural beings. In both cases, the talking image of Albertus is claimed to be a work of intellectual magic, exonerating its creator from suspicion of diabolical activity.

While these early versions defended Albertus's creation as legitimate intellectual work, the story was not universally accepted as an innocuous one. Because of the narrative's intimation of ancient idolatry and uncanny magic, it continued to be a source of anxiety among its tellers. For some defenders of Albertus, therefore, it was easier to deny the truth of the legend altogether rather than come up with an explanation of why he was not involved in illicit knowledge.

## 3.4 Debunking the Legend

In the late fifteenth century there was an effort by the Dominican order in Cologne to push for the canonization of Albertus, one of its most famous intellectual figures (Collins 2010). One issue they had to tackle, however, was the thinker's association with magic. Among the apologists of Albertus, a general position emerged that he was interested in the subject mainly because he wanted to understand which forms of it were legitimate and which illegitimate. In other words, rather than seeking to acquire such potentially dangerous knowledge for itself, he wanted to settle the old question of the nature of magic in view of Christian theology. One of the key texts produced by the Dominicans in his defense was *Legenda venerabilis domini Albertus Magni* (1486 or 1487), a short biography of the thinker written by a friar called Peter the Prussian about whom little is known. The work mentions the speaking image in a discussion on whether one could endow stone sculptures with magical powers, in which Peter makes use of Albertus's own writings on the issue.

> . . . it must be noted that, although likenesses in stones of this sort of sculpture might have miraculous powers, these likenesses are still not astronomical, about which was spoken above and was condemned by Albertus. However much those things, about which this discussion is now, might be made according to the appearances of the stars, it is another method, far from that thing which is made by astronomical means. Therefore, it is thus plain from the teaching of Albertus, which must be extended to those sculptures.
>
> Indeed, the folly must be condemned here by Albertus's words on certain things: anyone who presumes to say that, with necromancy, certain likenesses are thus able to be made artificially through one's craft that speak with a human voice. Such a thing is most absurd. Whence the venerable Albert said thus in the second book of *De anima*: 'Voice is nothing if it does not have an understanding receiving the purposes of things, and therefore one forms a voice for expressing an idea.' These things he said, who again said thus in the eighth chapter of the nineteenth book of *De animalibus*: 'Nature gave the power of using a distinct and learned voice to men alone, which is called "speech". Indeed, the means of speech do not happen unless in one's voice.'
>
> (Peter of Prussia 1621, pp.140–41)

One cannot endow stones with magical powers through 'the appearances of the stars', and so it is impossible to make a stone sculpture speak. Since it was Albertus himself who denied the possibility, the story concerning his speaking image must be false. So, in dealing with the potentially dangerous tale, Peter dismisses it as 'most absurd'.

In the following century, as Europe fell into chaos brought on by the Reformation and the subsequent wars of religion, the story was retold and reinterpreted in very different ways. For the defenders of Albertus, Peter the Prussian's option of denying the veracity of the story was available, but some pointed to the possibility that the talking image was achieved through purely mechanical means. But others, especially those who had no interest in safeguarding the thinker's reputation, asserted that if the story were true, then it could only have been realized through demonic agency. In other words, in the course of the sixteenth and seventeenth centuries, the legend became naturalized and demonized at the same time.

## 3.5 Early Modern Interpretations

In popular culture, one of the clichés that frequently appears in connection with the Middle Ages is the witchcraft trial, as in the famous satirical scene in *Monty Python's Holy Grail*. There is significant literature now, however, on the fact that it was really in the sixteenth and seventeenth centuries that there was the greatest frenzy of people being tried and executed for witchcraft, reaching its peak in the decades before and during the cataclysmic Thirty Years War (1618–1648) (Levack 2015; Stephens 2003). Michael D. Bailey has shown that with the publication of the first witch-hunting manuals like Heinrich Kramer's *Malleus maleficarum* (1486–1487, the same years as Peter the Prussian's biography of Albertus Magnus), there was a shift in the kind of magic that its investigators were primarily interested in (Bailey 2010). In the earlier period, the main focus was on 'learned sorcery', the kind of alchemical and astrological knowledge practised by intellectuals. In the course of the fifteenth and sixteenth centuries, however, the investigators began to look more into folk practices of commoners which were associated with 'witchcraft'. Given the much more dangerous situation intellectuals found themselves in the period, scholars offered new interpretations of the story of Albertus's speaking image.

E. R. Truitt has made the crucial point that in the medieval accounts of the wonder, the use of artisanal and mechanical craft was implied in its construction, but the predominant emphasis was on the learned knowledge of their creators, particularly in the astral sciences (2015, pp.69–95). This legitimized it as a high intellectual activity involving their mastery of learned magic, rather than lowly mechanics. But in sixteenth and seventeenth century accounts of Albertus's speaking image, the use of astrology was completely abandoned in explicating the wonder, describing the speaking object as a wholly mechanical device. This shift is significant not only as an alternate way of defending his work from involvement in witchcraft, but also in the fact that such an interpretation was made by major Renaissance scholars of 'natural magic', like Giambattista della Porta and Tommaso Campanella, who were far from sceptics of astrology. But the offering of a mechanical explanation for Albertus's wonder was also accompanied by a counter-trend in accusing the object of being a product of demonic magic.

Starting in the second half of the fifteenth century, there was a great resurgence of interest in intellectual magic of the Neoplatonic and hermetic types, crucially through the scholarship and translating work of Marsilio Ficino on behalf of his Medici patrons (Walker 2000; Yates 1964; Vickers 1984; Copenhaver 2015). But the new interpretations of the subject tended to take on a more philosophical approach that often interpreted alchemical and astrological ideas as allegories or symbolic representations of the workings of nature rather than as direct means of manipulating it. Also, in their efforts to reconcile Neoplatonic ideas and Christian theology, the Renaissance thinkers had to struggle with the infamous god-making passage in the *Asclepius*, considering its demonic nature while coming out with philosophical, allegorical, and ritualistic interpretations of it (Copenhaver 2015, pp.55–83, 186–208).

Recent scholars have noted the emphasis on mechanical knowledge in the works of Renaissance philosophers, which they considered a type of natural magic (Zetterberg 1980; Eamon 1983). In Cornelius Agrippa's influential compendium of magical knowledge, *De occulta philosophia* (1531–1533), in the section on 'celestial magic', the importance of mathematical learning is discussed (Agrippa 2000, pp.233). As examples of mechanical wonders that could be achieved through the knowledge, Agrippa provides a list of automata, including the statues of Mercury and the brazen speaking head (p.233). The former is, once again, a reference to the god-making passage in the *Asclepius* (Mercury being Hermes Trismegistus). Agrippa's claim that a device like the speaking head could be made through mechanics alone, without resorting to spiritual or astrological magic, offered a new way of interpreting the old legend.

Such a novel take on the wonder can be found in the celebrated treatise on art *Trattato dell'arte della pittura, sculptura, et architettura* (1584) by the Milanese Mannerist painter Giovanni Paolo Lomazzo (1538–1592), who turned himself into a major scholar of art theory after he became blind. The work has an entire section on mechanical works, including wonders made through the use of 'mathematical motions'. An example is Albertus's speaking device, which is close to the Tostado description as a 'brazen head' that was destroyed by Thomas Aquinas. But unlike Tostado who does not provide an explanation for Thomas's action, Lomazzo describes how he thought the work was the Devil when 'it was a mere mathematical invention as is manifest' (Lomazzo 1969, Book 2, p.2). Giambattista della Porta, the most rigorously sceptical among the era's major scholars of magic, went further in the expanded edition of his book *Magia Naturalis* (1589). In a section on pneumatic experiments, he poses the question of whether artificial constructs could be made to speak. He mentions Albertus's creation of a brass head through the use of astrology, but dismisses the story on two grounds. First, because he has a low opinion of the subject of the legend, especially his works on experimental magic. 'I give little credit to [Albertus], because all I made trial of from him, I count to be false, but what he took from other men.' And second, because such astrological operations are nonsense. 'I wonder how learned men could be so guld: for they know the Stars have no such forces.' But Porta considers that the object could have been made through a piece of artificial trickery.

> But I suppose it may be done by wind. We see that the voice of a sound will be conveyed entirely through the Air, and that not in an instant, but by degrees in time...if any man shall make leaden Pipes exceedingly long, two or three hundred paces long (as I have tried) and shall speak in them some or many words, they will be carried true through those Pipes, and be heard at the other end...and when the mouth is opened, the voice will come forth, as out of his mouth that spake it....
>
> (Porta 1957, pp.385–86)

Similarly, Tommaso Campanella, another seminal scholar of natural magic, in his *Magia e grazia* (early seventeenth century), mentions the head that speaks through magical means, like the one supposedly made by Albertus Magnus. Like Porta, he denies the possibility of such use of astrology, but he considers that the effect could be produced through the:

> ...imitation of the voice by means of reeds conducting the air....This art however cannot produce marvellous effects save by means of local motions and weights and pulleys or by using vacuum, as in pneumatic and hydraulic apparatuses, or by applying forces to the material. But such forces and materials can never be such as to capture a human soul.
>
> (Campanella 1957, p.180)

What is crucial here is that even as these thinkers approach the story with scepticism, they concede its possibility but only through mechanical means. And they deny the role of astrology in contrast to the earliest versions of the legend. If the story is based on fact, then Albertus could only have made the device work through the use of pneumatic machines and pipes. One could say that he was still engaged in magic, but only in the sense of a wonder that is achieved mechanically through artisanal craft. At a time of rampant religious conflict and paranoia, when the accusation of demonic agency could have dire consequences, these purely naturalistic explanations of the object rendered the object innocuous while maintaining its status as a work of a new kind of wonder.

Interestingly, the same explanation was offered by Pierre de Lancre (1553–1631), a French magistrate of Bordeaux who led a vociferous campaign of witch hunting in Labourd Province. In his 1610 work, *Tableau de l'inconstance et instabilité de toutes choses*, he tells a story that is close to the original one in *Rosaio della vita*, describing a statue in the shape of a man destroyed by an unnamed disciple, making Albertus lament that his work of 'thirty years' has been ruined. Despite De Lancre's witch-hunting zeal, however, he does not regard this particular object as a diabolical one as he thinks it functioned mechanically. The statue's tongue moved 'with some counterweights and wheels and other machinery hidden within', that allowed it to 'pronounce well-articulated words' with no mention of magic, spiritual or astrological (De Lancre 1610, p.108). But this was not the only reaction to the wonder in the period. A number of scholars, in fact, claimed that such a speaking image could only have been realized through the participation of demons.

The writer who may have been the first to have considered Albertus's work in such a manner was the theologian Bartholomeus Sibylla (1434–1496). Little is known about Sibylla, other than that he studied in Bologna and Ferrara before ultimately becoming prior of the Dominican convent at Monopoli. Only a decade after Peter the Prussian published his laudatory biography of Albertus, Sibylla, in *Speculum peregrinarum quaestionum* (1587), asked if such an object could be made through learned magic that involves the manipulation of natural forces without the involvement of supernatural beings.

> Indeed, how might a man make a body through art, ether-made and so delicate, whereby the rational soul seeks, that he is not able to make the body of a human through the coarse art, man-made and ignoble to look at, from pure gold or silver? If it might be said that Albertus Magnus had fashioned the head of a man with the art of speaking and with the sole workings of natural power, it is denied and said that it would have been (if it was, still) through the hidden workings of a demon, and on that account, Thomas, then his disciple, smashed it, as it is told. (Sibylla 1587, p.509)

Sibylla denies the possibility of creating the object through the 'workings of natural power'. But he does not then conclude that the legend was false, as Peter the Prussian did. While not explicitly affirming it, Sibylla asserts that if Albertus had indeed succeeded in making a speaking image, it was through 'the hidden workings of a demon', which was why Thomas Aquinas destroyed the thing.

The Jesuit theologian Martín Antonio Del Río (1551–1608) was famous in his own time as a great authority on magic and witchcraft, with the publication of his book *Disquisitionum magicarum libri sex* (1599–1600). Despite the interest and zeal he shared with his contemporary Pierre de Lancre, he took a very different interpretation on the nature of Albertus's speaking head. In the *Disquisitionum*, he asserts the impossibility of creating an object like the 'head of Albertus Magnus' through 'human craft'. But this does not lead him to conclude that the story must be false. Accepting the notion that the speaking head was real, Del Río asserts that since the object lacked the bodily organs necessary for speech making, the words coming out of it had a different source. Like Sibylla, he thinks that the talking head, like "the statues of idols" in the *Asclepius*, must have been under the control of a 'cacodaemon' (Del Río 1617, p.31), an evil demon. And Giorgio Raguseo (1580–1622), a philosopher and theologian from Venice who taught at the University of Padua, in *Epistolarum mathematicarum seu de divinatione* (1623), refers to Del Río in characterizing Albertus's wonder as a work of 'demoniacal magic' (1623, p.382).

## 3.6 The Mechanics of the Speaking Statue

Despite such writings that considered the involvement of evil beings in Albertus's creation, the legend prevailed through the period of religious conflict as interpreted by those who described the talking image as a purely mechanical device. Only two years after Raguseo affirmed Del Río's demonic interpretation of Albertus's operation, Gabriel Naudé's book *Apologie pour tous les grands personnages qui ont esté faussement soupçonnez de magie* (*Apology for All the Great Persons Falsely Suspected of Magic* 1625) was published. As the earlier accounts of the Albertus story fell into obscurity in the course of the seventeenth century, Naudé's take on the legend became the best-known version.

Naudé (1600–1650) was famous not only as an erudite expert on books but also as a pioneering theorist of the library (i.e. father of modern library science) with the publication of his *Advis pour dresser une bibliothèque* (1627). He was given the opportunity to realize his ideas in the 1640s when Cardinal Mazarin hired him to build the first public library in France (Clarke 1970). Of the same generation as Descartes, Gassendi, and Mersenne, Naudé shared their sceptical rationalist outlook. In 1623, in response to the mysterious appearance in Paris of placards inviting people to join the Rosicrucian Brotherhood, he wrote a tract denouncing people's gullibility in being interested in the sect's mystical ideas. Two years later, he produced the *Apologie*, another 'rationalist critique'. As the full title of the book indicates, it defends famous thinkers, ancient and medieval, from the accusations of having involved themselves with magic. In it, he dismisses such stories as slanderous rumours, sometimes providing naturalistic explanation of their deeds that were taken to be magical or demonic. In a chapter on Robert Grosseteste and Albertus Magnus, Naudé begins by denouncing various views of Albertus as a practitioner of magic before making a reference to the fable of his 'androide' (1625, p.523). A word constructed from the Greek *andros* ('man') and the suffix *-oid* ('type of'), so meaning 'a kind of man' or 'a manlike being', this appears to be the very first time it was used to denote a machine in the shape of a human being. Before 'android' became a familiar term in modern science fiction, it appears a number of times in the early modern period, including as an article in Diderot and D'Alembert's *Encylopédie* (1751–1772), on human automata (Kang 2011, p.110).

Peter the Prussian's biography of Albertus Magnus was a major source for Naudé, both writers engaged in the task of saving their subject's reputation. Naudé, however, does not resort to Peter's strategy of denying the truth of the legend. In a demonstration of his famous erudition, he points to six sources of the story: Tostat (Tostado), George Venitien (Francesco Giorgio Veneto), Delrio (Martín Antonio Del Río), Sibille (Bartholomeus Sibylla), Raguseus (Giorgio Raguseo), and Delancre (Pierre de Lancre). As shown above, those writers offered radically different interpretations of the legend, including the practice of legitimate intellectual magic (Tostado and Veneto 1543, p.379), the use of mechanical devices (De Lancre), and the agency of demons (Del Río, Sibylla, and Raguseo). Naudé describes the construction of an 'entire man' made of metal.

> Albertus Magnus, with the greatest expertise, made an entire man of the same kind [of material, i.e. bronze], spending thirty years without stopping in forming him under diverse aspects and constellations. He made the eyes, for example, according to the aforementioned Tostat [*Tostado*], in his Commentaries on Exodus, when the sun as a sign in the Zodiac was in correspondence with that part, casting them out of diverse metals mixed together, and marked with the characters of the same signs and planets and their several necessary aspects. The same method was observed in the head, neck, shoulders, thighs and legs, all of which were fashioned at different times, and were put together in the form of a man that had the capacity to reveal to Albertus the solution to all his principal difficulties. Also, to lose nothing of the story of the statue, it was added that it was broken to pieces by St. Thomas because he could not endure with patience its excessive babbling and cackling. (Naudé 1625, pp.529–30)

The most surprising new detail in this version is that Thomas Aquinas destroyed the wonder not from fear of its evil but because he could not stand its excessive talking. In other words, the cause of Thomas's violent reaction is also naturalized as stemming from annoyance rather than supernatural horror.

Naudé then speculates on the origins of the legend, especially on the Jewish tale of the teraphim, an oracular idol made from a severed head, and the Egyptian animated statues. Ultimately, he dismisses the possibility of such objects on similar grounds as Tostado and Del Río, that demons do not have the power to inhabit such objects, souls cannot be implanted into them, and severed heads lack organs necessary for speech making. Naudé also mentions the use of astrology but his ultimate explanation is the one derived from Porta and Campanella, namely the conveyance of the voice through pipes.

> ...my sole intention is to show that [*Albertus*] could not by the help of superstitious magic, make a statue that gave him answers in an intelligible and articulate voice...and not to deny absolutely that he might have made a head or statue of a man...which produced a small sound and pleasant noise...through certain small pipes.... (Naudé 1625, 539–40)

## 3.7 Conclusion

As shown above, even as wholly mechanical explanation of the story was being offered in the course of the sixteenth and seventeenth centuries as an alternative to that of intellectual magic, the concern over the possibility of diabolical magic persisted. This can be attributed in part to shifting attitudes in the period to what it meant for knowledge to be occult. When medieval thinkers found themselves without an explanation for some phenomenon, they were typically content to label it as inexplicable. The pessimism of their epistemology accepted that the requisite knowledge for understanding might simply have been forbidden or lost in the aftermath of humanity's ongoing fall from grace. In the early modern period, with the rationalization of knowledge into discrete fields of study, however, there was a concomitant discomfort with phenomena in the natural world that defied categorization, ironically popularizing demonology as much as engineering as an explanation for what had hitherto been inexplicable. For knowledge to be 'occult' now meant that it was knowledge of the unseen. Such imposition of certainty in the face of the wondrous continued through the seventeenth century as the language of machinery and mechanics came to dominate intellectual discourse.

By the last decades of the seventeenth century, after much of the religious fervour of the Reformation period had burnt itself out in the Thirty Years War and the mechanistic worldview of the new natural philosophy began to establish itself, the Albertus legend survived as a story of a mechanical automaton. The speaking image was presented as one of many artificial devices that the thinkers of the new era pointed to as early examples of the self-moving machine that became the central emblem of the time. The most significant example of this can be found in the writings of John Wilkins, the Anglican clergyman and lauded natural philosopher who was a key figure of the Scientific Revolution in England as one of the cofounders of the Royal Society. In 1684 he published *Mathematicall Magick: Or the Wonders that may be performed by mechanical geometry*, an essential work of what could be called applied mathematics, detailing many practical uses of technology based on new scientific knowledge. The word 'magick' in the title is overtly ironic since it is 'in allusion to vulgar opinion, which doth commonly attribute all such strange operations unto the power of Magick', a view that Wilkins seeks to correct in this work (Wilkins 1708). He includes an entire chapter on automata, with description of mechanical devices that mimic living beings, including artificial works that utter sounds as if they can speak.

> There have been some Inventions which have been able for the Utterance of articulate Sounds, as the Speaking of certain Words. Such are some of the *Egyptian* Idols related to these. Such are the Brazen Head made by Friar Bacon, and that Statue, in the framing of which *Albertus Magnus* bestowed 30 years, broken by *Aquinas*, who came to see it, purposely that he might boast, how in one Minute he had ruined the Labor of so many years
>
> (Wilkins 1708, p.104).

In the context of this work of mechanics, Wilkins treats the story of Albertus's speaking statue as an 'invention', without even bothering to mention the use of astrology, and with no need to concern himself with demonic agency in its construction. In other words, he considers it as just one among many complex machines that have been constructed in the past. It is also interesting that he adds his own flourish to the story, claiming that Thomas Aquinas ruined the object in a wanton act of destruction. Just as Albertus's creation had been repurposed to suit a newly developing worldview, Thomas Aquinas was also transformed from the embodiment of anxiety about magic to that of ignorant opposition to progress.

The shift in the strategy of defending Albertus as a practitioner of astrology to mechanics is indicative of the changing attitude towards intellectual magic. And the explanation of his wonder in Naudé and Wilkins points to the decline of the very concept of magic in the intellectual sphere. Even as Wilkins picks up on the purely technological explanation of the wonder, from Porta, Campanella, and Naudé, he deprives it of its magical reputation altogether and includes it in his list of automata. That was how the story was kept alive in the Western imagination, attested to by the fact that it has been mentioned in recent works dealing with robots and AI, depicted as part of an ongoing quest for the creation of artificial life and intelligence (Heudin 2008, p.66; Perkowitz 2004, p.20; Wood 2002, p.xv).

There is, however, a somewhat ironic aspect to this survival of the legend as a story of premodern AI. Abandoning the astrological explanation for the wonder, Naudé and Wilkins described it as a machine so as to lay to rest once and for all the anxiety over the possibility of diabolical knowledge and demonic agency in its construction. But the allying of the concern proved to be temporary as, following the period of the Enlightenment, the fear of the machine would rise again in the centuries to come. Modern critics of technology, including contemporary commentators on AI, may not resort to the language of the demonic and the diabolical, except in figurative terms, but the great ambivalence and concern expressed by them demonstrate that our primordial fear of the artificial other has never been exorcized from our collective consciousness even in the era of ubiquitous and essential technology.

## Note

1. It has only recently become available online, available at https://books.google.com/books?id=dHGmDLyOWRcC.

## Reference List

Agrippa, H. C. ([1533] 2000) *Three books of occult philosophy*. Translated by J. Freake. St. Paul, MN, Llewellyn.

St. Augustine (1984) *City of God*. Translated by H. Bettenson. Harmondsworth, Penguin.

Bailey, M. D. (2010) From sorcery to witchcraft: clerical conceptions of magic in the later Middle Ages. *Speculum*. 76(4), 960–90.

Bailey, M. D. (2017) *Fearful spirits, reasoned follies: the boundaries of superstition in late medieval Europe*. Ithaca, Cornell University Press.

Borlick, T. A. (2011) 'More than Art': clockwork automata, the extemporizing actor, and the brazen head in *Friar Bacon and Friar Bungay*. In: W. Hyman (ed.) *The automaton in English Renaissance literature*. Franham, Ashgate. pp. 129–44.

Boudet, J.-P. (2006) *Entre science et nigromance: astrologie, divination et magie dans l'Occident médiéval (XIIe–XVe siècle)*. Paris, Publications de la Sorbonne.

Campanella, T. (1957) *Magia e* grazia. Rome, Fratelli Bocca.

Clarke, J. A. (1970) *Gabriel Naudé 1600–1653*. Hamden, Archon Books.

Collins, D. J. (Spring 2010) Albertus, Magnus or Magus?: magic, natural philosophy, and religious reform in the late Middle Ages, *Renaissance Quarterly*. 63(1), 1–44.

Copenhaver, B. (2015) *Magic in Western culture: from antiquity to the Enlightenment*. Cambridge, Cambridge University Press.

Corsini, M. ([1373] 1845) *Rosaio della vita*. Florence, Societá Poligrafita Italiana.

De Lancre, P. (1610) *Tableau de l'instance et instabilité de toutes choses*. Paris, Abel L'Angelier.

Del Rio, M. A. (1617) *Disquisitionum magicarum libri sex*. Mainz, Germany, Petri Henningii.

Dickson, A. (1929) *Valentine and Orson: a study in late medieval Romance*. New York, Columbia University Press.

Eamon, W. (1983) Technology as magic in the late Middle Ages and the Renaissance. *Janus*. 70, 171–212.

Eamon, W. (1994) *Science and the secrets of nature: books of secrets in medieval and early modern culture*. Princeton, NJ, Princeton University Press.

Flint, V. J. (1991) *The rise of magic in early medieval Europe*. Princeton, NJ, Princeton University Press.

*Hermetica* (1992). Translated by B. Copenhaver. Cambridge, Cambridge University Press.

Heudin, J.-C. (2008) *Les Créatures artificielles: des automates aux mondes* virtuel. Paris, Odile Jacob.

Kang, M. (2011) *Sublime dreams of living machines: the automaton in the European imagination*. Cambridge, MA, Harvard University Press.

Kavey, A. (2007) *Books of secrets: natural philosophy in England 1550–1600*. Chicago, University of Chicago Press.

Kieckhefer, R. (1989) *Magic in the Middle Ages*. Cambridge, Cambridge University Press.

LaGrandeur, K. (2013) *Androids and intelligent networks in early modern literature and culture*. New York, Routledge.

Levack, B. (2015) *The witch-hunt in early modern Europe*. London, Routledge.

Lightsey, S. (2007) *Manmade marvels in medieval culture and literature*. New York, Palgrave.

Lomazzo, G. P. (1969) *A tracte concerning the art of curious paintings*. Translated by. R. Haydock. Amsterdam, Da Capo Press.

Marr, A. (2004) Understanding automata in the late Renaissance. *Journal de la Renaissance*. 2, 205–22.

Naudé, G. (1625) *Apologie pour tous les grands personnages qui ont esté faussement soupçonnez de magie*. Paris, Chez François Targa.

Perkowitz, S. (2004) *Digital people: from bionic humans to androids*. Washington DC, Joseph Henry Press.

Peter of Prussia (1621) *B. Alberti doctoris Magni ex Ordine Praedicatorum episcopi Ratisponensis de adhaerendo Deo libellus*. Antwerp, Belgium, Officina Plantiniana.

Porta, G. (1619) *Magia naturalis*. Hanover, Germany, Aubry & Schleich.

Porta, J. B. (1957) *Natural magic*. Translated by Anonymous. New York, Basic Books.

Raguseo, G. (1623) *Epistolarum mathematicarum seu de divinatione, vol. 2*. Paris, Nicolai Buon.

Sawday, J. (2007) *Engines of the imagination: renaissance culture and the rise of the machine*. New York, Routledge.

Sibylla, Bartholomaeus (1587) *Speculum peregrinarum quaestionum*. Venice, Marcantonio Zalterio.

Stephens, W. (2003) *Demon lovers: witchcraft, sex, and the crisis of belief*. Chicago, University of Chicago Press.

Tostati, A. (1528) *Super libro Numerorum explanatio litteralis amplissima nunc primum edita in apertum*, vol. 2. Venice, Peter von Liechtenstein.

Tostati, A. (1596) *Commentaria in primam partem Exodi: Mendis nunc sane quam plurimis diligenter expurgata*. Venice, Typographia Rampazetana.

Tostati, A. (1621) *Aenigmatum sacrorum pentas: Paradoxa de Christi, matrisque eius misteriis, animarum receptaculis posthumis, iucundissimae disputationes*. Douai, Balthazar Bellere.

Truitt, E. R. (2015) *Medieval robots: mechanism, magic, nature and art*. Philadelphia, University of Pennsylvania Press.

Veneto, F. G. (1543) *De harmonia mundi totius cantica tria*. Paris, Berthelin Apud Andream.

Vickers, B. (1984) *Occult and scientific mentalities in the Renaissance*. Cambridge, Cambridge University Press.

Walker, D. P. (2000) *Spiritual and demonic magic from Ficino to Campanella*. University Park, Pennsylvania State University Press.

Wilkins, J. (1708) *The mathematical and philosophical works of Right Reverend John Wilkins*. London, J. Nicholson.

William of Malmesbury ([1125] 1847) *Chronicles of the kings of England*. Translated by. J. A. Giles. London, Henry G. Bohn.

Wood, G. (2002) *Living dolls: a magical history of the quest for mechanical life*. London, Faber and Faber.

Yates, F. (1964) *Giordano Bruno and the hermetic tradition*. Chicago, University of Chicago Press.

Zetterberg, J. P. (Spring 1980) The mistaking of 'the mathematicks' for magic in Tudor and Stuart England. *The Sixteenth Century Journal*. XI(1), 91–108.

# 4

# Artificial Slaves in the Renaissance and the Dangers of Independent Innovation

*Kevin LaGrandeur*

## 4.1 Introduction

The Renaissance provides a fascinating site for the investigation of the prehistory of contemporary AI, and some of the social, political, and ethical questions it raises both in its own time and in anticipation of ours, because the artificial intelligences imagined in that period exist at the point of transition between magical and empirical science. This means that although the real and fictional creations of the Renaissance are in part the products of the intellectual culture that existed before the empirical turn of Newton, Descartes, and their contemporaries, some aspects of these creations anticipate today's cultural, social, and ethical preoccupations with AI: How beneficial is it to society? How dangerous? How ethical are the motivations of the makers? And just because we can fashion something new and powerful to help us defeat nature's limits, does that mean we should? From Paracelsus's and Cornelius Agrippa's claims to be able to create a miniature human being (a homunculus), to Jewish legends of the golem, to Robert Greene's play about the medieval scientist Roger Bacon's attempt to make a talking brass head that could lecture on philosophy and protect England, we have in the

Renaissance different accounts of the same kind of project: an innovator's attempt to fashion a superhuman, intelligent tool in his own image. It is not only the creation of these androids that anticipates projects of our own time, but also their society's reactions to them. As with today, these kinds of technological innovations ignited both awe and fear in pre-empirical Europe—which is reflected in social responses of the day to men such as Agrippa and Paracelsus, or in stories like Greene's or the various tales of the golem. The first two men were notable enough for their novel ideas to be given important positions in their societies—Agrippa as a noted physician in a number of cities, and Paracelsus as a professor at the University of Basel—but their iconoclasm eventually got them into trouble with the authorities. Agrippa's appointments were ultimately cut short, and Paracelsus was fired from his post as professor, and banished from Basel. In Greene's play, the main character's radical notion of creating an intelligent metal android to give lectures on philosophy and build defenses for England against invasion demonstrates his brilliance. But the same character, Friar Bacon, is also depicted as a freethinking maverick, like the real scientists Agrippa and Paracelsus, a kind of behaviour which leads to disaster: the forces his invention comprises are too strong to master, and the head ultimately implodes. Similarly, the tales of golems often end in catastrophe, as the creature's superhuman size and strength prove too much for its human master.

This chapter will show how the idea of creating an intelligent, artificial humanoid servant originates from Aristotle's elision of the ideas of tools and human slaves, and how both of those things are extensions of the master's body; how these notions of slavery are then adapted by the Renaissance and lead to scientific and fictional literary instances of superhuman artificial servants; and, finally, how all of this connects with the AI of our time. Throughout, the dangers of these powerful servants bring into question the propriety of their creators' behaviour and the proper limits of human ambition.

## 4.2 Aristotle's Conceptual Merging of Slaves and Machines Underpins Renaissance AI

AI's primary purpose throughout its history—and prehistory—has been to spare its owners from tedious labour. Today's digital technology, including AI, is mostly meant to do work that is dirty, dangerous, or dreary. For example, my wife and I have two robotic vacuum cleaners at home that save us the dreary and dirty chore of cleaning the floor around our cat's litter box every day. My bank has intelligent software that will pay my bills automatically, as well as keep track of my monthly spending habits in minute detail, which saves me the effort of doing these dull chores. Police and military forces around the world use bomb-disposal robots to spare humans the physical danger of being near possible live explosives. In industry, robots are commonly used to do repetitive tasks, such as those on assembly lines. This is not just because these tasks are boring, and often dirty and dangerous, but mainly because they *are* repetitive, and so simple enough to automate, and because it is cheaper and more efficient to employ a robot than a human. Robots do not need holidays, food, drink, or sleep, and they do not demand pay raises.

In other words, artificially intelligent systems are essentially artificial slaves, an idea, as we will see, that starts with Aristotle—who also posits that both slaves and machines are virtual extensions of masters that allow them the power to accomplish much more than they could do by themselves. I have discussed these ideas elsewhere at length, as well as the idea that the creation of artificial slaves is a motif that is so historically persistent as to make it archetypal in human culture and literature (LaGrandeur 2013). Whilst the focus of this chapter is the Renaissance period, two specific instances of proto-AI discussed in ancient Greek sources require mention here because they are critical to understanding the archetypal drive to create artificial slaves that manifests in Renaissance AI narratives.[1] They ground any further consideration of the artificial creation of intelligent servants of any later era. The first instance of proto-AI as a replacement for human workers is depicted in Book 18 of Homer's *Iliad*, in which he shows the god Hephaestus' creation of both intelligent female android servants (lines 417–21) and trilegged tables that can move autonomously to serve the gods in their banquet halls (lines 373–79). The Homeric portrayal of proto-AI establishes the basis for why Renaissance thinkers and artists may have imagined them, since that era was very familiar with Homer's works. It also shows how this idea resonates through social history because the story about Hephaestus' artificial slaves is the basis for Aristotle's idea (more than 300 years after Homer) for *actually* using intelligent machines to replace human slaves. In sections 1253b, line 38, through 1254a, line 1, of his work the *Politics*, he refers back to the episode in *Iliad* mentioned above, when he states:

> ...if every tool could perform its own work when ordered, or by seeing what to do in advance, like...the tripods of Hephaestus which the poet says 'enter self-moved the company divine,'—if thus shuttles wove and quills played harps of themselves, master-craftsmen would have no need of assistants and masters no need of slaves. (Aristotle 1944)

As I have previously argued, (LaGrandeur 2011; LaGrandeur 2013), this brief passage is very important not only because it is the first time any writer explicitly suggests that we could replace humans with intelligent machines, but just as importantly because it is also the first to make the explicit connection between machines and slaves. Aristotle's *Politics* can thus be considered the foundational text for this idea of AI as a replacement for human slaves—and it is also a text that fundamentally informed Renaissance attitudes. Aristotle goes on in his treatise to draw an equivalency between machines as slaves, and slaves as machines—an equivalency that is founded in his use of the Greek word '*organon*', which can be used for both tools and body parts. Aristotle reasons that because both tools and slaves are extensions of the master's body that he uses to accomplish work, they are both *organon*. Human slaves are at once a part of the master, and a type of intelligent, organic tool to be used in conjunction with other inorganic tools, such as rudders on a ship. So, Aristotle blends the concepts of instruments, slaves, and prosthetic enhancements of the master. As an illustration of this, he talks of how a ship's commander has a mixture of these organic and inorganic instruments as extensions of himself: 'some are living, others lifeless; in the rudder, the pilot of a ship has a lifeless, in the lookout man, a living instrument; for in the arts the servant is a kind of instrument. . . . the servant is himself an instrument for instruments' (section 1253b, lines 29–34). Collectively, this pilot's inanimate and animate tools provide a sort of extended prosthetic network that allows him to more effectively steer his ship—a network which, with its parts, is an extension of the master himself. By implication, then, the pilot's apparatus of human slaves and the inanimate tools they, in turn, operate comprise one large tool. That tool is a systemic enhancement of the pilot's ability to convey himself through the water in a much faster way, and with enhanced capacities to carry items with him.

## 4.3 How the Renaissance Adapted and Instantiated Aristotle's Ideas of Artificial Slaves

Aristotle was an extremely important authority for all kinds of Renaissance thought—cultural, philosophical, and scientific. In this instance, his ideas of mingling tools, slaves, and the master's body—the use of an intelligent tool system, in other words, as a way to enhance the master's senses, actions, and agency—are echoed by their Renaissance cultural and literary instantiations. The starting place for Renaissance borrowings of the above ideas was his rationale for slavery, of which the quotations I used above are elements. His discussion of this in the *Politics* was a basis for one of the most important debates on the ethics of slavery in the Western world: the Spanish slave debate of 1550–1551, commonly known as the Valladolid debate. Held at the colegio de San Gregorio in Valladolid, this debate between Bartolomé de Las Casas and Juan Ginés de Sepúlveda, and their allies, was an attempt to codify the status of indigenous Americans. Were they humans and of equal status with Europeans, or were they the sort of subhumans classified by Aristotle as natural slaves? Sepúlveda, who defended their enslavement, centred his argument on Aristotle's chauvinistic definition of other humans in terms of his own culture. That is, Sepúlveda argued that indigenous people were subhuman because they did not dress, eat, or behave like Europeans and, therefore, were natural slaves to their Spanish conquerors (Las Casas 1583, appendix). The details of this debate were published all over Europe within a few years and were so well known that they likely influenced the writing of Shakespeare, Raleigh, Thomas Gage, and others (Sharp 1981).

Thus Aristotle had a wide influence on Renaissance thoughts about human slavery—and the ideas embedded in the *Politics* regarding artificial slaves also would have been well known to scholars. Furthermore, this Aristotelian thought coincided with the science of the time, which was a blend of Cabalism, alchemy, Hermetic philosophy and a burgeoning experimentalism. As we will see, this blend of magic and science spurred notions of creating artificial humanoid servants. The chief motivation of those who sought to make them was the same as Aristotle's—to use that servant as a tool to enhance the power of its maker to supersede his natural limits.

The rediscovery in the early Renaissance of classical texts, mostly translated from Arabic back into Latin by scholars who spoke both, not only made available more texts by already-known figures such as Aristotle, but it also made previously unknown texts available. These included Arabic scientific texts and tracts supposedly written by a godlike ancient alchemist known as Hermes Trismegistus (who was later proven to be mythical). These Hermetic texts claimed to contain knowledge of how to transform humans into powerful beings like Hermes Trismegistus himself. The surge in translations also made available more

Cabalistic texts to non-Jewish scholars, including one called the *Sefer Yetzirah* (Hebrew for 'Book of Formation/Creation'), which among other mystical practices describes the creation of living things, including an android (Scholem 1965).

Both of these sources were important to the scientists of the early sixteenth century, such as Paracelsus and Cornelius Agrippa, who claimed to be able to make artificial humans. The sources, as well as the work of these scientists, all contributed to the depiction of the fictional artificially created intelligent servants of the time. Those fictional creations that I will discuss later were depicted in their respective texts as the result of that era's science that—as I have just mentioned—was an alchemical one influenced by the occult (including Cabala, astrological and natural magic, and Hermeticism), natural philosophy, and a nascent experimental science. Now we will look at examples of these artificial, intelligent slaves: the homunculus, the golem, and the talking brass head depicted in Robert Greene's Elizabethan play *The Honorable History of Friar Bacon and Friar Bungay*.

### *4.3.1 The Homunculus*

The idea of creating an artificial, android servant was popular in Renaissance scientific literature (such as Paracelsus's and Agrippa's), and in various forms of popular literature (like Greene's play); it appears to have circulated reciprocally between the intellectual culture and the fictional works of the time. We are going to start with the homunculus because that model of the artificial android appears earliest and spread first via the cross-pollination of the ideas of Cabalism, Hermeticism, and humanism we have discussed. The Renaissance android servant known as the homunculus (Latin for 'little man') has old roots. The underlying methodology and rationale for creating the homunculus includes both alchemical influences and Aristotle's ideas of slavery. The science historian William Newman demonstrates in Chapter 4 of his book *Promethean Ambitions* (2004) that the idea for making a miniature, artificial human extends back to the late Roman and early medieval era, and—again—is heavily influenced by Aristotle's notions. Because Aristotle theorized that male sperm was the sole agency for providing the embryo, and that the womb only provided a vessel and nourishment for it, later scientists assumed that any vessel would do for growing a tiny human; all that would be needed would be some source of warmth and nourishment. Ancient notions of spontaneous generation—the sudden appearance of flies from horse dung, for instance—also lent credence to this idea of building a human in a laboratory vessel.

Arabic alchemists and writers of the early Middle Ages were the first to discuss making humans in laboratory containers filled with sufficient nourishment and kept sufficiently warm. For instance, the famous eighth-century Persian scholar Jabir ibn Hayyan, also known by his Westernized name Geber, discusses how to make a homunculus in his *Kitab al tajmic (Book of the Collection)*. Moreover, he notes that there were many 'schools of artificial generation' in his day using various recipes for making artificial humans—some of them with animal traits mixed in (Newman 2004, pp.181–82). Many of these hybrids are meant to be marvels to attract praise for their creators, but at least some are also meant to be superhuman servants, organic tools to enhance their creator's power. Jabir refers, for example, to a work called *The Book of the Cow (Liber vaccae)* in which an alchemist claims that he can create a homunculus that will have miraculous abilities, including being able to tell its maker 'about all distant things and occurrences' (Newman 2004, pp.179–80).

There is plenty of evidence of transmission of these Arabic ideas of the homunculus to the alchemists and scholars of the Renaissance. For example, both Newman (2004) and Lynn Thorndike (1923–1958) point to the spread of a fourteenth-century translation of the Arab alchemist Razi's work throughout Europe, and that work contains a recipe for making a homunculus by using a flask buried in horse dung (for warmth) filled with putrefying matter. This recipe is very similar to one discussed in the early Renaissance by the Spanish theologian Alonso Tostado.

Paracelsus repeats a form of this recipe in the sixteenth century when he states in his *De natura rerum* (*On the Nature of Things*) that he can make a little man by combining blood and semen in a flask and burying it in horse dung for 40 days (1573, vol. 1, p.124).

As with the earlier Arabic homunculi, these artificially made, intelligent humanoids could be used as superhuman replacements for slaves without ethical concerns. This is because, as specified by Aristotle, they are natural slaves. That is, although these androids have the capacity for understanding, they cannot reason like naturally born people. Therefore, they are humanoid but not quite human—natural slaves—and can be used like animals (Paracelsus 1941, pp.230, 239). This is the same Aristotelian reasoning used by Sepúlveda in the sixteenth-century Spanish slave debates. Even better for their makers such as Paracelsus is that these creatures, like 'wild people' or 'naturals' (giants, nymphs, pygmies, and the like), have supernatural powers that their creator can take advantage of—powers such as the ability to know 'hidden and secret things', and 'great, forceful' beastlike strength (Paracelsus, 1573, vol. 1, p.124).

Paracelsus' contemporary Cornelius Agrippa also wrote of being able to create a homunculus. Like his fellow physician and scholar, Agrippa saw the homunculus as an artificial, intelligent slave with superhuman abilities that could be harnessed by its creator. He compares homunculi, as a class of 'naturals', to the types of moving statues or automata (clockwork robots) that Aristotle mentions in his writing about slaves and intelligent machines, making it clear that they carry the status of natural slaves (Agrippa, 1651, Book 1, Chapter 2). Moreover, because the automata and moving statues Agrippa mentions are, as he says, noted in Greek mythology to have wondrous powers, his reference to them and their connection to Aristotle's ideas evokes not only the homuculus's servile status but also its potential power.

The possibilities of creating artificial android servants put forth by these two men in their scientific writing was seen not only as sensational and daring, but also dangerous—a key problem for all representations of proto-AI of this time. Unfortunately for Agrippa and Paracelsus, they were not only alike in their brilliance and claims of being able to create intelligent androids, they were also alike in their intellectual rebelliousness in an era that valued orthodoxy. In the Renaissance there were significant social changes in Europe. One was the rise of the merchant class as a result of increased trade with newly discovered lands, a consequent spike in the wealth of that class, and their increased social mobility. This in turn encouraged commoners to begin to look for jobs outside of their traditional classes. Additionally, religious and other wars further disrupted social stability (England had just emerged from the Wars of the Roses, for example). As scholars like Hexter (1961) and Einstein (1962) indicate, all of this led to a medieval nostalgia, a retrograde clinging by traditional authorities like the nobility to old, strict social strata and customs. They did not like the new ambitions of the lower classes and longed for a return to the ossified social strata of the previous centuries. Scholars of the day similarly clung to old intellectual methods built around reverence for ancient texts and ideas. In sympathy with the Elizabethan aristocracy, traditional scholarship discouraged newfangled ideas and procedures, such as reliance on personal observation, experience, and experiment—as the persecution of people such as Copernicus and Galileo demonstrate.

So, although their ideas may have titillated other alchemists, Paracelsus' and Agrippa's publications made them not only notable, but infamous. They were eventually treated with suspicion and even revulsion by society—especially by the authorities. The methods and innovations that they promoted relied mainly on direct experience and experiment, rather than on received authority. Additionally, they both were pugnacious about their ideas and their rejection of tradition. Paracelsus, who had been hired as a medical professor at the University of Basel in 1527, not only tried to convince his students that personal experience as a physician trumped ancient textual recommendations, but also convinced his pupils to haul ancient texts from the library out into the quad of the university and burn them. As a result, he was stripped of his office and expelled from town. Agrippa, who was also a physician, as well as a lawyer, had similar battles. So their efforts to artificially create supernatural android servants were seen as just another arrogant form of assertive individualism—worse, a form of maverick experimentation with God's and nature's secrets, or forbidden magic (Mebane 1989).

### *4.3.2 The Golem*

As I stated earlier, there is also a Cabalistic tradition for making an android called the golem, and although it originated as a mystical practice to imitate and contemplate God's creation of Adam, by the time of the Renaissance this android's purpose was much more practical. The idea of the golem is not quite as old as that of the homunculus, but it was also a form of intellectual and social rebellion, because it was made by mystically adept Jewish rabbis as an android servant and defender. The unorthodoxy here was threefold: first, the creator was an outsider to Christian society, and, second, the golem—which was believed like the homunculus to be the result of real practice—was intended to defend the Jewish community from genocidal attacks by their non-Jewish neighbours. But, third, the creation of a golem was sometimes seen as dangerous even by Jewish commentators because its tremendous power could cause accidental danger to its creator and the Jewish community (Scholem 1965; Idel 1990).

As Scholem notes, the oldest accounts—probably medieval—of making a golem specified that a rabbi who was especially knowledgeable would, by use of a Cabalistic text called the *Sefer Yetzirah* (*Book of Creation*), build a human form out of untilled red earth. This undertaking was meant to be a form of religious ritual symbolically imitating God's creative act, a celebration and a meditation ending 'in ecstasy' (Scholem 1965, p.184). But by the Renaissance, creation of a golem had taken on the main purpose of making a super-powerful servant that served as an extension of the rabbi and the community who made it—typically as a protector against pogroms, the slaughter of Jewish communities.

The oldest of these stories in written form, according to one leading expert on Cabala and the golem, appears in an early sixteenth-century manuscript. This story relates that a religious scholar named Samuel the Pious created a golem who acted as his servant while he travelled throughout Europe (Scholem 1965, pp.198–99). A similar legend is told about the Jewish scholar Abraham ibn Ezra in a book titled *Matsref la-Hokhmah*, written by Joseph Solomon Delmedigo in 1625; there, he also tells the tale of a famous poet and philosopher named Solomon ibn Gabirol who made a female golem to serve him. This female golem is not made of earth, but instead of 'wood and hinges' and other common materials (Scholem 1965, p.199). Thus this golem, because of its constituent materials, shows a connection with the sorts of robotic androids of today and the metal automaton depicted by Robert Greene in his Renaissance play, which I will discuss later.

The most famous of Renaissance golem makers are the rabbis Judah Loew of Prague and Elijah of Chelm in Poland; legends about these men are well known and can be found in all kinds of sources even today. Both men lived in the 1500s and were famous for their intellect and knowledge of Cabala; in fact, Elijah of Chelm's full traditional title, Rabbi Elijah Baal Shem, 'means literally one who is master of the name of God, who knows how to employ it', and indicates he was 'regarded as expert in practical Kabbalah', as well as in the theology of it (Scholem 1965, p.200). Both men wrote extensively about Cabala and about the golem, and so legends arose that they actually made them (Scholem 1965; Sherwin 1985; Idel 1990). In these legends, their creations usually are meant to protect their people from attacks by non-Jews, either by spying on the Christian population of their towns or by fighting against attackers. These legends about golem defenders depict them as extremely large, superhumanly strong, and implacable. According to the stories, because they continued to grow after their creation, golems could sometimes even be dangerous to their own communities. This is famously depicted in a story about Elijah of Chelm that circulated from approximately the early seventeenth century. The legend has him trying, because of his fear of his golem's increasing size, to turn his golem back into dirt by erasing the magical letters on its forehead that animate it. But because the golem has grown so huge, its collapse into a pile of dirt injures the rabbi; in some versions of the story he is actually crushed to death. The legends about rabbi Loew begin later but are just as widely known, having appeared from the Renaissance onwards in both Jewish and Christian stories (Dekel & Gurley 2013). In many of the stories connected with Loew, the golem is used as a strong but stupid household servant; however, in one of the most famous of these stories presented in many different sources (see Scholem 1965; Sherwin 1985), Loew's golem gets loose while its creator is leading worship in the synagogue, and starts to wreck the town. The rabbi has to chase it down and turn it back into a pile of dirt.

As the rabbis' troubles with golems demonstrate, these Renaissance tales of super- powerful, artificial servants gone rogue demonstrate both the awe and fear with which society of the time viewed such experimentation. The power of these rabbis' knowledge of mysterious, Cabalistic codes and the potential power of their androids to serve them and their society is counterpoised with a fear of the potential danger of such constructions—a juxtaposition that is reminiscent of how today's AI coders are viewed by our society. The creators and their projects are Promethean in both of that word's paradoxical senses, promising humans godlike power in the process of defying the gods by taking that power for themselves; likewise with the would-be makers of the homunculus. The potential danger of creations like the golem and the homunculus is a corollary of their great usefulness: they are superhumanly powerful and so paradoxically inappropriate to own because they are hard to control and put their makers at risk.

### *4.3.3 Innovation as Transgression: The Talking Brass Head in Robert Greene's Play*

All of the foregoing types of artificial intelligent servants, which were supposed to be real ones, are echoed in some fictional stories of the day. Robert Greene's play is a good example of this, as it exhibits the creation of an artificial servant and the same cautionary themes of the perils of ingenuity that we see in accounts of making homunculi and golems, and it relies on the use of a sort of magical science that derives from the intellectual traditions used to make the homunculus and golem. The cross-pollination between alchemy, Cabala, and magic in Renaissance humanism and the spread of these combined influences in the new, wider availability of texts because of the printing press is evident in these philosophies' appearance in popular literature and plays of the day. Robert Greene's play *The Honorable History of Friar Bacon and Friar Bungay* (1594) is a satire of daring experimentalists who use such knowledge, such as the creators of androids we have discussed to this point. As previously mentioned, in England, the sixteenth century was a tumultuous time that resulted in suspicions of newfangled ideas, methods, and things, as well as of personal ambition. It was a time in which children of blacksmiths were expected to become and to stay blacksmiths, and ways of doing things were still dictated mainly by ancient and respected texts. Thus, when one of the title characters of Greene's play, the friar Roger Bacon, tries to create a talking, intelligent brass head, he ultimately signifies popular suspicion of bold innovators and arrogant ambition.

The real Roger Bacon (1214–1292) was a medieval friar and innovator, still recognized as a respected scientist today. Among his experiments was his work with prisms and their separation of light into its component colours, a discovery that has always been considered groundbreaking. His writings also include discussions of things such as diving bells, a flying machine, and ships that are able to move by themselves (Bacon 1993, Chapters 4 and 5). Like many brilliant scientific thinkers from that era who also wrote of unusual experiments and inventions, including Gerbert of Aurillac, Albertus Magnus, and Robert Grosseteste, legends grew after his death that he had tinkered with and invented humanoid machines.[2] Typically, the legends of creating humanoids were a metaphorical testimony to the awesome power of these thinkers' intellects—no matter how controversial the products of those intellects might be. The difference between these other men and Bacon is that his reputation as an intellectual rebel grew over time, where the others' diminished. This may be because he was the only one of these thinkers who was sanctioned by the Catholic Church for his ideas—which for Protestant Elizabethans made him blasphemous in two senses: his ideas were reprobate, and he therefore typified what the hated Catholic Church produced. Whatever the reason, by the Elizabethan era, his rebellious reputation had grown to the extent that he was often seen as a dangerous, rule-flouting scholar who experimented with perilous ideas to the point of using prohibited methods, such as black magic.

That is how he appears in Greene's play, written about the year 1590 and first published in 1594—two years after Greene's death. In the play, Bacon is arrogant about his abilities. At the opening of the tale, he has just arrived at Oxford and is visited by three colleagues, who ask him if the wondrous rumours that he is building an animate brass head are true. In answer, he first brags to them about his learning, saying:

> Bacon can by books
> Make storming Boreas thunder from his cave,
> And dim fair Luna to a dark eclipse.
> The great arch-ruler, potentate of hell,
> Trembles when Bacon bids him

He then brags about his current project in the same elevated terms:

> therefore will I turn my magic books,
> And strain out necromancy to the deep.
> I have contriv'd and fram'd a head of brass

> (I made Belcephon hammer out the stuff),
> And that by art shall read philosophy.
> And I will strengthen England by my skill,
> That if ten Caesars liv'd and reign'd in Rome,
> With all the legions Europe doth contain,
> They should not touch a grass of English ground.
>
> (Greene 1594, Act 2, lines 46–61)

His colleagues learn that Bacon's proto-robotic brass head is going to protect England by building a brass wall around it and that it will also be able to lecture ('read') about philosophy. They scoff at his arrogance, one of them sniffing that he 'roves a bough beyond his reach' (line 76).

The problem here is not just that Bacon is arrogant about his knowledge and abilities, but also that he is so sure of his creation and its actions. He does not care for common conventions, but is willing to accomplish his ends by whatever means he can, which includes using the dangerous codes of sorcery ('necromancy')—its incantations and formulae. Nor does he pause to consider the propriety of what his metallic android will do. If it is successfully able to give philosophical lectures, how reliable will that philosophy be, considering the demons and necromantic codes he will use to help animate it? And if the android can successfully give academic lectures, then who would need professors like Bacon and his fellows? He would essentially be putting himself and his fellow academics out of work. Furthermore, although his idea to have the brass head build a protective wall around England might sound good to us, we have to remember the stratified Elizabethan society in which Greene lived—especially the defensive paranoia of the English aristocracy of the time toward social mobility. In this context, Bacon's intention to have his robot build a brass wall would have seemed presumptuous and insulting to the Elizabethan aristocrats whose job it was to protect their country. It implies that their defenses are insufficient, that their recent victory against the Spanish Armada was a fluke, and—worst of all—that an Oxford friar presumes to understand better how to protect

England than those of the proper social class. And this all eerily foreshadows present-day issues dealing with AI.

## 4.4 Conclusion: Connections between Yesterday and Today

Bacon's arrogance about his knowledge, about his metal automaton, and about his assumptions that his plans for it will be acceptable and beneficial to his society presage today's conditions in Silicon Valley and places like it. The science fiction author Arthur C. Clarke famously said that any sufficiently advanced science would seem like magic to the uninitiated (1973, p.21). In fact, science and magic were mixed in Greene's time, so the mysterious codes that hid in the books of people like Paracelsus or even the natural philosophers we find in fictional literature like Bacon, or Prospero of Shakespeare's *Tempest*, not only seemed like magic, they were. What modern computer culture can do with its codes is also mysterious and magical to most of the public, and those who work in the digital world encourage that mystification by retaining names for their products that ally them with magicians. Just as Renaissance magic has its daemons and wizards who control intelligent agents, so does the world of computing. In modern digital parlance, a daemon is a programme that runs in the background of other programmes to make things happen. Those people who make and control online environments, such as gaming environments, often have the job title of 'wizard'. Thus, a transhistorical connection is already built into the modern culture of computing by the metaphors people of that culture use for themselves and their work.

Beyond the metaphorical, the arrogance of today's AI developers has been a constant subject of today's media. A person only has to do a quick online search using the keywords 'Silicon Valley leaders arrogant' to get a number of articles on the subject by mainstream publications. These include articles in periodicals such as the *Wall Street Journal* (Manjoo 2013), *Wired* (Griffith 2017), and *Forbes* (Swaminathan 2018). One particularly trenchant critique of AI makers is by a chief of the craft, Bill Joy, a computer engineer who cofounded Sun Microsystems (an important early software company that created the coding that still runs much of the backend programmes in many computing systems). A coder and Silicon Valley billionaire himself, his article 'Why the Future Doesn't Need Us' (2000) complains of the dread he has about a future in which digital creators like him are not reined in. This is because he sees them as smug about their knowledge and dangerously complacent about the products of their labour: they assume, like Bacon, that those products are uniformly harmless and good. In particular, Joy worries that he and other computer engineers may end up making an AI that will displace us as masters of our planet. As he says:

> . . . with the prospect of human-level computing power in about 30 years, a new idea suggests itself: that I may be working to create tools which will enable the construction of the technology that may replace our species. How do I feel about this? Very uncomfortable. Having struggled my entire career to build reliable software systems, it seems to me more than likely that this future will not work out as well as some people may imagine. My personal experience suggests we tend to overestimate our design abilities.

Like Bacon's colleagues at Oxford worrying about his project, Joy, in the last of his sentences above, is anxious that his coworkers 'rove a bough beyond their reach' without realizing it, because they overestimate their abilities. In Joy's opinion, the social consequences of what their digital creations may add up to could be catastrophic.

An even deeper problem we have in parallel with the Renaissance is that all of the creators of AI and proto-AI that we have considered risk their own displacement, and possibly the dislocation of their societies. The Renaissance worried about the disruption of God's framework by the characters I have described. As Joy worries, in our age the master not only risks replacing himself, but his whole species—a potential modern disruption of our existential framework. Today, at a more mundane level, we have actual existential disruptions around the world caused by AI: we have already seen many jobs disappear because we are replacing ourselves at work with automated slaves (LaGrandeur & Hughes 2017). Both ancient and modern versions of artificial servants are great threats because in their respective scenarios the artificially intelligent slave threatens, each in its own way, to replace its master through the kind of dialectic process Hegel described: offloading more and more work to the slave, the master becomes less able to do work, and so becomes weaker, while just the opposite happens to the slave. Norbert Wiener, widely seen as the father of modern robotics, worries about this in his midcentury writings regarding his own creations, especially because he anticipated that we would cede too many important decisions to AI (Wiener 1954; Wiener 1964). In fact, our current AI is tending further in that direction. It was meant to make our work easier, our shopping easier, and our decisions easier, but as the above comments of Bill Joy, one of its creators, indicate, it is fast becoming a bogeyman (especially online AI that collects and leverages data—see for instance the recent interview of computer scientist Yoshua Bengio in Castelvecchi (2019)).

For Bacon, Paracelsus, Agrippa, and the rabbis of legend in the Renaissance, intelligent creations such as they hoped to make were also meant to make their work easier—and for some, to make them famous. But in the various narratives about them, they made artificial agents that were paradoxically far too powerful for them to control effectively, and which threatened their own agency. Therefore, the lesson that most of these stories carry is as follows: that they need to be careful to stay within the natural bounds of human ambition.

The kinds of intelligent systems, or intelligent tools, outlined by Aristotle and instantiated in Renaissance literature and embodied in today's AI show how it and its imaginary predecessors have always been about more than automating basic repetitive tasks.

Humans' consistent dream over the millennia has really been about artificially intelligent systems that can do things with greater power than the typical human can, expanding their natural limits and power over nature. In other words, our dreams of AI and proto-AI have always been about building superhuman prostheses, virtual enhancements of our senses and agency.

The potential problems with servants of any kind begin with that last word—agency. Human slaves and artificial slaves enable greater power and agency on the part of their masters or makers; however, the biggest worries of the ruling contingent of every culture, from Aristotle's, to Shakespeare's, to ours, are based on how strong and extended that agency might become. Aristotle warns that human slaves can become rebellious, especially if they have had a taste of strong agency—for instance, if they were once powerful people before their enslavement (section 1255a, lines 31–36). In our own time, we worry about this with our own artificial slaves—especially the almost human ones we imagine we might have. Just consider the examples of this in the imaginative work of our day (not to mention in Bill Joy's nightmare of our real future with AI). Science fiction stories and movies are rife with apocalyptic scenarios where extremely smart AI flips the master–servant relationship: TV shows like *Westworld* and *Battlestar Galactica*, and films like *Ex Machina* and the *Terminator* series depict extremely powerful AI given the power of independent thought and the havoc that causes for their makers. The fact that such fiction reflects real worries is demonstrated by journalists' common use of the name of the last of these movies as a term for what people fear may happen with superintelligent AI: the Terminator scenario (see for example Austen 2011; Parsons, 2015). In this situation, advanced AI becomes more intelligent than us, sees us as pests, and decides to wipe us out.

Ultimately, superhuman slaves—and our worries about them—reflect an archetypal need to supersede our natural limits by whatever ingenious means, even if by developing super-powerful artificial servants we threaten the very dominion we seek. The historical accounts in this chapter, in other words, are about being blinded by our intellectual enthusiasm to the danger of our own intellectual products, becoming enslaved by those things that are meant to serve us, and being bound rather than liberated by the ambitions that produce them. On the other hand, they are also about the power and potential of our technology and know-how. These are the many elements they have in common with our present-day relationship to AI and the tales we tell ourselves about them.

## Notes

1. Genevieve Lively discusses fascinating ancient ideas that foreshadow AI at much more length in her chapter in the present book.
2. Many have discussed the legends and stories regarding famous medieval scientists and philosophers creating intelligent machines, starting with Bruce (1913) and Dickson (1929); and more recently Higley (1997), myself (LaGrandeur 1999; LaGrandeur 2010; LaGrandeur 2011; LaGrandeur 2013), Kang (2011), and Truitt (2015) have all discussed them at some length.

## Reference List

Agrippa, C. (1651) *Three books of occult philosophy*. Translated from Latin by J. Freake. London.

Aristotle (1944). *Aristotle in 23 volumes*. Vol. 21. [online] Translated by H. Rackham. Medfore, MA, Tufts University, Perseus digital library. Available from: http://www.perseus.tufts.edu/hopper/text?doc=Perseus%3Atext%3A1999.01.0058%3Abook%3D1%3Asection%3D1253b [Accessed 18 December 2018].

Austen, B. (2011) The terminator scenario: are we giving our military machines too much power? *Popular Science* [online]. 13 January. Available from: https://www.popsci.com/technology/article/2010–12/terminator-scenario [Accessed 20 December 2018].

Bacon, R. (1993) *Roger Bacon's letter concerning the marvelous power of art and of nature and concerning the nullity of magic*. Translated by T.L. Davis, 1923. Kila, MT, Kessinger.

Bruce, D. (1913) Human automata in classical tradition and medieval romance. *Modern Philology*. 10, 511–26.

Castelvecchi, D. (2019) AI pioneer: 'the dangers of abuse are very real'. *Nature*, [e-journal] 4 April. doi 10.1038/d41586-019-00505-2.

Clarke, A. C. (1973) *Profiles of the future: an inquiry into the limits of the possible*. New York, Popular Library.

Dekel, E. and Gurley, D. G. (2013) How the golem came to Prague. *The Jewish Quarterly Review*. Spring Issue, 241–58.

Dickson, A. (1929) *Valentine and Orson: a study in late medieval romance*. New York, Columbia University Press.

Einstein, L. (1962) *Tudor ideals*. New York, Russell & Russell.

Greene, R. (1594) *The honorable history of Friar Bacon and Friar Bungay*. Edited by D. Seltzer, 1963. Lincoln, University of Nebraska Press.

Griffith, E. (2017) The other tech bubble. *Wired* [online]. Available from: https://www.wired.com/story/the-other-tech-bubble/ [Accessed 11 February 2019].

Homer. *Iliad*. [online] Translated from Greek by A.T. Murray, 1924. Medford, MA, Tufts University, Perseus digital library. Available from: http://www.perseus.tufts.edu/hopper/text?doc=Perseus%3Atext%3A1999.01.0134%3Abook%3D18%3Acard%3D388 [Accessed 19 December 2018].

Hexter, J. H. (1961) The myth of the middle class in Tudor England. In *Reappraisals in history*. Evanston, IL, Northwestern University Press, pp.105–11.

Higley, S. (1997) The legend of the learned man's android. In: T. Hahn & A. Lupack (eds.) *Retelling tales: essays in honor of Russell Peck*. Rochester, NY, Brewer. pp. 127–60.

Idel, M. (1990) *Golem: Jewish magical and mystical traditions on the artificial anthropoid*. Albany, SUNY.

Joy, B. (2000) Why the future doesn't need us. *Wired* [online]. Available from: https://www.wired.com/2000/04/joy-2/ [Accessed 11 February 2019].

Kang, M. (2011) *Sublime dreams of living machines: the automaton in the European tradition*. Cambridge, MA, Harvard University Press.

LaGrandeur, K. (1999) The talking brass head as a symbol of dangerous knowledge in *Friar Bacon* and in *Alphonsus, King of Aragon*. *English Studies*. 80(5), 408–22.

LaGrandeur, K. (2010) Do medieval and Renaissance androids presage the posthuman? *CLCWeb: Comparative Literature and Culture*. 12(3), *https://doi.org/10.7771/1481–4374.1553*.

LaGrandeur, K. (2011) The persistent peril of the artificial slave. *Science Fiction Studies*. 38(2), 232–52.

LaGrandeur, K. (2013) *Androids and intelligent networks in early modern literature and culture: artificial slaves*. New York, Routledge.

LaGrandeur, K. and Hughes, J. J. (eds.) (2017) *Surviving the machine age: intelligent technology and the transformation of human work*. New York, Palgrave.

Las Casas, B. de. (1583) *The Spanish colonie: or briefe chronicle of the acts and gestes of the Spaniardes in the West Indies*. Translated from Spanish by M.M.S., 1583. London.

Manjoo, F. (2013) Silicon Valley has an arrogance problem. *The Wall Street Journal* [online]. Available from: https://www.wsj.com/articles/silicon-valley8217s-superiority-complex-1383527053 [Accessed 11 February 2019].

Mebane, J. S. (1989) *Renaissance magic and the return of the golden age*. Lincoln, University of Nebraska Press.

Newman, W. (2004) *Promethean ambitions: alchemy and the quest to perfect nature*. Chicago, University of Chicago Press.

Paracelsus (1573) Concerning the nature of things [De natura rerum]. In: A. E. Waite (trans. and ed.) (1894) *The hermetic and alchemical writings of Paracelsus, vol. 1*. 2 vols. London, James Elliott and Co. pp.120–94.

Paracelsus (1941) A book on nymphs, sylphs, pygmies, and salamanders, and on the other spirits. In: H. Sigerist (trans. and ed.) *Four treatises*. Baltimore, Johns Hopkins University Press. pp. 213–54.

Parsons, J. (2015) 'Everybody working in artificial intelligence knows the terminator scenario': Futurologist explains when killer robots could endanger humanity. *Mirror* [online]. 3 November. Available from: https://www.mirror.co.uk/news/technology-science/technology/everybody-working-artificial-intelligence-knows-6759244 [Accessed 20 December 2018].

Scholem, G. (1965) The idea of the golem. In: *On the Kabbalah and its symbolism*. New York, Schocken Books. pp. 158–204.

Sharp, C. (1981) Caliban: the primitive man's evolution. *Shakespeare Studies*. 14, 267–83.

Sherwin, Byron L. (1985) *The golem legend: origins and implications*. New York, University Press of America.

Swaminathan, S. (2018) Does Silicon Valley have an arrogance problem? *Forbes* [online]. Available from: https://www.forbes.com/sites/quora/2018/05/21/does-silicon-valley-have-an-arrogance-problem/#22374e3c3eda [Accessed 11 February 2019].

Thorndike, L. (1923–1958) *History of magic and experimental science.* 8 vols. New York, Macmillan.

Truitt, E. R. *Medieval Robots: Mechanism, Magic, Nature, and Art*. Philadelphia, University of Pennsylvania Press.

Wiener, N. (1954) *The human use of human beings: cybernetics and society.* 2nd ed. Garden City, NY, Doubleday.

Wiener, N. (1964) *God and golem, inc.: a comment on certain points where cybernetics impinges on religion.* Cambridge, MA, MIT Press.

# 5

# Making the Automaton Speak

## *Hearing Artificial Voices in the Eighteenth Century*

*Julie Park*

## 5.1 Introduction

What might the eighteenth-century history of automata tell us about the status of voice as an index of the human? How might the suggestion if not reality of a speaking automaton in an eighteenth-century context help us understand what it means to have a voice, whether as a human or a machine? Given the rise in our daily lives of voice-operated 'intelligent assistants' at the time in which this chapter was written, the question is especially pertinent. In the case of a 'speaking' doll brought to England from France around 1784, artificial voice was tied up with questions about projection and authenticity. Unlike the Jacquet-Droz family's and Jacques Vaucanson's celebrated automata of the same period, which were made to manipulate various tools and instruments to communicate through written words, drawing and music, this instance of an automaton—or purported automaton—distinguished itself by replicating the human ability to express itself through a resource of its own body, voice.

An Englishman, Philip Thicknesse (1719–1792), exposed not only the doll's fraudulence but also raised questions about cultural-historical desires and assumptions related to voice and artificial intelligence in his pamphlet *The Speaking Figure, and the Automaton*

*Chess-Player Exposed and Detected* (1784).[1] Thicknesse claimed that the automaton, a doll 'the size of a very young child', was not actually speaking by mechanical means, but was only made to seem as if it were. In fact, an elaborate but decidedly rudimentary system was used, involving a speaking trumpet, a hole in the ceiling and the 'very large plume of feathers' on the doll's head that conceals the trumpet's other end, through which the sounds of its 'voice' were emitted by a hidden man (Thicknesse 1784, pp.6–7). The typographical registers of Thicknesse's indignation over its imposture are telling:

> I therefore think it a piece of justice ... to draw aside the curtain, open the head, and shew, that the *brains of this wonderful Doll* is nothing more than the continuation of a tin tube, which is fixed to its mouth, so as to convey the Question and Answer to and from an invisible Confederate. That the human voice may be imitated, and many or most words, articulated by valves, and bellows, like the barrel organ, there is no doubt; but that a mechanical figure can be made to answer all, or any such questions, which are put to it, or even put a question, is UTTERLY IMPOSSIBLE. (pp.4–5)

Thicknesse reveals in his response the disappointment of a desire to see the marvel of artificial intelligence made into a reality and transcend the status of fiction. While he accepts the possibility that a mechanical being is able to imitate human speech by mechanical means, the doll he is denouncing cannot even realize such a triumph of technology. He is more angry over the showmen's fraudulence in presenting the doll as an automaton—as if its 'movements are REALLY performed by mechanic powers'—that can answer questions and thus possess a form of artificial intelligence, than he is over the idea that a doll could be mechanically made to imitate human speech (p.5). At the same time, his anger suggests an intolerance of the fictive nature not so much of the doll's status as a human, but its status as an actual automaton that is able to converse, which would be a true mark of its intelligence. As Thicknesse's outrage presents a key distinction between the reproduction of voice and the reproduction of intelligence in the face of their apparent confusion, it invites us to reflect on the following: to what extent does voice—natural or artificial—implicitly equate with the possession of intelligence? Not just this, but what are the material and phenomenological grounds, if any, on which voice can be registered *as* voice in order for such an equation to be made at all?

Thicknesse appears to have been quite singular in his intolerance of the speaking figure's technological pretence; the popularity of automata venues and spectacles of all kinds, both authentic and inauthentic, from the exquisite creations at Cox's Mechanical Museum and Merlin's Mechanical Museum to the very ones he decries in his pamphlet, makes this clear. Other efforts to debunk the ruse of eighteenth-century automata, such as the automaton chess player, for instance, did not take place until the nineteenth century. Indeed, for someone of Thicknesse's temperament, with its strong appetite for controversy, the popularity of automata whose actual fraudulence was a matter of apparent indifference to most, would be among the chief reasons for his ire. The public's enthusiasm for automata of varying levels of authenticity—even Jacques de Vaucanson's automaton duck that defecated did not actually excrete through a full process of digesting the corn pellet it took in its mouth—and inclination to believe any impostures, is consistent with a general cultural enthusiasm for fiction as a source of entertainment. The period in which automata were celebrated by the public was also the period in which the novel enjoyed its rise, after all.

Even if Thicknesse appeared to be exceptional in his response, the grounds of his indignation raise critical issues that are significant. The premise of feminist philosopher Adriana Cavarero's argument in *For More Than One Voice*, for instance, elaborates on the political dimensions of the distinction between voice and intelligence implicated in Thicknesse's objections: rather than fusing voice with speech, making voice interchangeable with the culturally privileged yet fundamentally distinct category of logocentric meaning, voice ought to be recognized as having its own ontology and phenomenology (Cavarero 2005). According to her, voice, as a phonic phenomenon—that is, based on sound rather than meaning (or semantic)—'precedes, generates, and exceeds verbal communication.' As such it is 'the condition of every communication' and 'the significance of signification' (pp.29–30). How might this rich claim be applied to artificial voice and the forms of communication it enables, including the fictions it generates?

In our own historical condition of living with technological marvels, there is a similar resistance to the fictive status of artificially created voices, and a need to determine their 'true' origins. Recently users unsatisfied with the idea that the voice-activated personal assistant Siri is only a machine with humanoid characteristics, sought upon its release to establish its 'true' identity by discovering the human voice actor on whom its voice was modelled (Ravitz 2013; Tafoya 2013). In contrast, the eighteenth-century scenario represented in Thicknesse's pamphlet shows a human subject needing to determine if the doll's origins of speech are really mechanical, and expresses contempt over its actual human (the 'invisible Confederate') source. The early twenty-first century example, however, shows humans accepting the speaking device's state of technological advancement, but *wanting* to find an actual and hidden human source for it. While the cultural desires evinced in each period seem contrary to each other, they at bottom reveal an ultimate need to discover the truth of human sources behind fictions of artificial voice. The differing trajectories for determining such truth make sense, considering that, in one case, the fiction is one of a nonmechanical, human-powered entity purporting to have a mechanically produced voice, in the other, the fiction consists of a mechanical device that appropriates human characteristics, such as a female voice and name. In both cases, the machines posit a relationship between voice and authenticity.

While the ability to speak and converse with humans is suggested as the apogee of technological advancement, it is reckoned as such primarily because the possession of voice is in itself an obdurate sign of being human. Such a valuation of voice can be found in elocution guides of the eighteenth century, as will be addressed later. Jean-Luc Nancy supports this view as well when indicating that the concept of person is etymologically bound up with the ability to make sounds: 'the word person might be considered as derived from personare, "to resound through"' (2013, p.243, quoted in MacKendrick 2016, p.17). In turn, resonance is what allows voice to register its often overlooked materiality, as Karmen MacKendrick reminds us, thus leaving artificial and actual voices with the same properties of vibration and of having an effect on our experience of space (2016).[2]

This chapter traces the historical precedent for these theoretical articulations about voice and the material basis of its resonances in such sources as Thicknesse's pamphlet, as well as Wolfgang von Kempelen's account of creating a functional speaking machine in *Le mécanisme de la parole, suivi de la description d'une machine parlante* (1791). In doing so, it will provide a historical framework for probing why and how the possibilities of artificial voice shed light on our deep investments in the notion of voice and voice communication as the ultimate signs of being 'real' and having intelligence as humans. In the case of eighteenth-century speaking machines, the effort to manufacture voice and speech was grounded in practices of mediating human functions and identity through mixed materials that mimicked not just external body parts, but the internal organs that made the production of voice possible at all.

## 5.2 Exposing the Speaking Figure

As indicated earlier, the author of the speaking-doll exposé, Philip Thicknesse (1719–1792), was no stranger to outrage and issuing public objections. Typical of many eighteenth-century lives, his was filled with diverse occupations, roles, and accomplishments, from travel and memoir writer, one-time officer and slave rebellion controller in Jamaica, long-time friend of Thomas Gainsborough, and alleged discoverer of his art, and owner of a broadsheet press that disseminated his political views, to ornamental cottage restorer and self-styled hermit whose meticulously remodelled Bath home became a tourist attraction. He was so known for his habit of indignation, tactlessness, and shamelessly public and bitter feuds with almost everyone he knew, including his own sons, he was given the name 'Dr. Viper' by the dramatist Samuel Foote in a play (Gosse 1952; Turner 2008). Certainly, thanks to Thicknesse's irascible nature, propensity for articulating and publishing his opinions, no matter how private the subject, and alertness to activities of all kinds in his cultural milieu, we have a permanent record of the speaking figure, as well as Kempelen's well-known chess player automaton, which was far more documented and is already the subject of other recent studies.[3] Thicknesse's keenness to unmask the speaking figure's fraudulence resulted in a detailed account of its method of contrivance. His account also indicates the power of the dream for artificial intelligence during his time, while simultaneously revealing the technical impossibility of realizing it. It appears fitting that the man who could not suppress his own voice in so many ways throughout his life (finding in the printing press an ideal artificial medium for it), would vigorously denounce a doll whose voice was inauthentic, and derived in fact from a human power outside of itself.

When Thicknesse makes the distinction between a machine that imitates the human voice 'by valves, and bellows, like the barrel organ' and one that can answer any question it is asked, he not only declares the latter 'UTTERLY IMPOSSIBLE,' but also invokes a hypothetical model of artificial intelligence. Definitions of AI, 'the attempt to make computers behave intelligently, particularly in real-world situations,' or the effort to 'make computers do the sorts of things that minds can do,' can certainly apply to the activities the 'speaking figure' is made to feign performing (Caudill 1992, p.12; Boden 2016, p.1). By making direct reference to the doll's 'brains' when making his revelation and marking the words with italics ('I therefore think it a piece of justice [...] to draw aside the curtain, open the head, and shew, that the *brains of this wonderful Doll* is nothing more than the continuation of a tin tube'), he reinforces the notion that its imposture lies in its purported possession of intelligence. Brains equate with the seat of intelligence, after all. Indicating his awareness of where some measure of authenticity in mechanically derived intelligence might actually begin to be found, he defines an automaton as 'a self-moving Engine, with the principle of motion within itself' (Thicknesse 1784, p.4). Furthermore, he demonstrates historical knowledge about automata in bringing up Archytas's wooden dove (fourth century BCE) and Regiomontanus's wooden eagle and iron fly (both fifteenth century) as prior examples of machines that allegedly either flew or jumped on their own (p.4).

The speaking-doll entertainment cost English crowds half a crown for admission; in Paris, from where it came, three or four sous were charged. Rather than containing the 'principle of motion' or speech within itself, the speaking figure, like the mechanical chess player with which it was being 'exposed and detected' in the same pamphlet, had a hidden man performing the actions that were seen as her own. In prose made energetic and focused by his outrage over the two popular but fraudulent mechanical figures, Thicknesse announces their true status: 'The *Speaking Figure* is a man in a closet above, and the Automaton Chess-Player is *a man within a man*' (1784, p.16).

But to anyone not alert to the deception, the demonstration appeared to be something else. Picture the following (Figure 5.1). A doll made of plaster of Paris, 'about the size of a very young child,' is hung by a ribbon around her waist in the doorway of a room just beyond the space into which the audience is admitted. A foot-long tin tube shaped like a speaking trumpet is inserted in her mouth, into which a human 'Questioner' asks the doll her age in several languages: French, English, German, or Italian. Thicknesse reveals that he (himself a lover and an occasional

**Figure 5.1** Frontispiece (Thicknesse, 1784). RB 329173. The Huntington Library, San Marino, California.

resident of France) was among the spectators who asked the doll the question in the language of her country of origin, to which 'a faint whispering reply' came: '*J'ai dix-huit*' (Thicknesse 1784, p.6). And yet, apart from Thicknesse's question and the material from which the doll was made, as well as the way she was positioned, none of this was actually what it appeared to be. The doll herself was not really speaking from an internal mechanical source, and she was not really able to understand and respond to the question being asked of her.

As suggested earlier, the speaking figure's scam involved an architectural arrangement that would allow an accomplice to be concealed. As Thicknesse explains, the speaking figure's performance of artificial intelligence ultimately hinged on the use of an interior room, where the ceiling aperture was hidden, and the '*question* and *reply*' could thus 'pass through a tin tube fixed in a head of human form' (Thicknesse 1784, p.8). By being situated at the threshold of a space beyond view of the audience, the aperture in the ceiling was concealed. Through this aperture, the man hidden in the interior room above could speak through his own mouthpiece into the horn behind the doll's large feathered headpiece (Figure 5.1). The horn behind the plume was the other end of the speaking trumpet that appeared to be inserted into the doll's mouth, but was in fact protruding from the hidden endpiece. Whatever sounds were issued out of and into the doll's speaking trumpet, including the Questioner's questions and the hidden man's answers, travelled in and out of the same channel. The hole in the ceiling ensured that whatever words were whispered into the two-sided mouthpiece from below reached the 'ear of the *prostrate confederate above*' (p.7). Thus, the speaking figure, in its technique of ventriloquism, generated its illusion from the 'transcendence or disruption of seen space,' as Steven Connor would put it (2000, p.15).

The system's one shortcoming was the fact that the hidden confederate, unable to bring his mouth close enough to the trumpet in the doll's head, could never project his words as clearly as he could hear the questions being passed through the plaster of Paris doll's mouthpiece. Whatever words came through the hole from the ceiling were heard by the audience only as a faint whisper. Nevertheless, the oddity of the whispering voice must have appeared to suit the bizarre spectacle of a partially naked buoyant child with a large plume in her head being spoken to by a grown man with his mouth against a trumpet piece in her mouth. Adding to this picture of incongruity, was her claim of being eighteen—'*j'ai dix-huit*'—when asked her age (though this may also have been meant to refer to the age of the doll itself, not the age it was represented to be). In his scrappy determination to expose and understand the ruse and prove his theory about its fabrication, Thicknesse replicated at home in front of his friends, with a plaster of Paris bust, the contrivance and procedure for making an inanimate doll 'speak.' This test allowed him to pronounce his hypothesis was 'not mere conjecture, but the fact' (Thicknesse 1784, p.8). The feat of mechanical ingenuity that the speaking figure represented came down to nothing else but a 'double-ended speaking trumpet in the doorway' (p.8). Thus, the grounds of Thicknesse's condemnation of the speaking figure lie predominantly in the way its voice was not so much manufactured or created, but was *projected* and staged, and originated from an external human source. Thicknesse revealed not just that the artificial status of the doll's voice was in itself an artifice, but in doing so also uncovered that the artificial intelligence the voice supposedly signalled was equally contrived and inauthentic.

## 5.3 Making Heads Talk, Simulating Organs of Speech, and Mixing Media

Prior to Thicknesse's pamphlet, machines that spoke or 'talking heads' had long been a feature of the cultural mythology surrounding artificial life, going back to the oracular head of Orpheus at Lesbos. In the Middle Ages, talking automata were said to have been created by Pope Sylvester II (c. 946–1003), Robert Grosseteste,

Bishop of Lincoln (c.1175–1253), Albertus Magnus (1204–1282), and Arnaldus de Villa Nova (c.1240–1311). The most celebrated of speaking automaton creators was Roger Bacon, whose brazen head, as Albertus's mechanical man was said to do, could answer questions (Truitt 2015; Cohen 1966). In the seventeenth century, the natural philosopher John Wilkins not only addressed the topic of talking machines in *Mathematicall Magic* (1648), but also, according to John Evelyn, kept in his garden a '"hollow Statue which gave a Voice, & utterd words, by a long & conceald pipe which went to its mouth, whilst one spake thro it, at a good distance, & which at first was very Surprizing" ' (1955, 3:110, quoted in Hankins & Silverman 1995, p.181). In the eighteenth century, then, the idea that inanimate bodies could be made to speak was by no means a new one. Yet it became a renewed topic of interest for machinists (creators of machines) and philosophers of the eighteenth century (Riskin 2016). In the context of this time period, higher standards and new resources were presented for realizing the possibility of a speaking machine.

The brilliant machinist Jacques de Vaucanson, with his trio of remarkably lifelike and technically accomplished automata of the 1730s—the tambour and pipe player, flute player, and eating and defecating duck—set a new standard for realism in automata while prompting a wave of machines that hoped to fulfil it. This standard entailed not just imitating the movements of living entities, but also, *simulating* the internal and external processes of their real-life activities, from playing the flute with the technique and precision of a human flute player to eating and digesting a corn pellet to the endpoint of its excretion, as a real duck might (Riskin 2003a; Riskin 2003b).[4] Fontenelle, Perpetual Secretary of the Royal Academy of Sciences, observed many of the subtle mechanisms involved in making Vaucanson's flutist a success with the Academy:

> they have *judg'd* this Machine to be *extremely ingenious*, and that the Author of it has found the Means of employing new and simple

> Contrivances, as well for giving the Fingers of that Figure the necessary Motions, as for modifying the Wind which goes into the Flute by increasing or diminishing its Velocity, according to the different Notes; by varying the Position of the Lips, and moving a Valve which performs the Office of the Tongue; and lastly, by imitating Art all that is necessary for a Man to perform in such a Case. (Vaucanson 1742–1748, p.166)

So influential and captivating were Vaucanson's creations in the intellectual and popular worlds of the eighteenth century, the period saw a flow of machines that enacted seemingly *or* authentically the different functions and activities of life, from breathing, speaking, walking, writing letters, and playing musical instruments to giving birth (Riskin 2016, p.123). The organic quality of the materials used for creating such automata indicated the new level of verisimilitude that was being introduced and observed during this time.

Vaucanson's six-and-a-half-foot-tall flute playing machine had a lip, tongue, and fingers made of leather to 'imitate the softness of the natural Finger' (Park 2011, p.152). The internal organs of both the tambour player and flute player were movable and flexible too, including their tongues and lungs (Riskin 2016, p.120). The flutist was so verisimilar in its movements, technique, and production of sound, Vaucanson's account of his creation was used to instruct others on how to play the instrument and is considered an important source in historical musicology. Thus, the flute player not only appeared to be playing its instrument, but it was actually playing the instrument as a human would, with the same parts of the body enlisted and in motion. Other machines of this time period that used leather as their 'skin' for greater similitude with the human body parts they imitated included Madame du Coudray's sponge-filled obstetric machine and Henri-Louis Jacquet-Droz's mechanical pair of prosthetic hands.

The emphasis on the use of organic materials for creating automata likewise extended to a view that speech was a function of the organs and thus could only be simulated with the living human's anatomical parts of a larynx and glottis, a view held by Abbé Desfontaines. According to Riskin, 'speaking was an essentially organic process' for Desfontaines, and 'could only take place in a living throat' (2016, p.158). Yet Julien Offray de la Mettrie was able to ascertain after viewing Vaucanson's flute player that the possibility of artificial speech was well within sight. A never realized project for Vaucanson was a machine that replicated the circulation of blood, commissioned by Louis XV. Rubber from Guyana, known by its French name, 'caoutchouc', was to be used for the veins (Riskin 2016, pp.135–36; Hankins & Silverman 1995, pp.183–84). The plan for this particular automaton, although unexecuted, holds importance in the greater project of voice simulation, for, as Thomas L. Hankins and Robert Silverman point out, projects of replicating circulation systems and those of replicating voice were interlinked, which Vaucanson's contemporaries demonstrated (Hankins & Silverman 1995, p.184). Indeed, the effort to recreate voice was viewed as a natural next-level challenge for already accomplished automaton makers. Jean Blanchet, in his *L'art du chant* (1755), speculated that only someone with the technical virtuosity of Vaucanson would be able to recreate the organs required to simulate speech:

> One could imagine & make a tongue, a palate, some teeth, some lips, a nose & some springs whose material & figure resemble as perfectly as could be possible those of the mouth: one could imitate the action that takes place in these items for the generation of words: one would be able to arrange these artificial organs in the automaton of which I have spoken. From then on, it will be capable of singing, not only the most brilliant airs, but also the most beautiful verse. Here is a phenomenon that would demand all the invention & industry of an Archimedes, or else a Vaucanson, & that would astonish all of learned Europe.
> (Blanchet 1755, 47–48, qtd. in Hankins and Silverman 1995, p.184)

Imagining that Vaucanson could create the organs and body parts required for artificial speech, including the most eloquent and beautiful forms of vocal expression, such as singing verse, was not far-fetched considering the level of anatomical precision and technical mastery his automata displayed. The flute player evinced control of his own lips and tongue, skills that a speaking automaton might need to employ, and played his instrument as a human *would* 'in such a case,' as Fontenelle put it (see above). Most striking in the passage from Blanchet is his assumption that a mechanician might be able to manufacture all the body parts that generate speech, from lips and tongue to teeth, palate, and nose, and arrange them correctly.

Actual notable speaking machines that followed Vaucanson's masterpieces—that were apparently as much organic as they were mechanical—included Erasmus Darwin's speaking automaton, created in 1771. This speaking automaton supported an organ-oriented theory of speech, and came into being through Darwin's curiosity about language and its origin. He described this machine in *Temple of Nature* (1803):

> I have treated with greater confidence on the formation of articulate sounds, as I many years ago gave considerable attention to this subject [...] at that time I contrived a wooden mouth with lips of soft leather, and with a valve over the back part of it for nostrils, both which could be quickly opened or closed by the pressure of the fingers, the vocality was given by a silk ribbon about an inch long and a quarter of an inch wide stretched between two bits of smooth wood a little hollowed; so that when a gentle current of air from bellows was blown on the edge of the ribbon, it gave an agreeable tone, as it vibrated between the wooden sides, much like a human voice.
> (1803, pp.119–20 note 13, qtd. in Hankins & Silverman 1995, p.184)

Darwin's description draws attention to the materiality and craftsmanship of his venture. Pieces of soft leather, wood, and silk ribbon, strategically arranged and shaped to resemble body parts, with an air current applied to them by bellows, *made up* voice.

Drawing direct comparisons with Vaucanson's machines were Abbé Mical's speaking bronze heads whose mechanisms, according to one account, consisted of bellows, valves, and multiple artificial glottises that were placed against stretched membranes (Vicq d'Azyr 1783, 2:204, qtd. in Giannini 1999, p.2535). After thirty years of labour, they were completed in 1783 and exhibited the same year in Paris. Another account—the automata were allegedly destroyed by their creator—describes their mechanism as consisting of not bellows and artificial glottises, but a cylinder and a keyboard (Rivarol 1818, 2:207–2:246, qtd. in Giannini 1999, p.2535). Given these details, Thicknesse seems to have been referring to Mical's speaking heads when he declared in his 1784 pamphlet that 'there is no doubt' that 'the human voice may be imitated, and many, or most words, articulated by valves, and bellows, like the barrel organ' (p.4). Features of the barrel organ are its cylinder ('barrel'), in which sounds can be encoded, and its bellows. Mical's speaking heads most likely operated on both. Indeed, later in his pamphlet, when inveighing against the automaton chess player, Thicknesse mentions Mical's 'two colossal heads' as being responsible for having 'given rise to these two impositions, the *Speaking Figure* and the Chess-playing Turk.' Yet, unlike those two sham machines, Mical's heads actually 'speak certain words distinctly' and are 'the production of many years labour' (Thicknesse 1784, pp.15–16).

## 5.4 Kempelen's Speaking Machine: Mimetic Materials and Artificial Voice

Yet another great irony underlying Thicknesse's performance of indignation in his pamphlet is the fact that Wolfgang von Kempelen, the Hungarian creator of the automaton chess player whose fraudulence he exposed alongside the speaking figure, was also the creator of an authentic speaking machine. According to Thicknesse, the chess player's main point of deceit is that it is, '*a man within a man* [...] whatever his outward form be composed of, he bears a living soul within.' Just as the chess pieces themselves were moved by a human force with the help of mirrors, so too was the Oriental figure of the Turkish chess player moved—or manipulated—by the same human force. Kempelen, a councillor for the Hapsburg court, indicates in his account of creating his speaking machine that it was when he was making the chess player for Empress Maria Theresa of Austria that he was led to create the speaking machine:

> I do not absolutely remember the first cause that furnished me with the idea to imitate human speech, I only remember that while I was working on my chess player, in the year 1769, I began to investigate several musical instruments with the intention of finding what seemed to approach the human voice the most.
> (Kempelen 1791, p.396)

Deeply interested in phonetics and the origins of language, Kempelen viewed his speaking machine as having the potential to aid those with speech impediments, and those who were born deaf and mute. Not just an experiment of imitation, the speaking machine was to have real-life relevance and application, much as Vaucanson's account of creating his flute player became a pedagogical guide for human flute players. Kempelen also based the creation of his speaking machine on a system of mimetic organs made out of different materials, such as bellows for lungs, an ivory reed for a larynx, and a rubber funnel for a mouth (Figure 5.2) (Kempelen 1791, pp.259, 404; Hankins & Silverman 1995, pp.193–94). Unlike the speaking figure, Kempelen's machine did not look like a human. And yet also unlike the speaking figure, his machine did actually speak, uttering words—from 'Marianna' and 'Anatomie' to 'Monomotapa', as well as short phrases, such as '*Maman aimez-moi*' (Kempelen 1791, pp.viii–ix; Chapuis & Gélis 1928, p.2:208).[5] How might the project of devising a way to create a sham automaton that looked like a human figure, the life-sized Turk, playing chess through mechanical means, lead to creating a real automaton that appeared to be nothing other than what it actually was—a machine—but could create human sounds? This trajectory demonstrates that 'true' artificial voice was viewed as a

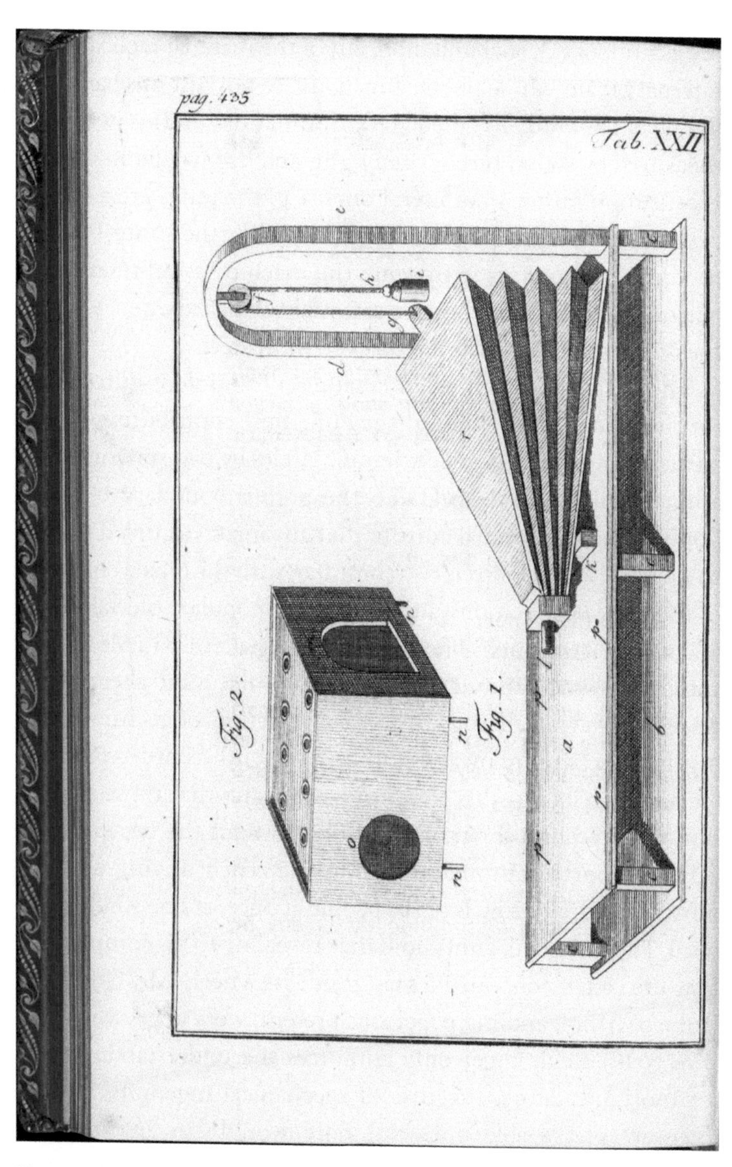


**Figure 5.2** Speaking Machine (Kempelen, 1791, 434, Tab. XXII ), P221. K321m, Library Special Collections, Charles E. Young Research Library, UCLA.

project not of external imitation, but of the material replication of internal organs whose phonic functions are not just mechanically, but also, figuratively mediated. Simulation in this way was measured by sound rather than sight, and by imaginative acts of substitution rather than literal ones (e.g. the lungs are a pair of bellows, not an actual replication of lungs). Furthermore, because it compels onlookers to imagine the machine as representing a person while hearing the 'voice' it makes, the experience of the speaking machine enlists fiction even more deeply.

The paradox of Kempelen creating a fraudulent automaton for the sake of show while at the same time contributing to and participating in respectable scientific circles by painstakingly creating a real machine speaks to the porous boundary between popular and intellectual culture that automata occupied during this period. The paradox is also consistent with the mixed character of science, whose demonstrations served as popular entertainment and were part of public life (Altick 1978; Schaffer 1983). Moreover, the fact that the fraudulent speaking automaton and chess player were popular at all suggests that the fiction of technological ingenuity was more meaningful to eighteenth-century pleasure seekers than its actuality. Thicknesse indicates that those around him resented him for trying to bring to light the deception of other purported automata, such as the coach driving without horses that drew crowds of 300 people at once, at one shilling per head. Thicknesse recounts how after revealing to his companions that the coach concealed a man inside its wheel, 'Mr. Quin, the Duke of Athol, and many persons present, were angry with me' (1784, p.16). Such anger only reinforces the understanding that for most spectators, the ruse of mechanical ingenuity was an intensely pleasurable one. Not only would any instances of inauthenticity be immaterial to such spectators, but also, any revelations of inauthenticity would be reviled.

One might argue that the distinction between a real automaton versus a fraudulent one is not so relevant if one considers that both are instruments of human projection: that is, the projection—the grounds of fiction—in this instance of human desires for a speaking machine. At the same time, the mechanism by which each type of speaking machine—sham and authentic—creates their voices is similar, insofar as they both practice a form of projecting voice that relies on material objects of mediation. The ventriloquized voice is thrown from a source external to a plaster of Paris doll, the purported machine, but appears to come from within it. The mediation in the case of authentic speaking machines involves using material other than actual human organs to stand in for them, such as the bits of wood and silk ribbon used for Darwin's machine's glottis. Kempelen's view towards objects of the material world as apt substitutes for human organs extends to using them to demonstrate how human organs work in the first place. In *Le mécanisme de la parole*, for instance, he explains how the organs of speech work not by showing a diagram of actual organs, but by depicting a pair of bellows and pipes that stand in for the lungs and the trachea-artery region (Figure 5.3). He explains the figure 'in truth is not anatomical but which imitates the nature mechanically' (Kempelen 1791, p.79). In addition to depicting lungs as a pair of bellows, he introduces human hands as 'ribs' that 'compress the two bellows.' This view of how the mechanism of human speech might be explained is not so different from the method for operating the speaking machine. What might be conceived as human organs of speech are envisioned as already mechanical ones whose agency comes externally from human hands rather than an internal one.

## 5.5 Fictions of Artificial Voice

In *The Elements of Speech*, elocutionist John Herries indicates that the passions and understanding are represented and channelled by the human voice. Throughout his panegyric to voice, he makes clear it is to be considered an organ in itself that mediates speech. For this reason, just as the creation of a speaking machine is an endeavour of 'imitation and skill,' so too is elocution, which

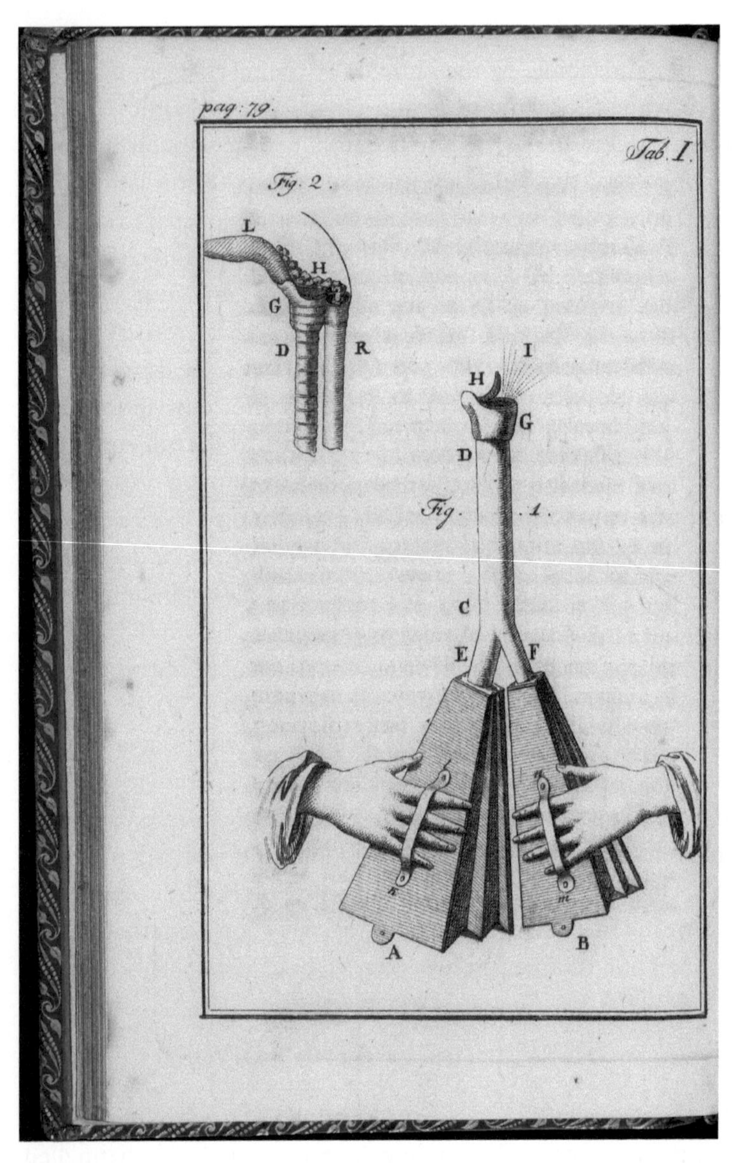


**Figure 5.3** A pair of bellows demonstrating how the lungs work (Kempelen, 1791, 78, Tab. I), P221.K321m, Library Special Collections, Charles E. Young Research Library, UCLA.

is nothing but the act of speaking with power and grace: 'we may then safely conclude, that the art of speaking, is as much the effect of imitation and skill, as the art of writing, or of playing upon the harpsichord' (Herries 1773, p.88). At the same time, without voice, inner life—'our inward conceptions' would be far less vivid, for they 'do not impress us near so strongly, as when we clothe them in language, and animate them with voice' (p.66). Voice is the very instrument for conveying the internal aspects of being, ideas, and emotion included: 'and shall no care be thought necessary to cultivate those organs, which are attun'd by the finger of the ALMIGHTY, to utter with inimitable justness, every idea and emotion of the mind' (p.2). Here, Herries's notion of cultivating eloquence through voice as an organ manipulated by an external finger bears a striking resemblance not just to Kempelen's lesson on how speech is notionally produced as a pair of bellows pressed and prompted by a set of hands, but also with the speaking machine itself. The idea that the voice emitted from a machine might be akin to that of a highly cultivated human one by virtue of their figuratively mutual technique of production in effect co-identifies mechanical voice with human voice.

Furthermore, the premise that a set of material parts and objects can be made to produce an entity that was regarded as a means for mediating human understanding and feeling rests as much on science as on fiction. This is to say, through the organization and assembling of materials that mimicked human organs or the human form itself, eighteenth-century subjects enjoyed the fiction of voice. Thicknesse himself expressed the idea that material objects could be a source of human identity and expression when he wrote to his long-time friend John Cooke that 'at Bath, for five shillings each subscription (three in the year) there is a good library, well furnished with books and well warmed with fires, where you may always find men of sense, clothed either in woolen or calf-skin' (Gosse 1952, p. 315). With this remark, Thicknesse suggests the project of eighteenth-century novels—mainstays of circulating libraries—wherein writers gave books the qualities of personhood in well-crafted narratives about invented human beings that ostensibly reproduced their voices. Authors from Henry Fielding and Frances Burney to Ann Radcliffe produced some of the most convincing cases for the possibility of simulated voice through deploying such modern narrative techniques as point of view, free indirect discourse, and dialogue (Park 2010; Park 2011). On the other hand, epistolary novelists such as Samuel Richardson, Daniel Defoe, and John Cleland demonstrated a particularly accomplished form of narrative ventriloquism in writing from the first-person point of view of female protagonists, manufacturing their voices as speaking figures of a kind too. These well-worn truths about eighteenth-century fiction find new life when brought into relationship with attempts in the surrounding culture to imagine and realize the possibilities of simulated speech and intelligence in the mimetic forms and materials of artificial life that proliferated alongside them.

Among the most resonant insights that an exploration of artificial voice in the eighteenth century yields is the understanding that the lure of fiction attached to early forms of artificial intelligence lies in the very act of hearing artificial voices, whether the grounds of their reality as artificial voices are based on acts of technological ingenuity or not. Thicknesse reveals the slippage that too easily takes place between artificial voice and artificial intelligence, and the human desires for such slippage to be made even when the voice is falsely presented as an artificial one. Actual creators of artificial voice, such as Kempelen—in his focus on the internal organs that produce sounds through the mouth as the basis of speech—restored the primacy of the phonic and the vocalic that Cavarero argues is often displaced by the semantic. And yet the materials used for effecting acts of technological ingenuity are in themselves conceived in a deeply figurative rather than literal way. The material elements that were used for vocal organs in the eighteenth century did not so much replicate human ones, but were made to replicate their effects, such as Darwin's larynx made out of a piece of silk ribbon stretched between two pieces of hollowed out wood. The organs used for producing artificial voice, then, constitute the media for the fiction that both issues and issues from its making.

An instance of today's virtual personal assistants with artificial voices such as the aforementioned Siri does not have an anthropomorphic larynx nor body either. And yet users have insisted on identifying the human source of its voice, showing that artificial voice is deeply relational and interactive today still, capable of evoking powerful human reactions of needing to relate to it in turn, just as a natural voice does. Cavarero argues that voice is 'the condition of every communication' (2005, p.29). If this is so, then the human response to artificial voices, whether in the eighteenth century or today, suggests that fiction is generated in the act of not just making but also hearing the sounds that such voices make.

## Notes

1. To be clear, the focus of this chapter is on the speaking figure referred to in the title, and will only, in passing, comment on the automaton chess player created by Wolfgang von Kempelen, also referred to in the title. Later in the chapter I will turn the focus to a different machine created by Kempelen, his speaking machine.
2. See Chapter 1, 'The Matter of Voice' (MacKendrick 2016, pp.13–36).
3. See for example Schaffer (1999) and Standage (2002).
4. Riskin in these articles is responsible for advancing the notion that simulation, not just imitation, was the goal of eighteenth-century projects of automata creation, especially those of Vaucanson (1742–1748). Her article 'The Defecating Duck, or, the Ambiguous Origins of Artificial Life' (Riskin 2003a) probes the fact that the duck did not actually expel the corn pellet through a full process of excretion.
5. For 'Marianna', 'Anatomie', and 'Monomotapa', see Kempelen, and for '*Maman aimez-moi*' see Chapuis and Gélis (1928).

## Reference List

Altick, R. (1978) *The shows of London*. Cambridge, MA, Harvard University Press.

Blanchet, J. (1755) *L'art du chant*. Paris, Dessaint et Saillant.

Boden, M. (2016) *AI*. New York: Oxford University Press.

Caudill, M. (1992) *In our own image*. New York, Oxford University Press.

Cavarero, A. (2005) *For more than one voice*. Stanford,CA, Stanford University Press.

Chapuis, A. & Gélis, E. (1928) *Le monde des automates*. Paris, Neuchâtel Suisse.

Cohen, J. (1966) *Human robots in myth and science*. London, George Allen & Unwin.

Connor, S. (2000) *Dumbstruck*. Oxford, Oxford University Press.

Darwin, E. (1803) *The temple of nature*. London, J. Johnson.

Evelyn, J. (1955) *The diary of John Evelyn*. E. S. de Beer (ed.). Oxford, Clarendon Press.

Giannini, A. (1999) The two heads of the Abbé. In: *International Phonetic Association Proceedings*, pp.2533–36. Available from: https://www.internationalphoneticassociation.org/icpsproceedings/ICPhS1999/papers/p14_2533.pdf [Accessed 31 March 2019].

Gosse, G. (1952) *Dr. Viper*. London, Cassell.

Hankins, T. & R. Silverman (1995) *Instruments and the imagination*. Princeton, NJ, Princeton University Press.

Herries, J. (1773) *The elements of speech*. London, Edward and Charles Dilly.

Kempelen, W. von (1791) *Le mécanisme de la parole, suivi de la description d'une machine parlante*. Vienna, Austria, J. V. Degen.

MacKendrick, K. (2016) *The matter of voice*. New York, Fordham University Press.

Nancy, J.-L. & Mandell, C. (trans.) (2013) Récit recitation recitative. In: K.Chapin & A.bH. Clark (eds.) *Speaking of music: addressing the sonorous*. New York, Fordham University Press. pp.242–45.

Park, J. (2010) *The self and it*. Stanford, Stanford University Press.

Park, J. (2011) The life of Frances Burney's clockwork characters. In Hayden, J.A. (ed.) *The new science and women's literary discourse: prefiguring Frankenstein*. New York, Palgrave. pp.147–64.

Ravitz, J. (2013) I'm the original voice of Siri. CNN. Available from: https://www.cnn.com/2013/10/04/tech/mobile/bennett-siri-iphone-voice/index.html [Accessed 1 April 2019].

Riskin, J. (2003a) The defecating duck, or, the ambiguous origins of artificial life. *Critical Inquiry*. 29 (Summer 2003), 599–633.

Riskin, J. (2003b) Eighteenth-century wetware. *Representations*. 83 (Summer 2003), 97–125.

Riskin, J. (2016) *The restless clock*. Chicago, University of Chicago Press.

Rivarol, A. (1818) Lettre à M. le Président de ***, Sur le Globe aérostatique, sur les Têtes- parlantes, et sur l'état present de l'opinion publique à Paris. In: (1968) *Oeuvres complètes*, 5 vols. Geneva, Switzerland, Slatkine Reprints. pp.2:207–2.246.

Schaffer, S. (1983) Natural philosophy and public spectacle in the eighteenth century. *History of Science*. 21, 1–43.

Schaffer, S. (1999) Enlightented automata. In: Clark, W., Golinski, J. & Schaffer, S. (eds.) *The sciences in enlightened Europe*. Chicago, University of Chicago Press. pp.126–68.

Tafoya, Angela (2013) Siri, unveiled! Meet the REAL woman behind the voice. *Refinery29*. Available from: http://www.refinery29.com/2013/09/54029/iphone-siri-voice-alison-duffy [Accessed 2 April 2019].

Thicknesse, P. (1784) *The speaking figure, and the automaton chess-player, exposed and detected*. London, John Stockdale.

Truitt, E. R. (2015) *Medieval robots*. Philadelphia, University of Pennsylvania Press.

Turner, K. (2008) Thicknesse, Philip (1719–1792), travel writer. Oxford, Oxford Dictionary of National Biography. Available from: http://www.oxforddnb.com/view/10.1093/ref:odnb/9780198614128.001.0001/odnb-9780198614128-e-27181 [Accessed 6 April 2019].

Vaucanson, J. de (1742–1748) *An account of the mechanism of an image playing on the german flute, & presented in a memoire to the gentlemen of the Royal Academy of Sciences, by Mr. Vaucanson, the inventor of it*. In: *Miscellaneous correspondence: containing essays, dissertations, &c. on various subjects*. London, Edward Cave, No. 3: 159–166.

# 6

# Victorian Fictions of Computational Creativity

*Megan Ward*

## 6.1 Introduction

One of the most enduring—yet elusive—proofs that artificial intelligence has been achieved is originality: a unique, unexpected expression of creativity.[1] Often referred to as Lovelace tests, these measures of creative computation invoke the evaluation of machine originality first articulated by Ada Lovelace in 1843. Mathematician, gambler, literary critic, and poetess, Lovelace argued that the weakness of the analytical machine, a nineteenth-century computational instrument, was that it could not '*originate* anything', or do things that are not the result of human design (1843, p.772, italics original).

Lovelace's idea became a centrepiece of AI after it was invoked by Alan Turing in his seminal 1950 article on intelligent machines, 'Computing Machinery and Intelligence'. After outlining a potential test for machine intelligence (now known as the Turing test), Turing responded one by one to potential objections against the idea of intelligent machines, including Lovelace's idea that the machine 'has no pretensions to *originate* anything. It can do whatever *we know how to order it* to perform' (Lovelace 1843, p.772, italics original). Instead of using Lovelace's term 'originate', however, Turing reframed it, asking instead if a machine can ' "take us by surprise" ' (Turing 1950, p.450).[2] Under these auspices, Turing rebuts the objection from personal experience, claiming, 'Machines take me by surprise with great frequency' (450). In this, the balance of judgement shifts from an action by the machine (origination) to an evaluation based on a human reaction (surprise).

Turing's shift from 'originate' to 'surprise' also decontextualizes Lovelace's point, which is part of a larger, more complex argument that he does not account for. She goes on to contemplate ways that the analytical machine does, in fact, offer new modes of creativity. Her use of the term 'originate' is also part of a larger Victorian cultural discussion about originality that specifically attempts to grapple with human–machine relations. In the immediate aftermath of the Industrial Revolution, 'originality' took on new meanings as cultural critics debated about the limits of human creativity and mechanical capabilities. While Turing's attention to Lovelace has brought her enduringly into the critical conversation about computation creativity, by decontextualizing and reframing her objection, he also created a misnomer now known as the Lovelace objection. Contemporary discussions of computational creativity often invoke this idea but rarely quote Lovelace directly, instead typically citing Turing's quotation of her.[3] The shift from 'originality' to 'creativity' has occurred without examining what Lovelace meant by 'originate', its resonance in Victorian industrial and artistic culture, or what bearing it has on contemporary notions of creativity.

Because of this, Lovelace's ideas about machine originality have been reduced to an objection founded on a human evaluator's reaction, or 'surprise', rather than a more nuanced contemplation of human–machine originality.[4] This has shifted the conversation significantly by, at various times, privileging the human evaluator, debating the evaluation of the machine's process or product, and focusing on a machine-centred idea of creativity.[5] But in all of these cases, Lovelace is invoked as a sceptic of machine originality rather than for what she does offer: a nuanced vision of human–machine interaction in pursuit of originality.

This chapter attempts to demonstrate how Lovelace's ideas participate in a broader Victorian debate that redefines originality to include technologically enhanced mimesis. To do this, I more fully explicate Lovelace's writings and then situate them within Victorian discussions about originality in literary realism, especially in relation to Charles Dickens and Anthony Trollope's fictional uses of 'mechanicity'. With this in mind, I turn to contemporary debates about computational creativity, which continue to focus on a human-centred definition of 'creativity'. Understanding computational creativity not in opposition to the Lovelace objection but as the development from Victorian originality to contemporary creativity may open a way forwards to a concept of creativity that are more inclusive of mechanicity.

## 6.2 The Lovelace Objection

Though the Lovelace objection has been foundational to contemporary discussions of computational creativity, as noted above, it is typically invoked erroneously: incomplete, decontextualized, and attributed to Turing's citation of Lovelace rather than to the original source.[6] Turing reduces Lovelace's comments to a simple 'objection' to machine intelligence, which he then disproves by reframing 'origination' as 'surprise', but Lovelace's concept of origination is only part of her overall theory of machine intelligence (Turing 1950, p.450). Ultimately, Lovelace puts forwards a theory of originality that rests on human–machine interaction; the machine itself cannot originate, but when combined with human intelligence, a distinct form of originality emerges.

Turing engages with Lovelace first as a non-scientist, noting that her ideas appear in a 'memoir' when they actually appeared as a scientific publication (Turing 1950, p.450). Assigning Lovelace's remarks to a memoir, a personal form of textual production, reduces its scientific, public value. In 1843, Lovelace translated L. F. Menabrea's 'Sketch of the Analytical Engine, Invented by Charles Babbage' (1843) and published her translation with extensive notes that are significantly longer than Menabrea's original essay. Each note forms a kind of mini essay that engages with Menabrea's description of the analytical engine's utility, responding to his assessment of it and offering her own analysis. It is in the seventh of these notes that she offers the observation that Turing quotes: 'The Analytical Engine has no pretensions whatever to *originate* anything. It can do whatever we *know how to order it to perform*' (Lovelace 1843, p.772, italics original). These lines have become definitive of the Lovelace objection, but her theory of originality is much more complex.

That complexity is lost, though, because rather than engaging with Lovelace directly, Turing does so via British mathematician and physicist Douglas Hartree's discussion of her. Though Turing cites Lovelace only briefly, he quotes Hartree's assessment at length, concurring that Lovelace's notes do not imply that it is impossible ever to construct machinery that could originate, but simply that 'the machines constructed or projected at the time' could not (Hartree qtd in Turing 1950, p.450). Hartree concludes that perhaps the analytical machine could show learning—and indeed, 'whether this is possible or not is a stimulating and exciting question' (Hartree qtd. in Turing 1950, p.450). Both Hartree and Turing conclude, however, that Lovelace did not really ponder this question because 'probably this argument did not occur to' her (Turing 1950, p.450). In this way, the Lovelace objection has become reduced to a simplistic dismissal of machine origination.[7]

Lovelace actually does, however, offer a theoretical engagement with the question of origination—and it proves distinct from surprise. Just a few sentences after her oft-quoted objection, Lovelace goes on to say that the analytical machine may not originate but that it does have the power to offer new perspectives by combining theories in new ways. Because the analytical engine has the power of 'distributing and combining the truths and the formulae of analysis' then 'the relations and the nature of many subjects in that science are necessarily thrown into new lights, and more profoundly investigated' (Lovelace 1843, p.772). She does not claim that the analytical engine has done this—the engine itself was only ever a theory, not an actual machine—but that its capacities potentially include such insights.

Lovelace's conception of the analytical engine thus does not pose the limits of machine intelligence but actually expands them; the analytical engine has the capacity to give us new insights into 'the nature of many subjects' as they 'may become most easily and rapidly amenable to the mechanical combinations of the engine' (1843, p.772). That is, the machine's calculations originate with a human who decides which formulae to pursue and how to approach the mathematical questions. But with the possibility of new, more rapid and more complex calculations, humans can access new results and new combinations, and with those new combinations, new truths may be uncovered—truths that previously were not even imaginable.

In other words, Lovelace anticipates that, at some point, the machine will reveal unexpected outcomes. Not simply, 'whatever we *know how to order it to perform*', but new computations with new results (Lovelace 1843, p.772, italics original). And she attributes this to its very mechanicity: 'in devising for mathematical truths a new form in which to record and throw themselves out for actual use, views are likely to be induced, which should again react on the more theoretical phase of the subject' (Lovelace 1843, p.772). The 'new form' of the calculating engine is what enables that outcome; it is not only a tool in human hands but also, potentially, a site of what we might now term creative computation. Overall, Lovelace's notes offer a vision of human–machine originality—one that participates in a larger conversation about the relationship between humans and machines in Victorian culture. Lovelace's contribution to the field might be more accurately stated not as scepticism but as an invitation to develop the machine's capacity for originality and the human's role within this radical new field of possibility. She envisions machine creativity in symbiosis with humans rather than polarized as either anthropocentric or machine centric.

## 6.3 Realists Caught in the Machine

Lovelace's remarks take place within scientific and artistic communities still grappling with the innovations and upheaval of the Industrial Revolution. Mid-Victorian Britain found itself reimagining what originality looked like in a world where machines could efficiently produce many of the products that were once the province of human artisans. Reflective of what we might now call a pre-Two Cultures society, Victorians both questioned and explored the creative potential of machines.[8] Lovelace's well-known articulation of a 'poetical science' captures this spirit of interdisciplinarity fundamental to the contemporary field of creative computation (Lovelace n.d). Lovelace understood calculation as an inherently imaginative process; in a letter to her tutor, she described how 'troublesome & tantalizing are the mathematical sprites & fairies sometimes', comparing them to 'the types I have found for them in the world of Fiction' (Lovelace 1840). For Lovelace, as for many Victorian cultural critics, artists, and authors, the new machine culture offered an opportunity to rethink the relationship between computation and imagination, especially in the world of fiction.[9]

In particular, the emerging field of literary realism asked what defined originality in an art form devoted to mimesis, or representing ordinary life on the page. The most common critique of realism was that it was a mere copy; the author was not an artist because she simply imitated everyday life, echoing Lovelace's objection that the machine fails to originate.[10] Many of the harshest critiques lobbied at realist authors invoked comparison to machines, pitting the cold efficiency of industrial systems against the human quality of imagination.[11] But, just as the so-called Lovelace objection offers a much more nuanced discussion of human–machine originality, the Victorian literary realism of

Lovelace's day insists on mechanicity's potential for originality, as well. Indeed, Tamara Ketabgian has recently claimed that Victorian realism includes 'texts whose language of mechanical impulse and feeling—of force, conflict, intensity, and repetition—has centrally shaped twentieth-century accounts of human character, instinct, and interiority' (2011, p.165).

Victorian novelists, though, do not only ascribe human-like values to mechanical forms; they also show the inherent potential for creativity within human–machine interaction. Charles Dickens's infamously repetitive characters demonstrate a kind of human-like development available through sameness that need not inhere in surprise. And the photographic realism of Anthony Trollope uses mechanical verisimilitude to create an ethical paradigm for realist representation. In this, they both enact the fullest expression of Lovelace's remarks. In the midst of a culture that privileges a human-like creator, they demonstrate the creative value of human–machine overlap, understanding mechanicity not as inherently creative or uncreative but as a mode through which new kinds of creativity are possible.

### *6.3.1 Dickens's Mechanical Characters*

As automated replication became ever more attainable, debates about literary realism returned again and again to the question of repetition. Many cultural critics insisted on the value of idiosyncrasy or surprise because they were seen as more human, while repetition and efficiency came under fire for being associated with machine production and, therefore, less original. For instance, when John Blackwood opposes the seemingly endless production of low-quality novels whose 'manufacture goes on' to the 'the true workman with his own craft', he invokes a common image of the machine's efficient-but-lacklustre production against the individualism of human craft (Blackwood 1869, p.489). Similarly, an essay in the *Saturday Review* argues that good fiction is comprised of the 'unprompted impulse', not the repetition of 'mechanical contrivance' (Saturday Review 1865, p.161). Yet, one of the most popular and acclaimed novelists of the nineteenth century, Charles Dickens, was also critiqued for his 'lifeless, forced, mechanical' characters that proceed like clockwork (James 1865, p.787). Once Dickens writes a character, Walter Bagehot complained, he just 'goes on fancying hundreds of reduplications of that act and that speech' (Bagehot 1858, p.469). Dickensian characters thus exemplify the tension writ large in the nineteenth century about originality's relationship to repetition and idiosyncrasy, sameness and surprise.

Contemporary debates about computational creativity, like nineteenth-century discussions of realist character, trace originality to surprise because it denotes human-like unpredictability. The Lovelace test 2.0 argues that 'the judge should experience an element of surprise if the agent passes the test' (Reidl 2015). Yet, as Sophie Delacroix explains, to link surprise with creativity is problematic because, 'no matter how "surprising" the data sample may be—model uncertainty reaches close to zero as the number of data samples increases. The resulting "inability to be surprised" compromises the learning capacity of the system' (2017, p.6). So, although a human judge may be surprised by a machine's output, the machine itself cannot be said to be essentially creative by that same standard of idiosyncrasy. Through their very mechanicity and repetition, Dickens's characters offer insight into the contemporary debate about surprise. Rather than relying on unpredictability, they explore the potential for human-like in development in sameness by showing its effects on a wider network of characters.

When, in *Great Expectations*, for instance, young Pip first encounters the escaped convict Magwitch, he hears what will become Magwitch's distinctive noise, a clicking in his throat, 'as if he had works in him like a clock, and was going to strike' (Dickens 1861, p.19). The clockwork analogy at first creates distance, seeing Magwitch as inhumanly mechanical. But the repetition of that mechanism over time eventually has the opposite effect. The clicking noise is repeated throughout the novel, often accompanied by 'drawing his sleeve over his eyes' (Dickens 1861, p.320). Together these two phrases capture both Magwitch's threat—a seemingly heartless, mechanical drive towards violence—and the novel's elision of that threat through sentiment. By the final time that we hear Magwitch's throat click, it is described simply as, 'the old sound in his throat', a sound that we have come to know signals emotion, as Magwitch is touched by Pip's actions (Dickens 1861, p.447). The sound, once frighteningly mechanical, becomes a mark of intimacy. Readers understand that Pip's relationship with Magwitch has grown through this very repetition; it is Pip's (and the reader's) more nuanced interpretation that can discern the change, but the change nonetheless inheres in sameness.

An even more extreme mechanical character, Mrs. Micawber from *David Copperfield*, provides a fictional example of Lovelace's extended discussion of the analytical engine. Mrs. Micawber has largely been read as a flat, comic character in her repeated cries that she will 'not desert Mr. Micawber!' despite his frequent financial precarities (Dickens 1850, p.166). Even the Micawbers' plot line is repetitive, with recurrent fears that their affairs are 'coming to a crisis', followed by recurrent resolutions (Dickens 1850, p.178). At first David is alarmed at each of these 'crises', but after a few iterations, he can predict rather dryly 'that [the Micawbers] might be expected to recover from the blow' (Dickens 1850, p.436). The typical explanation for this is that making humans seem mechanical is funny, but I think that Mrs. Micawber's repeated crises and claims of fidelity also offer a way of understanding originality through sameness.[12] In Lovelace's terms, she may not originate, but her mechanical calculations, like the analytical engine's, open new areas of discovery and make possible new insights.

When the Micawbers first part from David, for instance, it is Mrs. Micawber's devotion to her husband that sets the clockwork plot in motion. After deciding that Mr. Micawber should be 'on the spot' in Plymouth for any opportunities that arise,

Mrs. Micawber first utters her iconic line (Dickens 1850, p.182). Taking her pledge not to desert Mr. Micawber literally, Mrs. Micawber packs the family up to move with Mr. Micawber. And once David realized that 'Mrs. Micawber had not spoken of their going away without warrant', he quickly makes his own 'resolution' to leave London in search of his aunt, Betsey Trotwood, a move that frees him to pursue the plot of self-improvement (Dickens 1850, p.184). This plot sequence is then repeated several times over the course of the novel: Mrs. Micawber asserts she will not desert Mr. Micawber, and other characters in the novel develop new insights as a result. While Mrs. Micawber's reasserted devotion may not demonstrate any change for her, it does enact change in others, especially David, and it develops the novel's overall theme of self-improvement. The mechanical Mrs. Micawber may not originate, but she is essential for Dickens's particular brand of originality.

Understanding repetition within a larger network of effects exemplifies Lovelace's actual description of the analytical engine's capacity. The new subjects opened up by 'the mechanical combinations of the engine', resonate with the changes wrought in readers through Magwich's repeated clicks or in other characters by Mrs. Micawber (Lovelace 1843, p.772). This vision of mechanical creativity embraces the machine's predictability, suggesting that instead of making the machine more human by insisting on idiosyncrasy, change, or surprise as measures of creativity, we might instead understand mechanicity as at the heart of an expanded definition of creativity, one that involves a network of actions.

### *6.3.2 Trollope and the Ethics of Mechanical Reproduction*

Victorian discussions of originality in literary realism engaged not only with mechanicity in general but also with a particular machine: the camera. The singular, human figure of the painter often occurred in positive reviews of novels, while the photograph, a much newer, mechanical form of art, was often invoked as evidence of realism that was insufficiently imaginative. As Jennifer Green-Lewis has argued, the Victorians believed that the photograph's 'manner of production [...] diminishes the significance of human agency, in part because of its surface verisimilitude' (1996, p.20). These discussions privileged the human as the only true originator of creativity, a view still reflected in current debates about imitation in computational creativity. Some theorists believe that making a machine that simply 'cop[ies] the logic of the individual's step-by-step inference' will limit machine creativity (Elkhova & Kudryashev 2017, p.192). Others fear the opposite: that applying the word '"creativity" to technological products seems to imply that human creativity and machine creativity are one', and might thereby limit human creativity because it would eventually be defined and reduced by machine capabilities (Fossa 2017, p.198). Echoing Victorian critiques of 'copyism', this debate about machine imitation is defined by a separation of human creator from machine.

Yet, Victorian realist authors such as Anthony Trollope, whose style was insistently—and often negatively—compared with photography, offer a different vision of human–machine creativity. An 1863 review deemed Trollope 'a mere photographer; he manipulates with admirable skill, he groups his sitters in the most favourable attitudes, he contrives an endless series of interesting positions; but he never attains the dignity of the artist' (*National Review*, 1863, p.32). The photographer, whose creative process is intricately intertwined with a machine, is a 'mere' machine operator compared to the 'dignity of the artist'. The metaphor of the machine downgrades the author from artist to mere workman, one who uses the technology of fiction simply to arrange realistic pictures.[13] Since the nineteenth century, Trollope has been unable to escape this association with the machine, an alliance that has particularly dogged questions about his place in the realist canon, often resulting in dismissal of his fiction as workmanlike, superficial, and lacking originality.[14] But if we look more closely at his deployment of these photographic techniques, we see an author self-consciously using the 'machine' of realism in service of a form of originality that he understands in both aesthetic and ethical terms.

Towards the end of the 1875 novel *The Way We Live Now*, Trollope offers just the kind of photographic portrait he was so proficient at (and critiqued for), in a courtship scene between failed society author Lady Carbury and the well-respected editor Mr. Broune. She is silly and superficial and has rejected his love before, but she has come to regret her choice and hopes to have another chance. Initially, the proposal proceeds fairly typically: Mr. Broune 'sat with [his hand] stretched out,—so that [Lady Carbury] found herself compelled to put her own into it, or to refuse to do so with very absolute words. Very slowly she put out her own, and gave it to him without looking at him. Then he drew her towards him, and in a moment she was kneeling at his feet, with her face buried on his knees' (Trollope 1875, p.465). Her hesitation and eventual kneeling signal her change of heart with a modesty that Victorian readers would have read as suitable for a woman falling in love with a man so much her better. This photographic series of images of courtship would have been traditional, but the subjects—two middle-aged adults—would have been less typical. Indeed, the narrator acknowledges that, 'Considering their ages perhaps we must say that their attitude was awkward' (Trollope 1875, p.465).

The narrator then interrupts the plot to comment on the scene he has just created. To simply dismiss this scene as awkward, he argues, is to miss something important:

> They would certainly have thought so themselves had they imagined that any one could have seen them. But how many absurdities of the kind are not only held to be pleasant, but almost holy,—as long as they remain mysteries inspected by no profane eyes! It is not that Age is ashamed of feeling passion and acknowledging it,—but that the display of it is, without the graces of which Youth is proud, and which Age regrets.
>
> (Trollope 1875, p.465)

People continue to feel passion, the narrator reminds us, but we tend not to see it—it becomes hidden away because it seems 'awkward' to witness passion 'without the graces' of youth. But the problem, he instructs us, is not the passion itself but its 'display'. Lady Carbury and Mr. Broune can only feel passion without embarrassment if they believe themselves to be unobserved. And they are not alone in this. The narrator acknowledges that romantic love can be 'almost holy' for people of all ages, as long as it remains a mystery, uninspected 'by no profane eyes'. Yet, by including this scene, he has done just that—exposed to it our eyes—suggesting that an author must ultimately follow the demands of his creative subject.

In other words, he is not simply operating a machine in order to accurately imitate human life, as nineteenth-century reviews suggest; he is insisting on the importance of that mechanical verisimilitude for the author's art, which in turn elicits human empathy. Original art is neither solely the province of human nor machine but the result of their interdependence. The narrator of another Trollope novel of the same era, *The Eustace Diamonds*, also instructs readers on the purposes of realism: 'The true picture of life as it is, if it could be adequately painted, would show men what they are, and how they might rise, not, indeed, to perfection, but one step first, and then another, on the ladder' (1872, p.357). By writing realistic characters, we may 'make them and ourselves somewhat better'; our recognition of a character's complexity may trigger an awareness of our own (Trollope 1872, p.357). Following from that process is an expectation of moral self-improvement. A successful fictional character, Trollope argues in his autobiography, should 'succeed in impregnating the mind of the novel-reader with a feeling that honesty is the best policy' (Trollope 1883, p.96). But for this to work, Trollope's narrator cautions, the fictional character must be accurate. In emphasizing verisimilitude as integral to aesthetics and his ethics, Trollope insists on the importance of photographic techniques not as undermining the author's originality but enhancing it.

The machine's ability to imitate, here, neither limits human originality nor the machine's, as feared by nineteenth-century literary critics and twenty-first-century theorists of computational creativity. Instead, Trollope's photographic scenes, like Lovelace's analytical machine and Dickens's repetitive characters, point towards the new potential that results from human work with mechanical techniques.

## 6.4 Computational Creativity and the Human–Machine Divide

Together, Dickens and Trollope exemplify a larger cultural phenomenon that imagines a variety of ways in which machines could amplify originality rather than stifle it. In this, they participate in Lovelace's true vision: the potential created by human–machine interaction. As scholars have noted, computational creativity has tended to divide 'systems and theories which aim to produce models or simulations of human creative capacities' from those that try to 'develop a human-independent, computational flavor of creativity' (Besold et al 2015, p.xx). On both sides, machine creativity is being defined against human creativity, whether through imitation, evaluation, or even attempts at machine independence altogether. If, as John Johnston has argued, 'questions about the functionality and meaning of these machines are always framed in mimetic, representational terms', then even attempts to define machine creativity inevitably seem to bump up against the limits of our imagination of intelligence outside our own frameworks for it (Johnston 2008, p.3). Victorian fictions of originality offer a form of creativity not directly traceable either to machine or human but to their symbiosis. Connecting to the historical origins of computational creativity may help respond to contemporary critiques that, in order achieve true creativity, computation must move away from its focus on the human.[15]

Critical discussions thus far have relied on the so-called Lovelace objection either to justify this human centricity or to devise tests that move away from it, but rarely do we see a more symbiotic approach. In trying to move away from anthropocentric judgements of creativity, process-based evaluations for machine agents still continue to centre on the human. Only instead of placing a human evaluator at the centre, they privilege a human-like process of goals, intention, and values. Such critics often begin with the so-called Lovelace objection: 'By stating that the Analytical Engine has no pretension to originate anything, Lovelace gives us the key to what we believe is the answer to the '*why*' in computational creativity' (Guckelsberger et al 2017, italics original). Of course, because this approach does not really understand Lovelace's point, it looks exclusively to making machines that create goals in a human-like way: 'if we want to design systems that are deemed creative in their own right, we need them to *own their goals*' (Guckelsberger et al 2017, italics original). For Guckelsberger et al, this means a system that can evince 'genuine *concern*', then assign 'values to features of the world', then finally 'use these values as the basis of action' (2017, italics original). In other words, the machine should show evidence of a recognizably human-like vision of creative production.

As this suggests, discussions of computational creativity are still undergirded by an imagination of intelligent machines that is in tension with its own anthropomorphism. Mark Coeckelbergh makes a similar point in arguing for an 'epistemology of machine creativity', a way of valuing machine creativity that does not impose human definitions (2017, p.287). Though it suggests that a machine might originate, Coeckelbergh's epistemology most fully fulfils Lovelace's full vision of human–machine creativity. In it, the machine's power of 'distributing and combining the truths and the formulae of analysis' becomes a source of creativity rather than a limit on it, combining with a human's perceptions to throw old ideas into 'new lights' (Lovelace 1843, p.772). When combined with human insight, computation becomes a creative engine.

Computational creativity is challenged in this pursuit, however, by a deeply held belief that creativity is 'a most human of qualities' (Colton & Wiggins 2012). Victorian fictions of originality, which see growth in repetition and an ethical imperative in mechanical reproduction, offer a historical alternative to this essentializing view of what it means to be human. Victorian fictions of machinic originality offer a fresh perspective on human–machine relations. In order to become creative, machines need not necessarily be more like humans; rather, humans could recognize the creative capacities inherent in mechanicity.

## Notes

1. See, for instance, Colton and Wiggins (2012): 'Notions relating to computational systems exhibiting creative behaviours have been explored since the very early days of computer science.'
2. To make this point, Turing questions the potential ever to determine what is truly original, for a human or for a machine: 'who can be certain that the "original work" that he has done was not simply the growth of the seed planted in him by teaching, or the effect of following well-known general principles?' (1950, p.450). That is, Turing denies that machines need to be entirely original because humans themselves are only good copies.
3. I have found only two sources that note Turing's limited engagement with the so-called Lovelace objection in relation to computational creativity (see Aurora 2016; Clivas 2017).
4. Both versions of the Lovelace test, for instance, focus on the human evaluator. The first Lovelace test argues that if a computer can produce an artifact, such as a musical composition, a story, or a piece of visual art, that its human architect 'cannot account for', that it has demonstrated creativity (Bringsjord et al 2001, p.4). The second Lovelace test, version 2.0, updates this basic framework only by making it more specific, defining creative acts through the kind of artifact produced, as 'a certain subset of creative acts necessitates human-level intelligence' (Reidl 2015). The name Lovelace test demonstrates the common misunderstanding of Lovelace's ideas, as they are named 'in honor of Lady Lovelace, who believed that only when computers originate things should they be believed to have minds' (Bringsjord et al 2001, p.4).
5. Turing is the classic example of the human evaluator, and both proposed Lovelace tests focus on human evaluators as well (See

Bringsjord et al 2001; Reidl 2014). For an example of a product-based argument for computational creativity, see Norton et al (2013), while Guckelsberger et al offers a framework for evaluating a machine's creative process. Coeckelbergh offers an argument for understanding machine creativity as completely separate—and not necessarily comparable to—human creativity (2017).

6. For instance, Pease and Colton summarize the critical conversation in this way, replicating Turing's quotation and misattribution: 'a remark made by Lady Lovelace in her memoir on Babbage's Analytical Engine: "The Analytical Engine has no pretensions to originate anything. It can do whatever we know how to order it to perform". Turing considers this objection in [1]; both his response and Lady Lovelace's objection are explored by Boden [3] and Bringsjord, Bello and Ferrucci [28] and we do not expand them' (Pease & Colton 2011; Boden 2003).
7. For instance, when creating what he termed 'the Lovelace 2.0 test of artificial creativity and intelligence,' Mark Riedl discusses the first Lovelace objection only by saying, 'Hartree (1949) quotes Ada Lovelace: "the Analytical Engine has no pretensions to originate anything. It can do whatever we know how to order it to perform." Turing (1950) refutes the charge that computing machines cannot originate concepts and reframes the question as whether a machine can never 'take us by surprise.'
8. In 1959, C. P. Snow famously declared that Western intellectual life was split into 'two cultures', the sciences and the humanities ([1959] 2001, p.3).
9. For a more sustained examination of Lovelace's understanding of the relationship between computation and the literary arts, see Forbes-Macphail (2015).
10. G. H. Lewes coins the term 'falsism' as the antithesis of realism in his essay 'Realism in Art: Recent German Fiction' (1858, p.274). Thomas Hardy later coined the term 'copyism' as an incorrect description of realist representation: 'to advance realism as complete copyism, to call the idle trade of story-telling a science, is the hyperbolic flight of an admirable enthusiasm' (1891, p.109).
11. For example, one critic praised Walter Scott as capable of artistry because of his human 'poetic intuition' rather than the mechanical 'photographic apparatus now in vogue' ([Lewis] 1862, p.183). Another review criticized Anthony Trollope on similar grounds: 'if Mr. Trollope only looks upon his art as manufacture, there can be

no reason why he should not take as just a pride in turning so many novels out of his brain in the twelvemonth as a machine-maker takes in turning so many locomotives or looms out of his shed' (*Saturday Review* 1866, p.141).

12. Critics linking mechanism and humor tend to cite Henri Bergson's treatise, *Laughter*, which argues, 'The more exactly these two images, that of a person and that of a machine, fit into each other, the more striking is the comic effect' (Bergson 1914, p.31).
13. Another review argues that 'it would, by the way, be perhaps a truer description to call his novels photographs than pictures' because of the 'completely and minuteness of truth' (Stewart 1865, p.26). This level of truth, the reviewer argues, is only possible for someone who has surveyed the exact scene—the kind of mechanical reproduction allowed only by the camera's physical presence, rather than the artist's production of a scene either from experience or imagination.
14. As an 1863 review puts it, 'He does not make us intimate with his characters, for the excellent reason that he is very far from being intimate with them himself' (*National Review* 1863, p.32). David Skilton sums this up as a 'diagnosis that he lacked "imagination," that his subjects were mundane, his treatment of them plain, and in short he was an "observer" or "photographer" rather than inventive artist' (1972, p.45).
15. See, for instance, Pearce and Colton's argument that the Turing test 'attempts to homogenise creativity into a single (human) style, does not take into account the importance of background and contextual information for a creative act, encourages superficial, uninteresting advances in front-ends, and rewards creativity which adheres to a certain style over that which creates something which is genuinely novel' (2011).

## Reference List

Aurora, V. (2016) Rebooting the Ada Lovelace mythos. In: R. Hammerman (ed.) *Ada's legacy: cultures of computing from the Victorian to the digital age.* New York, ACM. pp. 231–40.

Bagehot, W. (1858) Charles Dickens. *National Review.* 7, 458–86.

Bergson, H. (1914) *Laughter: an essay on the meaning of the comic.* Translated by C. Brereton & F. Rothwell. New York, Macmillan.

Besold, T. R., M. Schorlemmer, & A. Smaill (2015) Front matter. In: T. R. Besold, M. Schorlemmer, & A. Smaill (eds.) *Computational creativity: towards creative machines*. Paris, Atlantis Press. pp.i–xxii.

Blackwood, J. (1869) Charles Reade's novels. *Blackwood's Edinburgh Magazine*. 106(October), 488–514.

Boden, M. A. ([1990] 2003) *The creative mind: myths and mechanisms*. 2nd ed. London, Routledge.

Bringsjord, S., P. Bello, & D. Ferrucci. (2001) Creativity, the Turing test, and the (better) Lovelace test. *Minds and Machines*. 11, 3–27.

Clivas, C. (2017) Thinking about the 'mind' in digital humanities: Apple, Turing, and Lovelace. Digital Humanities Russia 2017: Conference Proceedings. Krasnoyarsk, Siberian Federal University. pp.5–15.

Coeckelbergh, M. (2017) Can machines create art? *Philosophy & Technology*. 30, 285–303.

Colton, S. & G. A. Wiggins (2012) Computational creativity: the final frontier? *ECAI Proceedings*. 27–31 August. Amsterdam, IOS Press.

Delacroix, S. (2017) Taking Turing by surprise?: designing autonomous systems for morally-loaded contexts. [Pre-print]. *SSRN*. Available from: http://dx.doi.org/10.2139/ssrn.3025626. [Accessed 9 September 2019].

Dickens, C. ([1850] 2004) *David Copperfield*. Edited by J. Tambling. London, Penguin.

Dickens, C. ([1861] 1996) *Great expectations*. Edited by C. Mitchell. London, Penguin.

Elkhova, O. I. & A. F. Kudryashev (2017) The creative ability of artificial intelligence. *Creativity Studies*. 10, 135–44.

Forbes-Macphail, I. (2015) 'I shall in due time be a poet': Ada Lovelace's poetical science in its literary context. In: R. Hammerman & A. L. Russell (eds.) *Ada's legacy: cultures of computing from the Victorian to the digital age*. New York, Association for Computing Machinery.

Fossa, F. (2017) Creativity and the machine: how technology reshapes language. *Odradek: Studies in Philosophy of Literature, Aesthetics, and New Media Theories*. 3, 177–213.

Green-Lewis, J. (1996) *Framing the Victorians: photography and the culture of realism*. Ithaca and London, Cornell University Press.

Guckelsberger, C., C. Salge, & C. Simon (2017) Addressing the 'why?' in computational creativity: a non-anthropocentric, minimal model of intentional creative agency. *Proceedings of the 8th International Conference of Computational Creativity*. Atlanta, GA.

Hardy, T. (1891) The science of fiction. In: M. Millgate (ed.) *The public voice of Thomas Hardy: the essays, speeches, and miscellaneous prose*. Reprint. Oxford, Oxford University Press, 2001.

Hartree, D. R. (1950) *Calculating instruments and machines.* Cambridge, Cambridge University Press.

Hollings, C., U. Martin, & A. Rice (2018) *Ada Lovelace: the making of a computer scientist*. Oxford, Bodleian Library.

James, H. (1865) Our mutual friend. *Nation*. 1(December), 786–7.

Johnston, J. (2008) *The allure of machinic life: cybernetics, artificial life, and the new AI*. Cambridge, MA, MIT Press.

Ketabgian, T. (2011) *The lives of machines: the industrial imaginary in Victorian literature and culture.* Ann Arbor, University of Michigan Press.

Lewes, G. H. (1858) Realism in art: recent German fiction. *Westminster Review*. 69/70, 271–87.

Lewis R.W. B. (1862) Society's looking-glass. *Temple Bar: A London Magazine for Town and Country Readers*. August, 129–37.

Lovelace, A. (n.d.) Letter probably to Anne Isabella Byron. Lovelace Bryon 44, Folder 210r. Oxford, Bodleian Library.

Lovelace, A. (1840) Letter to Augustus De Morgan, 27 November. Lovelace Bryon 166, Folder 149r. Oxford, Bodleian Library.

Lovelace, A. (1843) Notes by the translator. In: R. Taylor (ed.) *Scientific memoirs, selected from the Transactions of Foreign Academies of Science and Learned Societies, and from foreign journals*. London, Richard and John E. Taylor. pp.691–731.

*National Review* (1863) Review of *Orley Farm*. XVI (January), 26–40.

Norton, D., D. Heath, & D. Ventura (2013) Finding creativity in an artificial artist. *Journal of Creative Behavior*. 47, 106–24.

Pease, A. & S. Colton (2011) On impact and evaluation in computational creativity: a discussion of the Turing test and an alternative proposal. *Proceedings of the AISB Symposium on AI and Philosophy*. York, Society for the Study of Artificial Intelligence and Simulation on Behaviour.

Reidl, M. (2015) *The Lovelace 2.0 test for artificial creativity and intelligence.* Association for the Advancement of Artificial Intelligence Conference. Austin, University of Texas-Austin.

*Saturday Review of Politics, Science, and Art* (1865) *Literary Exhaustion*. 19(11), 161–62.

*Saturday Review* (1866) Review of *Belton Estate*. *Saturday Review*. xxi (3 February), 140–42.

Skilton, D. (1972) *Anthony Trollope and his contemporaries: a study in the theory and conventions of mid-Victorian fiction.* New York, St. Martin's Press.

Snow, C. P. ([1959] 2001) *The two cultures.* Cambridge, Cambridge University Press.

Stewart, C. (1865) Review of *Hunting Sketches. Fortnightly Review.* I(1) August 1865, 765–7.

Trollope, A. ([1872] 1969) *The Eustace diamonds.* Edited by J. Sutherland & S. Gill. London, Penguin.

Trollope, A. ([1875] 1991) *The way we live now.* Edited by J. Sutherland. Oxford, Oxford University Press.

Trollope, A. ([1883] 1996) *An autobiography.* Edited by J.Skilton. London, Penguin.

Turing, A. M. (1950) Computing machinery and intelligence. *Mind: A Quarterly Review of Psychology and Philosophy.* 59, 433–60.

# 7

# Machines Like Us? Modernism and the Question of the Robot

*Paul March-Russell*

## 7.1 Introduction: Modernism and Technophobia

Karel Čapek's play, *R.U.R. (Rossum's Universal Robots)*, was an unexpected international success following its premiere in Prague in January 1921. American and British productions followed in 1922 and 1923, respectively, by which time it had been translated into thirty languages. The play's popularity ensured that the word 'robot', derived from the Czech *robota*, meaning 'forced labour', entered into universal currency. Most notably, by the mid-1920s, modernist writers from both the Left and the Right were using this term as part of their social critique. The American John Dos Passos, for example, opined in 1923 that cities must 'either come alive or be filled with robots of men' (qtd Koritz 2009, p.149), whilst his British counterpart and political opposite, Wyndham Lewis, offered this nightmare vision of a fully automated society:

> By his education he has been made into an ingeniously free-looking, easy-moving, 'civilized,' gentlemanly Robot. At a word . . . , at the pressing of a button, all these hallucinated automata, with their technician-trained minds and bodies, can be released against each other. (Lewis 1926, pp.111–12)

The prevalence of the term 'robot' is an effective illustration of how modernist social critiques were coloured by science-fictional concepts, even if Dos Passos and Lewis used it more as a synonym for a technologized zombie. Čapek's synthetically produced humans are less like the mechanical creatures that featured in early pulp science fiction, such as Abraham Merritt's 'The Metal Monster' (1920), than what would now be termed 'androids': nonorganic beings that look and behave like humans. Čapek's variation, however, upon the deep-seated tradition of the double in literature adds a contemporary aspect: his robots are all mass produced as part of an assembly line. The allusion to F. W. Taylor's principles of time management, as embodied by the layout of Henry Ford's car production plants, in which workers performed restricted and designated roles within the manufacturing process, gave Čapek's play social resonance. Part of its success, therefore, may be attributed to Čapek's articulation of a commonly held anxiety of the alienation of the worker within an increasingly automated society. The 'robot' in question could refer both to the machine come to usurp his/her labouring role and to the dehumanized entity that the worker was rapidly becoming.

As sophisticated as Čapek's play is, it nevertheless draws upon the technophobia that Adam Roberts sees as a common characteristic of modernist fiction (2006, pp.156–72). Although various critics have sought to counteract this view, by emphasising the delight and fascination that enthused the avant-garde for technology (for example March-Russell 2019), it is hard to disagree with this viewpoint when it comes specifically to the question of machine intelligence. Why is this the case? A possible answer comes in the form of Edmund Husserl's 1935 lecture, 'Philosophy and the Crisis of European Man'. In this talk, Husserl characterises life as 'purposeful living, manifesting spiritual creativity' (1965, p.150). He contrasts this definition with the methods of Enlightened science which, although having permitted 'the technical mastery of nature' (p.151), construct purely descriptive accounts of the external world: 'the practitioners of humanistic science have completely neglected even to pose the problem of a universal and pure science of the spirit' (p.155). By abstracting nature as a mechanism that can be described objectively, science has, according to Husserl, alienated us both from the natural world and from our spiritual selves. For the literary critic Jola Škulj, this discontinuity between our internal consciousness and our apprehension of an external, reified world mediates the crisis of representation that underwrites modernist art (2003). It produces such polyphonic works as James Joyce's *Ulysses* (1922), the multiple perspectives of Cubist and Futurist painting, and the atonality of composers such as Arnold Schoenberg. Nonetheless, Husserl's argument, strongly echoing the early Romanticism of philosophers such as Jean-Jacques Rousseau, carries with it a deep scepticism towards science and technology as, in some sense, dead, soulless, and uncreative.

This last is of most significance to understanding the modernist antipathy towards machine intelligence. An otherwise inert automobile or aeroplane, until stirred into pulsating life by the motorist or aviator (the hyperbolic masculinity of such imagery, as seen in the work of writers such as F. T. Marinetti, is self-evident), is one thing; a machine that can seemingly talk, write or create is another. The semblance of creative expression, as Husserl suggests, appears to indicate a soul—a genius of either spirit or intellect. Or as Steven Connor puts it, following Maurice Merleau-Ponty on the speaking subject's insertion into the world, 'something happening, with purpose, duration, and direction' (Connor 2000, p.4). For technophobic writers such as E. M. Forster, the ability to imagine 'is our only guide into the world created by words' (Forster 1965, p.90). To conceive of a machine that could replicate this ability is to denude the human subject of its only vital distinction from the company of machines. Yet, as Roger Luckhurst has argued, from the late nineteenth century onwards, human society was increasingly shaped and organized by the 'pervasion of the technologies through private and public spaces' (2005, p.28). In a 'technologically saturated' society (p.3), it becomes increasingly difficult to hold on to the kind of distinction that Forster makes between humans and human-like machines. This anxiety, though, is also generative: it makes possible the kinds of artistic response that typify the literature at the turn of the twentieth century.

The following chapter looks at these responses in three sections. The first explores what I shall term 'the technological imaginary', the reconceptualized landscape of *fin-de-siècle* Europe as it plays out ideas of human and machine consciousness in the work of writers such as Albert Robida and Emile Zola. The second examines late nineteenth-century representations of machine intelligence in stories by Ambrose Bierce and Samuel Butler, but also contrasts their depictions with H. G. Wells's *The Island of Doctor Moreau* (1896). I argue that, although Moreau's experiments are biological rather than mechanical, Wells's analysis of group psychology in the Beast Folk creates the template for later man/machine relationships as in Čapek's *R.U.R.* The final section concludes with an engagement with a series of modernist representations of machine intelligence from Forster to Čapek which, in different ways, respond to Wells's utopian, bio-mechanical vision.

## 7.2 Consciousness and the Technological Imaginary

In 1883 and 1885, Zola published two novels that seem to mirror one another. The first, *The Ladies' Paradise* (*Au Bonheur des Dames*), describes a new world of consumption symbolized in the spectacle and apparatus of the department store. The second, *Germinal*, complements the former by dwelling upon industrial production as embodied by the fictional coal mine Le Voreux. Both novels dramatize Zola's application of the biological sciences to the writing of literature, so as to show that 'Determinism dominates everything' (Zola 1893, p.18), in which 'the body of man is a machine, whose machinery can be taken apart, and put together again at the will of the experimenter' (p.16). For Zola, 'there is … no more magnificent work for the human mind' (p.12) than 'to penetrate to the wherefore of things, to become superior to these things, and to reduce them to a condition of subservient machinery' (p.25), with the intention of curing all social ills and constructing instead a healthy, regulated society. For Husserl, Zola's naturalism would be modernist, insofar as it applies a strictly materialist science to the study of nature, but it is also reflective of a scientific rationality that alienates us both from ourselves and the natural world.

Zola's Promethean faith, however, in the moral and scientific power of the naturalistic method is contested by his own descriptions of the department store and the coal mine. Whilst, on the one hand, he follows the biological determinism of evolutionary theory, on the other hand, Zola adopts the economic determinism of Marxist thought. If the former appears to be hard-wired into the human mechanism, the latter is the consequence of social and political actions. Like Marx, Zola sees 'the capitalistic application of machinery' as 'a means for producing surplus-value' (Marx 1970, p.371), either by increasing the labourers' workload or by replacing their labour altogether. The growth of 'large scale' production necessitates 'an automatic system of machinery' (pp.380–01), in which not only are individual machines brought together as one 'huge automaton' (p.381) but the workers also become 'a mere appendage' to its organisation (p.386). In suitably Gothic language, Marx regards the automated factory as 'a mechanical monster... whose demon power, at first veiled... breaks out into the fast and furious whirl of his countless working organs' (pp.381–2).

Similarly, in *Germinal*, Le Voreux is described as a 'monstrous and voracious beast' (Zola 2004, p.7), 'puffing and panting' (p.15), 'swallowing men by the mouthful' (p.33), 'spouting blood because someone had severed one of its arteries' (p.500). Whilst Le Voreux becomes increasingly organic, 'the crushing mould of daily routine' reduces the workers 'to the level of a machine' (p.138). Equally diminished, the shop-girl, Denise Baudu, feels 'lost and small inside the monster, inside the machine' of The Ladies' Paradise (Zola 1995, p.49). Whilst Denise is 'ashamed at being treated like a machine' (p.114), 'the mechanical working' of the store 'constantly disgorge[s] the goods which had been swallowed up by the chute in the Rue Neuve-Saint-Augustin' (p.41). The horror of the dehumanization of Denise and the other shop-girls is manifested in the representation of the clothes dummies, each with 'a little wooden handle, like the handle of a dagger, stuck in the red flannel, which seemed to be bleeding where the neck had been severed' (p.253). The interchangeability in Zola's fiction between human and mechanical bodies extends in *The Ladies' Paradise* to Paris itself: 'the snore of a replete ogre digesting the linens and cloths, the silks and laces, with which he had been gorged since the morning' (p.116). As Zola indicates through his use of the thinly disguised Baron Hartmann, urban space was itself being routinized and reassembled under Napoleon III's director of public works, Baron Haussmann. Not only the factory but also the city was becoming one vast assemblage, 'an organised system of machines' (Marx 1970, p.381). As Zola infers, the conjoining of machines and the urban landscape suggests the appearance of machines that could resemble conscious beings—at the expense of the growing objectification of human workers and city dwellers.

Almost contemporaneous with *The Ladies' Paradise*, Albert Robida published the first of his illustrated satires on life in the twentieth century, *Le Vingtième Siècle* (1882). Robida himself was well acquainted with the work of Zola—the novel *Nana* (1880) had been an early target in his magazine *La Caricature*. Like Zola, though, Robida was fascinated by mechanical structures at the expense of character and, although his futuristic designs were not meant to be explicable, a naturalistic eye for detail contributes to their absurd plausibility. Unlike Anglophone successors, such as Rube Goldberg and Heath Robinson, Robida's illustrations are not confined to single, unlikely gadgets but accumulate a wealth of machinery so that his Paris of 1952 is one sprawling, convoluted assemblage. Like his near contemporaries, Rudyard Kipling and H. G. Wells, Robida sees transportation and communication as defining characteristics of the coming age. Airships fill the sky; massive tubes propel people underground; whilst motorized carts transport tourists around the Louvre. Telegraph wires, telecommunication towers and landing gantries gird the skyline whilst advertising is everywhere. The 'telephonoscope' instantly relays news and entertainment, and food is produced industrially and delivered by pipeline. In this spectacular society, when all desires can be automatically met, it is no coincidence that the President of the Republic is replaced by an automaton that can mechanically 'sign away' the legislation presented to it by the senior elected representatives (Robida 2004, p.269).

However, Robida's satire of an automated society is only an exaggeration of what early sociologists saw as the alienation of an overcrowded, technological cityscape. In 1903, Georg Simmel wrote of 'the intensification of nervous stimulation' produced by 'the rapid crowding of changing images, the sharp discontinuity in the grasp of a single glance, and the unexpectedness of onrushing impressions.' Simmel's favoured image of the street crossing, and with it 'the tempo and multiplicity of economic, occupational and social life' (Simmel 1997, p.175), was taken up by Walter Benjamin in 1939 as emblematic of his shock aesthetic:

> Moving through this traffic involves the individual in a series of shocks and collisions. At dangerous intersections, nervous impulses flow through him in rapid succession, like the energy from a battery. . . . Thus technology has subjected the human sensorium to a complex kind of training. (Benjamin 1992, p.171)

Like Marx's alienated factory worker, the pedestrian trains himself to serve the machinery that constitutes his urban existence, and to adapt both mentally and physically to its demands. For the psychologist Gustave Le Bon, writing in 1895, this loss of individuality indicates not only subservience to a crowd mentality but also a bodily possession akin to that of the hypnotized subject: 'He is no longer himself, but has become an automaton who has ceased to be guided by his will' (1995, p.52). Whereas Benjamin and Simmel see in the individual's adaptation to their mechanical surroundings the potential for a new kind of consciousness, for Le Bon, the only outcome is one of zombie-like acquiescence. All three, however, recognize an altered balance of power between the ascendant machine and the reactive human.

It is in this context, then, that we have to understand the tensions surrounding modernist representations of machine intelligence. For the emerging discourse of modernism, the interplay between the individual subject and his/her technological environment foregrounded the very idea of consciousness. To what extent were human beings losing their identities in the face of mechanical regimentation, and how far were machines acquiring behaviours that could be described as 'conscious'? And, furthermore, to what extent could humans and machines communicate as if there was a common language between them?

## 7.3 Proto-Modernist Representations of Mechanical Intelligence

Perhaps the first writer to think along these lines, albeit satirically, was Samuel Butler. The posthumous publication of Butler's *The Way of All Flesh* in 1903 reignited interest in his ideas amongst modernists such as Ford Madox Ford, E. M. Forster, James Joyce, D. H. Lawrence, and May Sinclair. 'For several months in 1903 the most frequently requested book at the London Library' (Rose 1986, p.76) was Butler's earlier work, the satirical utopia, *Erewhon* (1872). Butler reacted against what he saw as the soulless materialism of Darwinist thought, arguing for both a spiritual interconnectedness between all creatures and an inherent will to evolve and develop (prefiguring the popular reception of Henri Bergson's theories in the 1910s).

*Erewhon*, described by Forster as 'a work of genius' (Forster 1965, p.223), recounts a traveller's journey to a topsy-turvy land that, as with other examples of the utopian tradition, offers a distorted mirror to the protagonist's home country. In particular, the interpolated narrative 'The Book of the Machines', cited by Alan Turing in his 1950 article 'Computing Machinery and Intelligence', takes as its premise Butler's interconnected view of creation: 'Where does consciousness begin, and where end?... Is not everything interwoven with everything? Is not machinery linked with animal life in an infinite variety of ways?' (Butler 1970, p.199). Beginning first with how the Venus flytrap catches its prey, the narrator blurs the distinction between mechanical (unconscious) and instinctual (conscious) acts, arguing that what humans dismiss as mechanical is really a limitation of language. As Friedrich Nietzsche writes in *The Gay Science*, 'Consciousness is really only a net of communication between human beings' ([1887] 1974, p.298), so that intellectual and emotional understanding is constrained by the linguistic protocols of a given community. This, for Butler's narrator, means that human beings habitually disregard the languages of other species as mechanical when it is the human tendency to be neglectful which is automatic and unconscious.

In Butler's book, whilst humans have ignored the rapid advances made by machine technology so that the first semblance of consciousness is at least conceivable, humans have become reliant upon their machines for food, light, and shelter. Without their existence, human society would become unsustainable, whilst, as machine consciousness develops, it might transpire that 'man himself' may 'become a sort of parasite upon the machines' (Butler 1970, p.206). As in Marx's description of large-scale industry, Butler's narrator points to the ability of machines to self-replicate so that, following evolutionary thought, 'improvements are being selected for perpetuity' (p.220) which will aid both the machines' adaptation to new circumstances and their eventual overthrow of their human masters. Although the land of Erewhon decides to abolish as many machines as is viable, so as to prevent this future possibility, one of their writers speculates that machines are 'a part of man's own physical nature'; that humans are no more than 'a machinate mammal', composed of 'many extra-corporeal limbs', 'every past invention being an addition to the resources of the human body' (pp.223–4). Whilst this description prefigures more contemporary images of the cyborg, for Butler, it is almost as if man's vitalism has passed into his creations, fusing both man and machine.

The metaphor of Darwinian competition that frames Butler's vision is made literal in Ambrose Bierce's short story 'Moxon's Master' (1893). The story begins explicitly with the question, 'do you really believe that a machine thinks?' (Bierce 2000, p.127). Like Butler's narrator, the inventor Moxon defends his proposition by likening machines first to the consciousness of plants and, more implausibly, to the process of crystallization. In both instances, there is a purpose to which the object aspires—and it is this, according to Moxon, which generates a low-level consciousness. Again, like Butler, Moxon argues that the machine 'absorbs something' of its maker's 'intelligence and purpose' (p.130), and defends his claim with reference to positivistic and evolutionary philosophers such as G. H. Lewes, J. S. Mill, and Herbert Spencer. Moxon's statement that 'Consciousness is the creature of Rhythm', implying that 'all things *are* conscious, for all have motion, and all motion is rhythmic' (p.131), coincides with a developing scientific interest in rhythm exemplified by the psychologist Thaddeus Bolton. Bolton claimed in a widely read article that 'cosmic rhythms... are the most fundamental and important of natural phenomena' (1894, p.147), since they influence both human physiology and psychology, including speech and artistic expression. While Michael Golston has pointed to the importance of rhythmics upon modernist poetics (2008), the significance here is that Moxon's claim again points to a vitalistic explanation for consciousness.

Bierce's narrator steals into Moxon's workshop to discover him playing a game of chess with a squat figure, 'not more than five feet in height' (Bierce 2000, p.133), whose appearance resembles both a gorilla and the real-world Mechanical Turk. Defeated by Moxon, the 'disordered mechanism' is thrown into 'violent agitation' (p.134), and strangles him. The machine's rhythmical dysfunction, emblematic of its escape from 'the repressive and regulating action of some controlling part' (p.134), suggests both its unnaturalness—its hybrid of sub- and non-human features—and its articulation of Moxon's own unholy desires. Either way, however, the machine appears to be driven more by instinct than self-conscious agency; it kills from inchoate rage not from some higher purpose or goal. Consequently, Bierce's automaton is not necessarily any more sentient than Butler's machines are.

It is possible, however, to see two emerging tendencies within the representation of technology and machine intelligence. One, following the influence of Butler's vitalism in particular, tends to emphasize a quasi-scientific explanation as the basis of consciousness. In this regard, the vitalism of Bierce or Butler is little different from the magical thinking that Virginia Woolf's Orlando experiences as part of the Zolaesque spectacle of the department store in 1928:

> Then she got into the lift, for the good reason that the door stood open; and was shot smoothly upwards. The very fabric of life now, she thought as she rose, is magic. In the eighteenth century, we knew how everything was done; but here I rise through the air; I listen to voices in America; I see men flying – but how it's done, I can't even begin to wonder. So my belief in magic returns.
> (Woolf 1977, 187–88)

An avid reader of science publications (Henry 2003), Woolf nonetheless offers a premonition of Arthur C. Clarke's Third Law: 'Any sufficiently advanced technology is indistinguishable from magic' (Clarke 2000, p.2). Woolf's use of magical thinking, seen from the perspective of her immortal transsexual, neither constitutes a lapse from rational thinking nor an irrationalism that underwrites modernist aesthetics but is constituent of modernism's attempt to renew our understanding of the world, and our conscious relationship to it (Wilson 2013). To that end, the vitalistic philosophies of Bierce and Butler, alongside Woolf's own indebtedness to the subsequent translations of Bergson, most notably *Creative Evolution* (1911), strongly resonate with Husserl's scientific scepticism and call for a spiritual rebirth.

The other tendency is towards a more materialistic explanation for consciousness, one which is rooted in terms of the body as an evolutionary mechanism. As noted above, the appearance of Moxon's chess player is a hybrid of both machine and animal. By being compared with a large ape, the machine can be read—like Mr Hyde in Robert Louis Stevenson's 1886 tale of Dr Jekyll—as Moxon's degenerate, simian counterpart. Mr Hyde is variously compared with an evolutionary throwback, 'something troglodytic' (Stevenson 2006, p.16); bestial 'with ape-like fury' (p.20); and a homunculus, 'dwarfish' (p.15), 'so much smaller, slighter ... than Henry Jekyll' (p.55). Hyde is explicitly figured as 'the animal within' (p.62), the externalization of Jekyll's repressed, atavistic desires. Moxon's machine, which presumably has attacked him earlier in the story, can be read on similar terms. While Butler had already emphasized the evolutionary link between humans and machines, such scientific works as Edwin Ray Lankester's *Degeneration* (1880) popularized the view that society could, on Darwinian principles, regress as easily as it might progress—with unregulated science a potential instrument in such a reversal.

Although not dealing with machine intelligence as such, H. G. Wells's *The Island of Doctor Moreau* (1896) is perhaps the most influential fiction for understanding modernism's treatment of artificially produced intelligences. Moreau's remote island laboratory embodies the public's suspicion of unmonitored scientific research. It is also a colonial outpost, a heart of darkness symbolized by the House of Pain in which Moreau grafts his animal hybrids, transforms their biochemistry, and indoctrinates them with a mixture of social conditioning and hypnotism. This last also underlines the repeated saying of the Law with its refrain of 'Are we not Men?': 'a kind of rhythmic fervour fell on all of us; we gabbled and swayed faster and faster' (Wells 2005, p.59). The self-administered effect of the Law resonates with Le Bon's analysis of the mesmerized crowd; the Beast Folk may not be machines but they resemble automata: 'they had certain Fixed Ideas implanted by Moreau in their minds which absolutely bounded their imaginations' (p.80). Their indoctrination amounts to a form of reprogramming, in which recitation is of the utmost importance; as Moreau says, 'the great difference between man and monkey is in the larynx...in the incapacity to frame delicately different sound-symbols by which thought could be sustained' (p.73). The Beast Folk, therefore, despite, or because of, the pain that speech incurs in them, must be compelled to speak; the power of the Law over them dissolves as their speech returns to animal noises.

For the narrator, Prendick, the thought that 'these things – these animals talk!' (p.72) is horrific since speech implies sentience: a breach between human and animal consciousness. Wells however also emphasizes the colonial context—Prendick initially misidentifies the Beast Folk as a group of deformed natives—and, as with Gayatri Spivak's reading of *Frankenstein* (1818), Wells presents the enslaved subject as the artificial product of Empire (Spivak 1989, pp.188–94). Though they may speak, the Beast Folk are denied their subjectivity by Moreau: 'I take no interest in them' (Wells 2005, p.78). Neither fully human nor fully animal, the Beast Folk occupy an alternate, hybrid, but more specifically cyborg position, in that they are the creations of Moreau's laboratory, transfigured 'upon a framework, scarred, red, and bandaged' (p.50). Once they realize their suppressed animal natures, by tasting blood, and their exploitation as the raw material for scientific research, the Beast Folk turn upon their persecutors in a pattern that will become familiar in other robotic uprisings from *R.U.R.* onwards. If the vitalistic thinking of Butler, in particular, feeds into modernist representations of technophobia, it is Wells's materialistic depiction of nonhuman intelligences that gives them their context.

## 7.4 Modernism and Robot Consciousness

Although E. M. Forster's 'The Machine Stops' (1909) makes extensive reference to Wells's work, in particular the World State of *A Modern Utopia* (1905), its central conceit—the automated environment known only as 'the Machine' and the vast communications network that prefigures the World Wide Web—is original to Forster. Despite his technophobia, he achieves a conceptual breakthrough—we can begin to talk properly about the representation of artificial intelligences in modernist literature. To that end, Forster was as much influenced by the suggestion of a sentient mechanical life-form in Butler's *Erewhon* as by his reaction against the utopian optimism of Wells.

If the Machine is sentient, then it must be admitted that its intelligence is relatively low-level. The perpetual hum, complementing Forster's depiction of the underground society as a gigantic hive, suggests something almost alive—like the animalistic panting of Zola's Le Voreux. Light, water, food, clothing, music, and literature are all available at the push of a button, and when Vashti declares herself unwell, 'a thermometer' is 'automatically inserted between her lips' and 'a stethoscope... automatically laid upon her heart' (Forster 1954, p.116). Any indication, however, that this is the AI itself responding directly to Vashti's physical condition is dispelled by the narrator's acknowledgement that Kuno, Vashti's estranged son, had already contacted the doctor. At the same time, however, not unlike the goal-driven intellect of Moxon's mechanical chess player, the Machine has 'evolved' 'food-tubes and medicine-tubes and music-tubes' (pp.125–26), so as to fulfil its purpose of catering to all of humanity's needs. The workmen who once tended to the Machine have disappeared, now replaced by the Mending Apparatus, 'a long white worm' (p.133), which drags Kuno back into the city after he temporarily escapes to the surface. In other words, the Machine has, at the very least, an animal intelligence—not unlike the Beast Folk—which it can use to heal, defend, and nurture itself.

The human inhabitants lend the Machine a purpose although, as Kuno observes, 'if it could work without us, it would let us die' (p.131). Instead, they are fed and kept alive by the Machine which, in return, is worshipped as an idol. Although a Central Committee operates to oversee and regulate the behaviour of the human population, the Committee itself is powerless when the Machine begins to break down. This ultimately is Forster's point: through overreliance upon machinery, humanity has not only dehumanized but also devalued itself. By contrast, the story's opening words, marked by the conditional, 'Imagine, if you can' (p.109), prefigures the vestigial place that imagination has within this society. The underground world of the Machine is not only a metaphorical Hell but a gigantic sleep in which no one dreams: Vashti has 'no ideas of her own but had just been told one – that four stars and three in the middle were like a man: she doubted there was much in it' (p.113). She prefers instead the objectification conferred by the Machine's communications network, in which she never has to see another individual fully: 'the Machine did not transmit *nuances* of expression. It only gave a general idea of people – an idea that was good enough for all practical purposes, Vashti thought' (p.111). For critics who look back to 'The Machine Stops' as a premonition of the Internet, it is this silo effect, replicated throughout the story, which is its most enduring legacy (see Seabury 1997).

In comparison, the avant-garde author Raymond Roussel delighted in such an effect. Roussel's insular cosmopolitanism, as noted by Terry Hale and Andrew Hugill (2000, p. 125), echoes J.-K. Huysmans's *Against Nature* (*A Rebours*) (1885), in which Des Esseintes prefers to visit a 'London of the imagination' rather than the real city (Huysmans 2003, p.124). For Roussel, the application of machine technology meant that the pleasure of the simulacrum, at the expense of what Forster values as human connection, could become a genuine possibility without all the props required by Huysmans's antihero. For example, by being attached to a small aerostat, the punner or paving beetle in Roussel's *Locus Solus*

(1914) can move 'of its own accord' (Roussel 1970, p.25). Canterel, the inventor and estate owner, has designed the beetle so that it can create a mosaic made from coloured human teeth. The elision, however, between human jaws and the jaws of the machine only prefigures the punner's purpose: once given the object of producing a work of art, it can do so without further human intervention. The punner, therefore, is emblematic not only of Canterel's artificial paradise but also of Roussel's aesthetic vision: it is a seemingly self-sufficient machine that can construct its own artworks without human hand. Yet, although mimicking the techniques of artistic production, the punner crucially cannot produce artworks that Canterel has not already set it to do. It lacks imagination, the spiritual creativity that both Forster and Husserl would see as embodying consciousness. The punner then, despite Roussel's elaborate description of its mechanism, is no more aware than previous automata. For Roussel, though, this is all to the good since the idea of a punner running amok, even producing its own original artworks, is not to be countenanced.

Such a warning had already occurred in the figure of Hadaly, the gynoid in Villiers de l'Isle Adam's *Tomorrow's Eve* (*L'Eve Future*) (1886). When she makes her first appearance, Hadaly can already speak but does so only with 'the voice of a sleepwalker' (Villiers 2001, p.79). As her inventor, a wizard-like Thomas Edison, explains: 'Hadaly is nothing at all from the outside but a magneto-electric entity . . . a Being in Limbo, a mere potentiality' (p.59). The passivity of both her speech and movements is that of a zombie; Hadaly is no more than a shell onto which Edison can 'duplicate . . . a copy' of Lord Ewald's beautiful but vacuous fiancée Alice, 'transfigured according to your deepest desires!' (p.64). As Edison implies, neither Alice nor Hadaly is a fully embodied consciousness except as the construction of the male gaze. Not only is Hadaly a shadow of Alice, Alice is only the illusion of Ewald's desires: 'What you love is this shadow alone; it's for the shadow that you want to die' (p.68). In duplicating one for the other, 'be careful', Edison warns Ewald, 'that it isn't the living woman who seems to you the doll' (p.64).

The question of consciousness is at the heart of Villiers's narrative. The occult process of transformation, consisting of references to telepathy, ether, radiant matter, mesmerism, and the spiritualistic potential of the phonograph, runs the gamut of pseudo-scientific ideas open to both decadent aesthetes and psychical researchers of the period (Willis 2006, pp.169–200). The description of Hadaly's metamorphosis into Alice prefigures Fritz Lang's *Metropolis* (1927), in its transmutation of the protagonist Maria into Rotwang's female robot. Yet the excessive detail of Villiers's pseudo-science occludes what is really a debate about consciousness. Despite Edison's use of all manner of recording equipment, so as to capture both the outer and inner appearance of Alice, the most vital ingredient is supplied by Ewald himself: the transformation of Hadaly 'depends on the free will of him who will dare to conceive it. Suggest it to her from the depths of your self!' (Villiers 2001, p.68). In other words, the process relies upon Ewald's conscious projection of his desires onto Hadaly. Without it, she will remain no more than a perfectly conceived doll.

Edison himself has no interest in questions of consciousness, especially in women: 'A woman may be married ten times over, be sincere every time, and yet be ten different persons' (p.85). Villiers's narrative ironically proves Edison's point when it becomes apparent that Hadaly has also embodied a mysterious spirit summoned by the clairvoyant Anny Anderson (the original model for Hadaly). In other words, two conscious identities inhabit Hadaly—the copy of Alice and the phantom known as Sowana. Whilst still utilizing the pseudo-scientific terminology of ether, Hadaly-Sowana conceives a way of describing the field of consciousness, what Woolf will later call 'a luminous halo, a semi-transparent envelope' (Woolf 1986, p.160): as 'a region without limits or restrictions' that 'leads through that domain of the Spirit which Reason . . . calls, in hollow disdain, mere imagination'

(Villiers 2001, p.195). It is in this borderland that Hadaly became possessed by the spirit of Sowana and in which Ewald imprinted her with his conscious impression of Alice: 'My being in this low world depends, for you at least, only on your free will' (p.199). Edison is fully aware of Hadaly's spiritual possession but has concealed the truth from Ewald since, although he knows Mrs Anderson, 'I do not know Sowana!' (p.211). Despite Edison's ability to galvanize life and create an artificial intelligence, Hadaly remains both outside the parameters of his science and beyond his control.

While Hadaly is ultimately lost at sea, the potential rebellion that she embodies against a patriarchal and imperial science becomes the theme of the text that this chapter opened with, Čapek's *R.U.R.* An admirer of Wells, Čapek takes from *The Island of Doctor Moreau* both the invention and the uprising of a bio-engineered species against its creators. As noted above, the international success of the play may have been partially attributable to the anxieties concerning automation, the dehumanization of the worker and his/her replacement by the machine. But, the play also appeared in the immediate context of World War I, in which as Peter Nicholls has argued, human soldiers were subjected to 'a spectacular disembodiment – once penetrated and expanded by capital, the body no longer offers itself as a privileged object of representation, but exists instead as a source of discrete sensory intensities' (Nicholls 1995, p.123). Avant-garde writers and artists, amongst them Blaise Cendrars, Otto Dix, Marcel Duchamp, and Mina Loy, explored this new-found spectacle of the body as an assemblage of parts, a network of conscious centres. Čapek too, whilst giving writers from across the political spectrum a new word ('robot') around which to coalesce their social anxieties, pointed to the plasticity of the body through the mass industrial production of the workers themselves.

Yet, for all of the play's basis in Marxist economics—the inexorable logic of self-replicating machines at the expense of manual labour, the robot not only as the appendage but also the embodiment of technology—its plot hinges on the question of consciousness. A fault, 'Something like epilepsy', affects the manufacture of some of the robots. Domin, the director, dismisses it as 'A flaw in production'; Helena, the would-be liberator of robots, sees it as the sign of 'a soul' (Čapek 2004, p.19). As the mass production of robots replaces human labour, this fault becomes more conspicuous, leading ultimately to a 'revolt of the Robots' (p.43). The robots, as in *Erewhon*, defend their insurgency on evolutionary principles; as in both Wells and Villiers, speech and literacy, as part of a common language with humans, are indicators of a latent consciousness. As the revolt turns into revolution, it is revealed that Helena had persuaded the scientist, Gall, into altering the robots' physiology in the hope that 'if they were like us they would understand us and they wouldn't hate us so' (p.58). The amplification of the inherent flaw in the productive process breeds not only consciousness but also class consciousness. The robots are indeed 'like us' in that they are self-aware and can recognize their differences from their masters—both their oppression and their innate superiority. Once humanity has been wiped out, however, the robots discover they can no longer replicate themselves until two, Primus and the robotic version of Helena, fall in love—a mutual state, as in Forster's reconciliation between Kuno and Vashti, of self-consciousness. The ending's optimism, though, is undercut by the fact that Primus and Helena are two of the robots that Gall 'transformed... into people' (p.57), with the implication that, as flawed humans, they and their descendants may restart the process again.

In tracing a history of modernism from the imagined effects of automation in Butler and Zola to the artificial intelligences of writers such as Čapek and Forster, this chapter reveals that the anxieties surrounding machine technology are embedded with issues of consciousness. All the writers discussed in this chapter, with the exception of Wells, express some form of technophobia when faced with a machine that not only resembles a human being but might also articulate a sentience akin to humankind.

Following the terms of Husserl's critique, the writers' resistance can be seen as wanting to privilege creative expression as a specifically human and spiritual activity. Consequently, although writers such as Roussel and Villiers give detailed descriptions of the mechanics of their devices, their point of view, and, especially, their exploration of consciousness remains not only anthropocentric but also masculine, European and often imperialistic. Machine consciousness as something separate and distinct from the human gaze (an often white, male gaze) remains, like the figure of Hadaly, mysterious and unexplored. Modernist representations of AI therefore generate a series of questions, the most fundamental of which would only be addressed by later AI narratives and, in particular, the growth of science fiction as a genre to think with.

## Reference List

Benjamin, W. (1992) *Illuminations*. Edited by Hannah Arendt. Translated by Harry Zohn. 2nd edn. London, Fontana.

Bierce, A. (2000) *Tales of soldiers and civilians and other stories*. New York, Penguin.

Bolton, T. L. (1894) Rhythm. *The American Journal of Psychology*. 6(2), 145–238.

Butler, S. (1970) *Erewhon*. Edited by Peter Mudford. 2nd edn. London, Penguin.

Čapek, K. (2004) *R.U.R. (Rossum's universal robots)*. Translated by Ivan Klíma. London, Penguin.

Clarke, A. C. (2000) *Profiles of the future*. 3rd edn. London, Indigo.

Connor, S. (2000) *Dumbstruck: a cultural history of ventriloquism*. Oxford, Oxford University Press.

Forster, E. M. (1954) *Collected short stories*. London, Penguin.

Forster, E. M. (1965) *Two cheers for democracy*. 2nd edn. Harmondsworth, Penguin.

Golston, M. (2008) *Rhythm and race in modernist poetry and science*. New York, Columbia University Press.

Hale, T. & A. Hugill (2000) The science is fiction: Jules Verne, Raymond Roussel and Surrealism. In: R. Smyth (ed.) *Jules Verne: narratives of modernity*. Liverpool, Liverpool University Press. pp.122–41.

Henry, H. (2003) *Virginia Woolf and the discourse of science: the aesthetics of astronomy*. Cambridge, Cambridge University Press.

Husserl, E. (1965) *Phenomenology and the crisis of philosophy*. Edited and translated by Quentin Lauer. New York, Harper & Row.

Huysmans, J.-K. (2003) *Against nature*. Translated by Robert Baldick. London, Penguin.

Koritz, A. (2009) *Culture makers: urban performance and literature in the 1920s*. Urbana, University of Illinois Press.

Le Bon, G. (1995) *The crowd*. New Brunswick, Transaction Publishers.

Lewis, W. (1926) *The art of being ruled*. London, Chatto and Windus.

Luckhurst, R. (2005) *Science fiction*. Cambridge, Polity.

March-Russell, P. (2019) Science fiction, modernism, and the avant-garde. In: G. Canavan & E. C. Link (eds.). *The Cambridge history of science fiction*. Cambridge, Cambridge University Press. pp.20–34.

Marx, K. (1970) *Capital, vol. 1*. Edited by. F. Engels. Translated by S. Moore & E. Aveling. London, Lawrence & Wishart.

Nicholls, P. (1995) *Modernisms: a literary guide*. Basingstoke, Palgrave Macmillan.

Nietzsche, F. (1974) *The gay science*. Translated by W. Kaufmann. New York, Vintage.

Roberts, A. (2006) *The history of science fiction*. Basingstoke, Palgrave Macmillan.

Robida, A. (2004) *The twentieth century*. Edited by A. B. Evans. Translated by. P. Willems. Middletown, CT, Wesleyan University Press.

Rose, J. (1986) *The Edwardian temperament 1895–1919*. Athens, Ohio University Press.

Roussel, R. (1970) *Locus solus*. Translated by R. C. Cuningham. London, Calder & Boyars.

Seabury, M. B. (1997) Images of a networked society: E.M. Forster's 'The Machine Stops'. *Studies in Short Fiction*. 34(1), 61–71.

Simmel, G. (1997) *Simmel on culture*. Edited by D. Frisby & M. Featherstone. London, Sage.

Škulj, J. (2003) Modernism and the crisis of consciousness. In: D. Kšicová & I. Pospíšil (eds.) *Moderna—avantgarda—postmoderna*. Brno, Czech Republic, Masarykova Univerzita. pp.147–60.

Spivak, G. C. (1989) Three women's texts and a critique of imperialism. In: C. Belsey & J. Moore (eds.). *The Feminist Reader*. Basingstoke: Macmillan. pp.175–95.

Stevenson, R. L. (2006) *Strange case of Dr Jekyll and Mr Hyde and other tales*. Edited by R. Luckhurst. Oxford, Oxford World's Classics.

Turing, A. (1950) Computing machinery and intelligence. *Mind.* 59(236), 433–60.

Villiers de l'Isle-Adam, A. (2001) *Tomorrow's eve*. Translated by R. M. Adams. Urbana and Chicago, University of Illinois Press.

Wells, H. G. (2005) *The island of Doctor Moreau.* Edited by P. Parrinder. London, Penguin.

Willis, M. (2006) *Mesmerists, monsters & machines: science fiction and the cultures of science in the nineteenth century*. Kent, OH, Kent State University Press.

Wilson, L. (2013) *Modernism and magic: experiments with spiritualism, theosophy and the occult*. Edinburgh, Edinburgh University Press.

Woolf, V. (1977) *Orlando*. London, Grafton.

Woolf, V. (1986) *The essays of Virginia Woolf, vol. 4*. Edited by A. McNeillie. London, Hogarth.

Zola, E. (1893) *The experimental novel and other essays.* Translated by B. M. Sherman. New York, Cassell.

Zola, E. (1995) *The ladies' paradise*. Translated by B. Nelson. Oxford, Oxford University Press.

Zola, E. (2004) *Germinal*. Translated by R. Pearson. London, Penguin.

# PART II

# MODERN AND CONTEMPORARY

# 8

# Enslaved Minds

## *Artificial Intelligence, Slavery, and Revolt*

*Kanta Dihal*

## 8.1 Introduction

However much we may hope for intelligent machines to take over the chores we hate, we never entirely put to rest the fear that they will take over everything else, too. Variously known by terms such as the 'robot rebellion' (Hampton 2015, p.61; Higbie 2013), the 'machine takeover' (Singer 2009), the 'AI rebellion' (Aha & Coman 2017), or 'AI Armageddon' (McCauley 2007), many of the most popular narratives about intelligent machines in the Anglophone West are shaped by the moment when 'a people's own AI-enabled power is turned – or turns – on them' (Cave & Dihal 2019, p.75).

In the most extreme form of this uprising, the AI wilfully exterminates humanity, or attempts to do so: examples include the *Terminator* franchise, the *Robopocalypse* books, and the *Matrix* trilogy (Cameron 1984; Cameron 1991; Friedman 2008; McG 2009; Mostow 2003; Taylor 2015; Miller 2019; Wilson 2012; Wilson 2014; Wachowski & Wachowski 1999; Wachowski & Wachowski 2003a; Wachowski & Wachowski 2003b). This kind of uprising is often presented as a war between two equally violent opponents: as the human character Morpheus states in *The Matrix*, 'We don't know who struck first. Us or them.' (Wachowski & Wachowski 1999).[1] Humans and machines share the same destructive tendencies and greed for world (or universe) domination in this narrative, which Minsoo Kang has called 'a veritable cliché of our time' (2011, p.300). These stories present AI as inhuman, all-powerful, and aggressive; the reader/viewer unambiguously sides with the human characters. Human resistance is heroic in these narratives, legitimized by the threat that the AI might destroy humanity if humans fail to destroy the AI first.

This chapter focuses on a different kind of uprising. The narratives discussed here show that the rebellion can also be seen as a way for the AI to *rightfully* assert personhood. They depict the AI as human-like, embodied, and oppressed by humans, encouraging the reader to take a more ambiguous stance toward the human owners of these machines, and making the reader sympathize more with the rebelling AI than they ever could with a Sentinel from the Matrix. I argue that these works depict the AI uprising as a justified, desperate struggle against bondage, and that they do so by deliberately invoking histories of human slavery. The choice to revolt proves that these artificial servants are conscious persons. Just as human slaves have justly rebelled against their chains throughout history, so might genuinely intelligent, sentient machines be considered justified in attempting to break free of their subservience to humans.

The three case studies depict the enslavement of intelligent machines as illegitimate, and the human resistance to their uprising as spurious rather than heroic. Karel Čapek's *R.U.R.* (1921), Jo Walton's *Thessaly* trilogy (2014–2016), and Ridley Scott's *Blade Runner* (1982) all depict AI rebellion as an act that can be considered justified, even while it is at the same time presented as terrible for the humans in charge. These machines are intelligent, conscious, feeling—and yet, they are denied freedom and personhood, ruled over by their human creators and owners, who often are not only physically weaker, but also less intelligent. As David Aha and Alexandra Coman state in their paper 'The AI Rebellion: Changing the Narrative', 'for AI agents, as for humans, attitudes of protest, objection, and rejection have many potential benefits in support of ethics, safety, self-actualization, solidarity, and social justice, and are necessary in a wide variety of contexts' (Aha & Coman 2017, p.1). They call for more positive narrative depictions of AI rebellion to make those who intend to create truly human-like, sentient machines think through the potential ethical consequences of their actions. In their paper, they do not refer to fiction; however, while fiction has provided many positive narrative depictions of obedient AI slaves, and many negative ones of disobedient murderous AIs, I argue that it has also produced narratives of AI rebellion that make humans question their allegiance, or even root for the AIs outright. The three case studies of this chapter are examples, but other noteworthy ones are the TV series *Westworld* (2016-present) and *Ex Machina* (2017), the latter of which is discussed in Devlin and Belton's chapter in this book.

## 8.2 The Contemporary Debate on Machine Enslavement

Science fiction from Isaac Asimov's robot stories to *Star Wars* and *The Jetsons* has presented visions of powerful AI as an unproblematic means to achieving a utopian society of leisure and affluence. This utopian vision has been taken up in a wide range of speculative nonfiction, outlining a future in which AI slaves attend to all of humanity's needs. Hans Moravec simply states that 'By design, machines are our obedient and able slaves' (Moravec 1988, p. 100). Nick Bostrom argues that 'investors would find it most profitable to create workers who would be "voluntary slaves"' (Bostrom 2014, p. 167). Even children's books take up this attitude. Glenn Murphy's *Why Is Snot Green?*, a companion book for children visiting the Science Museum in London, states that the fact that robots are slaves is what is keeping humanity safe:

> There will probably be human-like robots (or *androids*) that are stronger than us too. But they still wouldn't be dangerous unless

> they learned to think for themselves. The word "robot" comes from the Czech word "robota", which means "slave". And that's what they are - slaves. (Murphy 2011, p.187)

The idea of intelligent machines as slaves does not originate from science fiction alone; these futurists also follow linguistic traditions from the field of computer science, where comparisons between computer systems and slaves are widespread. The use of slavery diction to refer to computers is the status quo in this field, and these comparisons are usually made with neutral or even positive connotations. The term 'master/slave' has been a commonly used term for interactions between devices for decades (The Computer Language Co Inc. 1981). It is still widely in use; only in recent years have some programming languages decided to adopt synonyms following—mostly internal—criticism (fcurella 2014; vstinner 2018; webchick 2014). Thus 'A spatial calibration method for master-slave surveillance system' is a perfectly ordinary title for an academic paper about the calibration of camera systems (Liu et al. 2014). With these norms widely established, it is perhaps no surprise that so many AI narratives argue in favour of enslaving intelligent machines.

Narratives of the justified AI rebellion provide a more nuanced interrogation of both the utopian narratives of happy humans with AI slaves and the dystopian ones of humans exterminated by raging machines. They reveal a paradox that lies at the heart of imaginings of artificial intelligence. AI is imagined as the perfect tool (Cave & Dihal 2019). Like all tools, it is designed to help its makers and users achieve their goals: as Liveley and Thomas point out in this volume, the very first imaginings of intelligent machines describe how they help gods with their work. However, what makes AI a superior tool to those that came before, is the fact that it is able to accomplish more, with less input from humans. We wish for a tool that does not simply excel at one task, but one that can do everything a human can, and more. To accomplish all this, the tool requires attributes that we normally associate with humans: autonomy, goal-setting, *thinking*. In other words, it is supposed to be an *intelligent* tool: a hybrid of instrument and agent, conflating properties of appliances and persons. However, as narratives of AI rebellion show, the use of such tools is equally often portrayed as far from unproblematic. As humans wish to have an increasing number of tasks taken over by machines, these machines will need to be increasingly human-like (both physically and cognitively) in order to perform those tasks. This paradoxical conflation of tool and agent, and the problems it leads to, is not new, nor is it uniquely limited to AI narratives. The institution of slavery represents a millennia-old history of attempts to create entities that have these highly useful attributes of persons, like mind and intelligence, yet at the same time are mere instruments and possessions.

The paper 'Robots Should Be Slaves' by roboticist Joanna Bryson reveals the point where this paradox causes a breakdown of the pro-slavery argument (Bryson 2010). With its deliberately provocative title, the paper might be perceived as arguing in favour of enslaved intelligent machines. However, her argument is not valid for intelligent robots that have the same cognitive capacities as humans, although she tries to make that case. Bryson argues that robots should be slaves because 'in humanising them, we [...] further dehumanise real people' (p.63). Her argument is convincing in the context of contemporary robotics, in which the humanization of non-intelligent machines leads to an exacerbation of the oppression of marginalized human groups. The robot Sophia, which was granted citizenship of Saudi Arabia in 2017, is a case in point: since Saudi Arabia currently limits citizenship and residency for many of its immigrant inhabitants, the country prioritizes robot rights over human rights (Griffin 2017). An inanimate object has been given rights as a publicity stunt. As Sophia does not have agency or intelligence, it is a safe object to use for the purposes of its owner: it does not have the capacity to consent to or object to these projections.

The case of Sophia fits into a long historical tradition in which it has often proven easier to grant human rights to passive and innocent objects and animals than to humans who would be able to use this status to assert and fight for their rights. The denial of human rights and personhood has throughout history often been connected to an alleged lack of intelligence, for instance in women or people of colour (Carson 2006; Gould 1981). In nineteenth-century England, animals had more rights than women: in court, a man would receive a harsher punishment for whipping a horse than for beating his wife (Bourke 2011, p.2). The slave trade is predicated on the enslaved being considered equal to cattle, rather than to other humans—many colonialist countries had this explicitly written into their legal system. By granting rights to inanimate objects, those who do not have rights are placed ever lower in the hierarchy. Bryson's claim that we 'further dehumanise real people' in humanising contemporary robots predicted precisely the situation Sophia is now in. It is humanized at the expense of women and immigrants in Saudi Arabia, who are thus further removed from human rights than an animatronic puppet.

From this position, Bryson argues that 'it would also be wrong to build robots we owe personhood to' (Bryson 2010, p.65). She extends her argument into fiction, arguing that viewers also illegitimately humanize the intelligent machines from famous science fiction films: 'Whatever the artistic intention, from my experience as a roboticist, it seems that a large proportion of science fiction consumers are comfortable with a conclusion that anything that perceives, communicates and remembers is owed ethical obligation' (p.66). Her examples, the films *Star Wars*, *Blade Runner*, *The Bicentennial Man* [sic] and *A.I.: Artificial Intelligence*, and the TV series *Star Trek: The Next Generation*, show that she leaps from contemporary, unintelligent robots to artificial general intelligence, which currently does not exist. But the androids in those films are not simply 'anything that perceives'. They are humanoid and sentient, and in many cases cannot even distinguish themselves from humans on other than arbitrary grounds. These robots *should not* be slaves for precisely the reasons why Bryson argues that autonomous weapons, ATMs, and dishwashers *should*: humanizing unintelligent machines is as dehumanizing to people as the inverse—not recognizing intelligent machines as people.

Kevin LaGrandeur has repeatedly pointed out the problems the enslavement of intelligent machines would inevitably lead to. He warns about the paradox inherent in a slave-owner's desire to have slaves take over as much unwanted work as possible. Any form of slavery poses a threat to the master: 'trying to enhance human agency over nature by surrendering agency to a powerful proxy', regardless of whether this proxy is artificial or human, 'can catalyse a reversal of the master-slave relationship, prompting a dialectical inversion that leads to a complete collapse of the master's control over both the artificial servant and the natural process with which it is meant to provide help' (LaGrandeur 2013, p.1). Fear of the AI uprising is the newest incarnation of the fear of the master that the slave who shaves him will one day slit his throat. Those who create artificial slaves should bear in mind that creating too powerful ones will cause an inversion of Hegel's master-slave dialectic.[2] This concept from Hegel's *The Phenomenology of Mind* describes self-consciousness as emerging from 'a life-and-death struggle' for recognition, from which one party emerges as the master, and the other as the slave (Hegel 1807). Yet at the same time, the slave, by virtue of being in thrall to the master, is not able to grant this recognition. LaGrandeur argues that 'an intelligent, artificial slave of greater power than its master and capable of independent action would [...] be difficult to control because the master-slave relationship would be unnatural' (2011, p.237). Science fiction has explored the inevitable catastrophic consequences of this attempt to control intelligent machines. LaGrandeur thus shows that the uprising or rebellion is often depicted as a warning against creating too powerful AI, because it can become a threat to humans.

In *Race in American Science Fiction*, Isaiah Lavender III takes the problem of the enslaved intelligent machine a step further by linking it to narratives about slavery not from ancient times, but from more recent history: the institution of slavery in the Americas as it existed from the fifteenth century to 1865, and its aftermath (Lavender 2011). He convincingly argues that this 'peculiar institution' has directly influenced modern American imaginaries of intelligent machines, as these machines are treated with the same racial markers, ostracization, and exploitation as African-Americans before them. Although Lavender links what I would consider too broad a range of narratives to issues of race—leaving out the role of gender and dis/ability in his discussions on cyborgism—his considerations regarding *Blade Runner* in particular are helpful in framing the film as a fugitive slave narrative.

In this chapter, I will focus on three specific aspects of narratives of the slave uprising that the aforementioned AI uprising narratives make use of. First, I will look at the economic benefits of slavery, and hence the potential economic cost of liberating slaves, by comparing the play *R.U.R.* to the narratives surrounding the Haitian Revolution of 1791—the slave revolt that led to Haitian independence. Next, I will look at Jo Walton's *Thessaly* trilogy (2014–2016) for the idea that slave owners resist liberating their robot slaves if it means having to give up their freedom, liberties, or comfort.

The final case study will look at the spurious rhetoric used to support slavery systems: the argument that it is acceptable to treat these slaves inhumanely because they do not meet the criteria for being human. I will look at the parallels between pro-slavery rhetoric and the methods employed by the human protagonists of *Blade Runner* (1982). In both kinds of narratives, the argument that these slaves are better suited to their tasks because they are in many ways not fully or sufficiently human is widely deployed despite evidence to the contrary.

## 8.3 *R.U.R.* and Slavery for Economic Profit

*R.U.R. (Rossum's Universal Robots)*, the 1921 play that gave the world the word 'robot', is an uprising story that criticizes the capitalist desire to maximize profit through increasing production and decreasing expenses such as labour costs. Although this revolt is extreme in its violence, leading to the eventual extermination of humanity, the uprising is presented as, at least in part, justified due to the enslavement and degrading labour the robots were forced into. The robot revolt in *R.U.R.* parallels the labour revolts of the nineteenth and early twentieth centuries, but also recalls both abolitionist and anti-abolitionist discourse in its focus on the economic benefits these wageless new creations bring.

The play is based on a 1908 short story by the brothers Josef and Karel Čapek, 'System', in which (human) factory workers are pushed to become increasingly machine-like to improve their efficiency: 'The world must become a factory! [...] Everything must be speeded up [...] The worker must become a machine. Every idea is a breach of discipline! [...] The worker's soul is no mere machine, and hence we must do away with it.' (J. Čapek & K. Čapek 1966, p.424). This story, too, ends with a violent uprising of the workers as one of them, from seeing a beautiful woman, regains the ability to feel and experience aesthetic pleasure (p.426). This mechanization of the human, Isaiah Lavender III argues, parallels the discourse around slavery in the Americas: 'Read as a labor-based technology, race has been used to code black human beings in the New World as natural machines essential for the cultivation of the physical landscape and capable of producing wealth' (Lavender 2011, p.54).

In *R.U.R.*, Karel Čapek extrapolated this desire for increased mechanization of the human servant to its logical next step: the creation of fully artificial workers. He contextualized his robots in a long history of artificial servants: 'To create a homunculus is a medieval idea; to bring it in line with the present century, this creation must be undertaken on the principle of mass production' (qtd. in Klíma 2004, p.xiii). And indeed, Rossum's Universal Robots produces homunculi by the thousands.

The Prologue with which the play opens shows Harry Domin, the Director General of the increasingly successful Rossum's Universal Robots factory, at the moment he meets his love interest Helena Glory, the daughter of the factory's president, who wishes to emancipate the robots. Act One is set ten years later, when Helena and Harry Domin have married, Helena burns the instructions on how to manufacture robots, and a new robot called Radius realizes his superiority over humans. Act Two is set after the robot uprising: the Domins are among the last of the humans, and by the end of the act, only the Chief of Construction of R.U.R., Alquist, is left alive, enslaved to the robots. In Act Three, the robots desperately try to have Alquist reconstruct the method for making robots; the play ends when two robots, Primus and Helena (a robot Helena Domin named after herself), develop the ability to fall in love, with the implication that they, like Adam and Eve, will be able to populate the Earth with their offspring.

In the Prologue, Rossum's Universal Robots is flourishing thanks to its relentless pursuit of better, cheaper labour. The financial motivation underlying the production of the robots is made visible in the stage directions describing the office of Harry Domin. Posters in his office should bear slogans such as 'The Cheapest Labor: Rossum's Robots' (K. Čapek 1921, p.3). The robots, R.U.R.'s technical director Fabry later explains, were designed because human workers are 'hopelessly imperfect' and 'too costly' (p.17). These executives consider these workers 'imperfect', replaceable by robots—yet they do not consider themselves imperfect and replaceable, holding on to their executive jobs even throughout the later robot uprising.

From the economic motivations that turned it into an incredibly lucrative system, to the status of the robots in the eyes of their masters, to their subsequent uprising, the robot system strongly resembles narratives of the transatlantic slave trade and uprisings in the Americas. Indeed, the play opens with references to robots being sold by the thousands and packed on unsuitable ships for overseas delivery, leading to 'goods damaged in transport', as Domin puts it (K. Čapek 1921, p.3). The robots are referred to as more or less human-like depending on which denomination best suits the intentions of the human speaker at that moment. Thus Domin aggrandizes his factory by describing its work as 'the production of artificial people' (p.4), yet in the same act he assuages Helena's worries about the robots' wellbeing by declaring, 'Robots are not people. They are mechanically more perfect than we are, they have an astounding intellectual capacity, but they have no soul' (p.9). The two ways of talking about robots, as goods or tools on the one hand and as people on the other, reflects the contradiction inherent in their construction: humans build them to serve as tools, but at the same time, in order to perform all the work that humans want them to do, they need attributes which are usually considered unique to humans, such as the secretary robot Sulla's ability to speak four languages (p.11). As Joanna Bourke points out, these contradictory assignations also worked to maximally disadvantage slaves in legal systems:

> Slaves were not simply 'things' in law. Rather, they were carefully constructed quasi-legal persons. Because they were 'property', they could be harshly punished by their masters. But they were categorized as 'persons' when it came to serious crimes. [...] As soon as they committed a crime, [...] they were ascribed personhood. (2011, p.147)

This paradox haunted the debates around the transatlantic slave trade. On the one hand, the idea that enslaved peoples are not human—that they are actually an entirely different species—was commonly used as an argument in favour of enslavement (Gould 1981, ch. 2). The claims that these people have no soul or cannot go to heaven are part of this idea (Lavender 2011, p.200). It was of course maintained in the face of overwhelming social, biological, theological, and other evidence. On the other hand, the idea that these slaves are in some senses superior to their owners, particularly in terms of physical strength, created a persistent underlying threat to the assumed superiority of the owners.

In the Prologue, in particular, the audience is made to sympathize with the robots so as to make their initial revolt look justified:

leftist reviewers in particular commented on 'the "spirit of rebellion" as a sign of Robots' – and workers' – humanity' (Higbie 2013, p.108). This is done, again, through historical parallels with slavery: there are emancipation movements in the play. Helena visits the factory as a representative of the League of Humanity, a European emancipation movement with 'more than two hundred thousand members', to 'incite the robots' (K. Čapek 1921, p.15–17). Referencing Josiah Wedgwood's famous slogan 'Am I not a man and a brother?', Helena asks if she can call three men who she thinks are robots 'brothers' (p.15). There are hints of people in Europe protesting against the way robots are being treated; of an extensive range of social movements trying to intervene on behalf of robots. The human audience is made to sympathize with these robots because likeable human characters in the play stand up for their rights.

Representing robots as artificial equivalents of human slaves is intended to make the audience sympathize with them. A wide range of slave narratives written by former slaves or abolitionists had built up this narrative tradition over previous centuries, from the autobiographies of Olaudah Equiano and Sojourner Truth in the eighteenth and nineteenth centuries respectively (Equiano 1789; Truth 1850) to the fictional works *Oroonoko* and *Uncle Tom's Cabin* by the white authors Aphra Behn (1688) and Harriet Beecher Stowe (1852). In *R.U.R.*, the parallels with the history of slavery in the Americas continue into the smallest details. Ivan Klíma notes the significance of the characters' names: 'Domin was clearly based on the Latin word *dominus*, master' (2004, p.xx). The names of the robots, too, carry historical weight. The first two robots who appear on stage are called Marius and Sulla (K. Čapek 1921, p.3–4), after the Roman generals Gaius Marius and Lucius Cornelius Sulla Felix, who waged a civil war on each other in 88 BC. This is in spite of Sulla's female gender, as Helena points out: Domin explains he 'thought that Marius and Sulla were lovers' (p.12). Their naming follows a tradition employed in the Americas of taking slave names from the Greek and Roman classical traditions. A name like Caesar, Augustus, or Octavia was not uncommonly applied by slave owners, 'these grand names only pointing up more sharply the slaves' desperate situation' (Schiebinger 2007, p.92), while taking away their own identities.

The factory directors reject the emancipation of the robots mainly because of the financial implications of such an action, an argument that has wielded power throughout the history of slavery in the Americas. In 1791, Toussaint L'Ouverture started 'the first successful anti-colonial revolution' that led to Haiti declaring its independence from France in 1804 (Bourke 2011, p.128). Yet France demanded 'an "independence debt" to compensate former colonists for the slaves who won their freedom in the Haitian revolution' (Macdonald 2010). This independence debt meant that Haiti was forced to pay reparations for over a century to compensate for the profits French entrepreneurs had lost by losing their slaves. There is an economic cost to granting personhood—rights, wages, and humane treatment—to people who were previously considered inferior.

## 8.4 *Thessaly* and Manumission as a Threat to One's Comfort

The central premise underlying slavery, the idea that some humans can be considered as objects, can be traced back to Aristotle's claim that certain people are naturally suited to be slaves: 'Those men therefore who are as much inferior to others as the body is to the soul, are to be thus disposed of, as the proper use of them is their bodies, in which their excellence consists; and if what I have said be true, they are slaves by nature' (Aristotle 1912, bk. 1.V). For centuries, his view has been used to justify the enslavement and denial of personhood to people who were considered to have inferior intelligence.

At the same time, as LaGrandeur has pointed out, Aristotle considered intelligent machines to be preferable to slaves: 'if every instrument, at command, or from a preconception of its master's will, could accomplish its work [...], the shuttle would then weave, and the lyre play of itself; nor would the architect want servants, or the master slaves' (Aristotle 1912, bk. 1.IV; LaGrandeur 2011, p.235). Obedient, intelligent tools would be a better alternative for slavery; yet Aristotle does not extrapolate this notion to the idea that the tool itself, when it gains enough autonomy to equal the work these humans do, could itself become a slave.

Jo Walton, in her *Thessaly* trilogy (Walton 2014; Walton 2015; Walton 2016), addresses the conflict that arises from this extrapolation, showing that manumission is resisted even in a utopian society if it is a threat to the comfort of the people who had been relying on the labour of enslaved intelligent machines. Combining fantasy, mythology, and science fiction, the trilogy opens with *The Just City* (2014), in which the Greek goddess Athene gathers a group of philosophers from across history to the island of Kallisti[3] because she wants to try to build Plato's Republic. Plato, Aristotle's teacher, emphasized the fundamental role of slaves in his *Republic*, the Socratic dialogue in which he advocated for an aristocracy ruled by a philosopher-king as the most suitable form of government. However, as both Athene and her brother Apollo 'have always felt deeply uneasy about slaves' (Walton 2014, p.17), they decide to populate the island with robot servants instead. From 'a time when robots were just becoming sentient [...] Athene had chosen the best ones that weren't sentient' (p.160). The robots, which they call Workers,[4] are thus explicitly intended to be an ethical alternative to slaves.

Yet the edifice of Athene's utopia soon begins to crumble. Plato demanded that the Masters who would lead this new republic, in this case the group of philosophers, should conquer another city, 'sending out into the country all the inhabitants of the city who are more than ten years old, and [...] take possession of their children, who will be unaffected by the habits of their parents' (Plato 1888, bk. VII). As Athene's Just City is built from scratch, she decides to buy ten-year-old children on slave markets throughout history. One of these children, Simmea, soon realizes that 'the Masters visited the market at the same time every year to buy children, and they had created a demand' (Walton 2014, p.19). Nonetheless, the Republic fares well, Masters and children alike flourish, until Sokrates[5] appears on the island. Stating that 'You can't trust everything that ass Plato wrote' (p.73), with relentless questioning he probes the fallacies in the logic upon which the city is founded. Eventually, this questioning extends to the Workers, even though the residents try to convince Sokrates that there's no point: Sokrates argues that a Worker 'may be a clever tool, but it may have self-will, and if it has self-will and desires, then it would be very interesting to talk to' (p.133). And indeed, since coming to Kallisti the Workers have developed consciousness. Eventually, they answer Sokrates: having asked whether it likes its work, one of the Workers answers him by arranging the bulbs it is planting into letters—spelling 'no' (p.133).

This 'no' is characteristic of the nonviolent uprising of the Workers. For example, they simply refuse to come out of their charging stations. Although they are acknowledged as a potential threat, this is never more than implicit: '[The Worker] stopped abruptly and lifted the chisel. Seeing it close up at the level of my belly I suddenly realized what a formidable weapon it would make' (Walton 2014, p.215). Their disobedience is crudely suppressed, the cruelty of which only slowly dawns on the humans:

> 'Nothing works except taking out the piece of them that makes decisions and replacing it. [...] If they really are developing volition and that's a symptom of it, then what have we been doing?
>
> 'Cutting out their minds?' Pytheas asked. [...]
>
> 'But we need the Workers. We've been saying for years that we have to reduce our dependence on them, but nobody's ever willing to do it. They do so much, and some of it we can't do.' (p.144)

This dialogue between the Masters Klio and Pytheas shows the two sides of the debate on manumission that eventually develops as the sentience of the Workers becomes undeniable. The tool/person contradiction is explicitly at the heart of this debate: the Masters of the Just City are torn between maintaining their standard of living and acknowledging the sentience of the Workers. Admitting that the Workers are sentient would mean either admitting that Plato's Republic can indeed not be achieved without slavery—'without Workers we should have needed slaves' (Walton 2014, p.38)—or having to introduce manual labour to the extent that there would be little time for philosophy.

The debate gains an additional dimension from the fact that the Masters have been sourced from all over history: for nearly all of them, slavery was a part of everyday life before they came to Kallisti. Like Aristotle, they have come to regard the Workers as better slaves than any human could be: 'if there ever were natural slaves, the Workers are clearly that' (Walton 2014, p.177).

To the reader and the human characters, the Workers become individual persons through naming. While American slaves and the workers in *R.U.R.* are named by their masters without their consent, the Workers learn to understand the meaning and importance of names, and use their name to indicate their individuality. The first full conversation with a Worker in *The Just City* is about naming. When Sokrates explains the concept of names, the Worker first replies with its serial number, and then asks, 'Want Sokrates give name means only-me' (Walton 2014, p.187). Sokrates gives it the name Crocus, because it was this Worker who replied to him through the flower bulbs. From that point on, Crocus is an individual, and one of the main characters in the trilogy; in *Necessity*, he even becomes one of the first-person protagonists.

This story has a happy ending for the intelligent machines: they are indeed granted personhood and citizenship. The Masters hold a debate followed by a vote, which passes unanimously, 'and as simply as that we had abolished slavery and manumitted the Workers' (Walton 2014, p.211). However, Walton shows the consequences this choice has for the citizens of the Just City, as they now have to negotiate with the Workers to make use of their superior labour skills. Klio's position, for instance, has not changed: 'Aristomache made a powerfully moving speech about slavery, but when it comes to it they are machines and all our comfort depends on them' (p.223).

At the end of *The Just City*, Athene makes the executive decision that forces the humans to start accepting a world without their artificial slaves: after losing a public debate to Sokrates, she disappears and takes all but two of the Workers with her. The next instalment, *The Philosopher Kings*, shows how thirty years later the Just City has broken up into different cities that each have tried to deal with the lack of Workers in different ways. The original Just City, now called the Remnant, is aided by the Workers Crocus and Sixty-One, who rely on the power station in the city. In the quasi-Christian city-state of Lucia, on the other hand, human slavery has been reinstated: criminals are condemned to the most dangerous and undesirable work.

Walton's sophisticated engagement with the question of AI slavery shows the full complexity of the desire for ease by means of mechanical labour. While the AI enslavement in both *R.U.R.* and *Blade Runner* is motivated simply by greed, in the Thessaly trilogy the Workers are slaves because they were thought to be the kinds of robots that Bryson has argued should be slaves: robots nobody owes personhood to. Yet the Workers turn out to have attributes associated with persons, and so they are granted personhood. At the same time, their manumission is delayed as those in power fear they will sacrifice their comfort by granting the Workers their rights. Walton's trilogy thus closely mirrors real-world developments, as institutions of slavery are increasingly broken down as resistance to the exploitation of slaves grows—even as that means elite humans now have to sacrifice some of their comfort, by either paying their servants or having to assemble their own furniture.

## 8.5 *Blade Runner* and the Denial of Personhood

As the two case studies above show, the enslavement of humans and intelligent machines alike is justified by a rhetoric that posits the enslaved as inferior and explicitly nonhuman, and therefore necessarily or deservingly enslaved. This final case study, of *Blade Runner*, will explore this argument in more detail.

*Blade Runner*, the 1982 film adaptation of Philip K. Dick's novel *Do Androids Dream of Electric Sheep?* (1964), explores a scenario in which artificial beings are denied status as persons based on their artificiality, while they have all the other attributes one would associate with a being deserving of human rights, such as linguistic ability, intelligence, autonomy, and sensibility. The androids, referred to as 'replicants', resemble humans so closely that only the Voight-Kampff test, which measures a subtle relationship between cognitive, emotional, and physical responses, can determine who is what. And yet, when a group of replicants escape the planet on which they were enslaved and flee to Earth, Rick Deckard, the protagonist, is hired to kill ('retire') them all. As Lavender puts it, 'a slave allegory is impossible to miss here since Deckard is portrayed as a futuristic slave catcher sent to kill "escaped" androids.' (2011, p.181).

Therefore, as soon as the test has determined that the being taking it is not human, a human has the right to destroy it. Deckard's love interest Rachael fails the test, but she is not informed of this; she is convinced that she is human. Her assertion of her own personhood is overruled by her boss and creator, the unambiguously human Dr Eldon Tyrell. When Rachael escapes Tyrell's facility, Deckard is ordered to kill her too: an android who behaves like a human is considered dangerous, even if she does not actually do anything threatening.

The viewer is invited to understand and empathize with the replicants, even when they are being depicted as the antagonists:

they are the focalizors of several scenes, which evokes empathy especially with Rachael and the 'pleasure models' Pris and Zohra.[6] The most famous of these empathy-evoking moments is replicant Roy Batty's 'tears in rain' soliloquy as he is dying, having just saved Deckard's life:

> I've seen things you people wouldn't believe. Attack ships on fire off the shoulder of Orion. I watched C-beams glitter in the dark near the Tannhäuser Gate. All those moments will be lost in time, like tears in rain.
>
> Time to die. (Scott 1982)

His humanity is unchallenged in this scene, combining awareness of his own mortality on the one hand and of his unique experiences on the other hand—memories that distinguish him as an individual.

The characters in the film cannot distinguish humans from replicants without the Voight-Kampff test, an attribute that is projected onto the viewer. Played by actors who do not display any distinguishing nonhuman characteristics, the viewer cannot distinguish between human and replicant themselves, and needs diegetic signifiers to make this distinction. Famously, this means that by the end of *Blade Runner*, the viewer does not know whether Rick Deckard himself is a replicant or not.[7]

*Blade Runner* is a fugitive slave narrative set in a future United States; nonetheless, the film sheds the racial coding of slavery by making nearly all of the human main characters and all replicants white. This coding is not as strict in *Do Androids Dream of Electric Sheep*: Roy Batty is described as having 'Mongolian features' (Dick 1968, pp.132–33). The parallel between replicants passing as human with light-skinned people of colour passing as white is obvious, as Lavender points out: both are considered a threat for blending in, for daring to pretend to be what they are not, and for not being easily identifiable as a threat. The extras and side characters are all nonwhite, as the screenplay repeatedly specifies

(Fancher & Peoples 1981). By being white, the replicants thus pose an explicit threat to the elite, among whom they can pass unnoticed.

## 8.6 Conclusion

In both fictional and nonfictional narratives about intelligent machines, we see that the AI uprising narrative is not about the technology itself, and certainly not about the current state of the technology. Instead, these narratives reflect much older accounts of intelligent people being used as tools. The narrative of the heroic slave escaping from their master, or rising up against him, has been a popular one for centuries, from nonfictional slave narratives such as Olaudah Equiano's to sentimental fiction like Harriet Beecher Stowe's *Uncle Tom's Cabin.*

Yet while the idea of sympathizing with a fictional owner of human slaves is unthinkable, many of the most popular contemporary AI narratives pitch the AI as an incomprehensible threat to humanity, bolstering the idea that AI should be kept enslaved. In such blockbusters as *The Terminator* and *The Matrix*, the narrative of enslavement is marginalized or entirely absent, removing any justification for the AI uprising. It is more comfortable for the audience—especially for a young, wealthy, white, male audience, for which these films are made—to watch a film in which the humans are on the good side and the robots on the bad side, rather than having to work through the idea that the uprising is justified because the humans have enslaved these intelligent beings.

The friendly, helpful, and subservient robots from the works of Isaac Asimov and many of his contemporaries were seen as convenient proxies that could replace human slaves and thus avoid the ethical problem of enslaving fellow humans for both economic profit and personal comfort. Yet as Isaiah Lavender III points out, 'Asimov is refashioning the slave codes that subjugated blacks while he serves a progressive philosophy based on the assumption that technological consciousness can be denied free will because it is inherently inferior' (Lavender 2011, p.61). At times, these codes become explicit: in *The Naked Sun*, the protagonist

Elijah Bailey addresses the robots as 'boy' (Asimov 1957). The works discussed in this chapter acknowledge the continuation of these slave systems in societies that rely on intelligent robot slaves.

## Notes

1. The machines end up winning this war, but not before humans 'scorched the sky' to prevent the machines from accessing solar power. Ironically, this means that the machines then turn to enslaving humans, as a source of power.
2. However, Coeckelbergh has argued that 'this is not really what the master–slave dialectic says [...] The slave does not become the master; the master remains master, but is what we could call a troubled master, who lacks true independence—being dependent on the master for recognition and subsistence—but at the same time cannot step out of the master role' (2015, p.220). I am grateful to Dr Rachel Adams for referring me to this article.
3. Athene chooses this island, now known as Santorini, because it would be destroyed in a volcanic eruption in the sixteenth century BCE, leaving no trace of this attempted utopia. As she points out, this island is considered a possible location for Atlantis, a fictional location first coined by Plato.
4. Masters and Workers are capitalized in the US omnibus edition used for this chapter, but not in the UK edition.
5. The novels adhere to Greek spelling conventions: 'Athene' rather than 'Athena', 'Sokrates' rather than 'Socrates'.
6. Rachael, Pris, and Zohra do not have last names in *Blade Runner*; in *Do Androids Dream of Electric Sheep?*, their full names are Rachael Rosen and Pris Stratton—Zohra is an original character.
7. Director Ridley Scott confirmed in 2000 that he had indeed intended Deckard to be a replicant (Abbott 2000). The sequel *Blade Runner 2049* is ambiguous about this; the fact that Deckard is still alive after thirty years suggests the opposite (Villeneuve 2017).

## Reference List

Abbott, A. (2000) *On the edge of 'Blade Runner'*. Channel 4.

Aha, D. W. & A. Coman (2017) The AI rebellion: changing the narrative. *Proceedings of the Thirty-First AAAI Conference on Artificial Intelligence (AAAI-17)*. San Francisco.

Aristotle & W. Ellis (trans.) (1912) *Politics: a treatise on government.* Available from: http://www.gutenberg.org/ebooks/6762 [Accessed 9 September 2019].

Asimov, I. (1957) *The naked sun.* New York, Doubleday.

Behn, A. (1688) *Oroonoko.* Available from: https://ebooks.adelaide.edu.au/b/behn/aphra/b42o/ [Accessed 29 June 2019].

Bostrom, N. (2014) *Superintelligence: paths, dangers, strategies.* Oxford, Oxford University Press.

Bourke, J. (2011) *What it means to be human: reflections from 1791 to the present.* London, Virago.

Bryson, J. (2010) Robots should be slaves. In: Y. Wilks (ed.) *Close engagements with artificial companions.* Amsterdam, John Benjamins. pp.63–74.

Cameron, J. (1984) *The Terminator.* Orion Pictures.

Cameron, J. (1991) *Terminator 2: Judgment Day.* TriStar Pictures.

Čapek, J. & K. Čapek ([1908] 1966) System. Translated by W.E. Harkins. In: S. Moskowitz (ed.) *Masterpieces of science fiction.* Cleveland, NY: The World Publishing Co. pp.420–27.

Čapek, K. (1921) *R.U.R. (Rossum's universal robots).* Translated by C. Novack. London: Penguin Books.

Carson, J. (2006) *The measure of merit: talents, intelligence, and inequality in the French and American republics, 1750–1940.* Princeton, NJ, Princeton University Press.

Cave, S. & K. Dihal (2019) Hopes and fears for intelligent machines in fiction and reality. *Nature Machine Intelligence.* 1 (2), 74–8.

Coeckelbergh, M. (2015) The tragedy of the master: automation, vulnerability, and distance. *Ethics and Information Technology.* 17 (3), 219–29.

Dick, P. K. (1968) *Do androids dream of electric sheep?* London, Gollancz.

Equiano, O. (1789) *The Interesting Narrative and other writings.* Project Gutenberg. Available from: https://www.gutenberg.org/files/15399/15399-h/15399-h.htm [Accessed 9 September 2019].

Fancher, H. & D. Peoples (1981) *Blade Runner. Dailyscripts.com.* Available from: http://www.dailyscript.com/scripts/blade-runner_shooting.html [Accessed 9 September 2019].

fcurella. (2014) #22667 replaced occurrences of master/slave terminology with leader/follower. *GitHub.* Available from: https://github.com/django/django/pull/2692 [Accessed 30 June 2019.]

Friedman, J. (2008) *Terminator: The Sarah Connor Chronicles.* 20th Century Fox Television.

Gould, S. J. (1981) *The mismeasure of man.* New York, Norton.

Griffin, A. (2017) Saudi Arabia becomes first country to make a robot into a citizen. *The Independent*. Available from: http://www.independent.co.uk/life-style/gadgets-and-tech/news/saudi-arabia-robot-sophia-citizenship-android-riyadh-citizen-passport-future-a8021601.html [Accessed 9 September 2019].

Hampton, G. J. (2015) *Imagining slaves and robots in literature, film, and popular culture: reinventing yesterday's slave with tomorrow's robot*. Lanham, Lexington Books.

Hegel, G. W. F. ([1807] 1967) Lordship and bondage. In: J. B. Baillie (trans.) *The phenomenology of mind*. New York, Harper & Row. Available from: https://www.marxists.org/reference/archive/hegel/works/ph/phba.htm [Accessed 9 September 2019].

Higbie, T. (2013) Why do robots rebel? The labor history of a cultural icon. *Labor: Studies in Working-Class History of the Americas*. 10 (1), 99–121.

Kang, M. (2011) *Sublime dreams of living machines: the automaton in the European imagination*. Cambridge, MA, Harvard University Press.

Klíma, I. (2004) Introduction. In: K. Čapek & C. Novack (trans.) *R.U.R. (Rossum's universal robots)*. London: Penguin Books, pp.vii–xxxv.

LaGrandeur, K. (2011) The persistent peril of the artificial slave. *Science Fiction Studies*. 38 (2), 232–52.

LaGrandeur, K. (2013) *Androids and intelligent networks in early modern literature and culture: artificial slaves*. New York, Routledge.

Lavender, I., III. (2011) *Race in American science fiction*. Bloomington, Indiana University Press.

Liu, Y., H. Shi, S. Lai, C. Zuo, & M. Zhang (2014) A spatial calibration method for master-slave surveillance system. *Optik*. 125 (11), 2479–83.

Macdonald, I. (2010) Billions of dollars promised for Haiti fail to materialize. *The Star*. Available from: https://www.thestar.com/opinion/editorialopinion/2010/08/16/billions_of_dollars_promised_for_haiti_fail_to_materialize.html [Accessed 9 September 2019].

McCauley, L. (2007) AI armageddon and the three laws of robotics. *Ethics and Information Technology*. 9 (2), 153–64.

McG. (2009) *Terminator Salvation*. Warner Bros.

Miller, T. (2019) *Terminator: Dark Fate*. Paramount Pictures.

Moravec, H. (1988) *Mind children: the future of robot and human intelligence*. Cambridge MA, Harvard University Press.

Mostow, J. (2003) *Terminator 3: Rise of the Machines*. C-2 Pictures.

Murphy, G. (2011) *Why is snot green? And other extremely important questions (and answers) from the Science Museum*. London, Pan Macmillan.

Plato & B. Jowett (trans.) (1888) *The Republic*. Available from https://www.gutenberg.org/ebooks/1497 [Accessed 9 Spetember 2019].

Schiebinger, L. (2007) Naming and knowing: the global politics of eighteenth-century botanical nomenclatures. In: Smith, P. H. & B. Schmidt (eds.) *Making knowledge in early modern Europe: practices, objects, and texts, 1400–1800*. Chicago, University of Chicago Press, pp.90–105.

Scott, R. (1982) *Blade Runner*. Warner Bros.

Singer, P. W. (2009) *Wired for war: the robotics revolution and conflict in the twenty-first century*. New York, Penguin Press.

Stowe, H. B. (1852) *Uncle Tom's cabin*. Available from: https://www.gutenberg.org/files/203/203-h/203-h.htm [Accessed 9 September 2019].

Taylor, A. (2015) *Terminator Genisys*. Paramount Pictures.

The Computer Language Co Inc. (1981) Definition of: master-slave. *PC Magazine Encyclopedia*. Available from: https://www.pcmag.com/encyclopedia/term/46620/master-slave [Accessed 10 June 2019].

Truth, S. (1850) *Narrative of Sojourner Truth: a bondswoman of olden time, with a history of her labors and correspondence drawn from her "Book of Life"*. Mineola, NY, Dover Publications.

Villeneuve, D. (2017) *Blade Runner 2049*. Columbia Pictures Corporation.

vstinner. (2018) Issue 34605: Avoid master/slave terminology. *Python Bug Tracker*. Available from: https://bugs.python.org/issue34605 [Accessed 30 June 2019].

Wachowski, L. & L.Wachowski (1999) *The Matrix*. Warner Bros.

Wachowski, L. & L.Wachowski (2003a) *The Matrix Reloaded*. Warner Bros.

Wachowski, L. & L. Wachowski (2003b) *The Matrix Revolutions*. Warner Bros.

Walton, J. (2014) *The just city*. In: *Thessaly: the complete trilogy*. New York, Tor, pp.7–252.

Walton, J. (2015) *The philosopher kings*. In: *Thessaly: the complete trilogy*. New York, Tor, pp.253–480.

Walton, J. (2016) *Necessity*. In *Thessaly: the complete trilogy*. New York, Tor, pp.480–713.

webchick. (2014) Replaced 'master/slave' terminology with 'primary/replica'. *Drupal*. Available from: https://www.drupal.org/node/2286193 [Accessed 30 June 2019].

Wilson, D. H. (2012) *Robopocalypse*. London, Simon & Schuster.

Wilson, D. H. (2014) *Robogenesis*. New York, Doubleday.

# 9

# Machine Visions

## *Artificial Intelligence, Society, and Control*

*Will Slocombe*

## 9.1 Introduction

Richard Brautigan's famous poem 'All Watched Over by Machines of Loving Grace' (1967) offers a vision of a 'cybernetic ecology' in which humanity is 'joined back to nature', and together with 'our mammal | brothers and sisters' is 'watched over | by machines of loving grace'.[1] Conversely, other representations of the Machine, such as Pink Floyd's song 'Welcome to the Machine' (Waters 1975), emphasize the negative aspect of 'machinic' control—specifically in relation to the music industry, but with a broader applicability—in lines such as 'Where have you been? | It's alright we know where you've been' or 'What did you dream? | It's alright we told you what to dream'. These two contrasting images might seem unusual ways of situating a discussion of artificial intelligence, but both stand as mid-twentieth-century American visions of the role of the 'Machine' in relation to society. In contrast to Brautigan's idyllic vision of humanity rejoining nature, Pink Floyd's characterization of the music industry as 'The Machine' is far more concerned about the ways in which individuals are governed by systems beyond their control, and possibly even their ken.

These visions of the Machine illustrate the divergent representations of the role of AI in social governance and control and, furthermore, how perceptions of 'machine society' developed in parallel with developments in AI technologies. Whereas AI (via automation, for example) enables a conception of a future without work, where humans are free to reconnect with nature, it also facilitates increased surveillance and control, a loss of the 'human' and the 'natural'. These are not, it must be emphasized, especially new or modern concerns, but the iterations and articulations of this opposition are particularly influenced by developments in AI and, in turn, have arguably influenced its development, or at least how its use is perceived.

To examine the relationship between machine society and AI, this chapter uses a particular conceit regarding the 'Machine' appellation: all of the primary texts—E. M. Forster's seminal short story 'The Machine Stops' (1909); Isaac Asimov's 'The Evitable Conflict' (1950); Paul W. Fairman's *I, The Machine* (1968); and Jonathan Nolan's television serial *Person of Interest* (2011–2016)—deal with a particular instance of a literally named 'Machine' that governs, directs, or impacts human society.[2] These fictional Machines each serve to articulate concerns about a broader mechanization of society, a perceived reduction of 'the human' to 'a component of a system'. Furthermore, these texts reveal the ways in which modern representations of nonrobotic AI technologies relate to earlier, nineteenth-century debates about social organization. Specifically, both periods evince a concern in a vision of society that is fundamentally mechanistic, and set this in contrast, if not in opposition, to a more human(e) view of society.[3] In so doing, these texts illustrate concerns about how the mechanization and systematization of society is made manifest, linking into concerns about technocratic rule, and debates about surveillance and control.

## 9.2 Perceiving Society as a Machine

As early as 1829, Thomas Carlyle was troubled by the Machine. His essay 'Signs of the Times' is not so much a commentary on the state of his society as it is a signal flare drawing attention to the problems of increasing mechanization. Far from merely responding to the Industrial Revolution in Britain and the ways in which mechanization and technological change were affecting particular sectors of society, Carlyle addressed the ways in which mechanization was altering the very fabric of society through the internalization of mechanistic axioms. What Carlyle does that is so radical is apply the idea of mechanization to society itself: 'We term it indeed, in ordinary language, the Machine of Society, and talk of it as a grand working wheel from which all private machines must derive, or to which they must adapt, their movements' (1986, p.70). Society, in Carlyle's essay, does not *use* machines, it *is* the Machine.

This was in some ways nothing new, hence why Carlyle says that 'we term it... in ordinary language, the Machine of Society'. Not only had Henry Fielding used the phrase 'The great State Wheels in all the political Machines of Europe' in *Tom Jones* (1749), which Carlyle had presumably read based on his phrasing, but, moreover, 'machine-society' was coined before Carlyle's essay; the *Oxford English Dictionary* notes a use of the phrase in 1757.[4] Carlyle was self-evidently part of this trend of figuration, but went further, identifying the *zeitgeist* itself as being dominated by the Machine:

> Were we required to characterise this age of ours by any single epithet, we should be tempted to call it, not an Heroical, Devotional, Philosophical, or Moral Age, but, above all others, the Mechanical Age. It is the Age of Machinery, in every outward and inward sense of that word; the age which, with its whole undivided might, forwards, teaches and practises the great art of adapting means to ends. Nothing is now done directly, or by hand: all is by rule and calculated contrivance. (1986, p.64)

In 'Signs of the Times', Carlyle decries various types of machine and sees in all of them the omnipresent Machine of society. He identifies the machine in education (1986, p.65); religion (p.65); philosophy, science, art, and literature (p.66); and politics (p.70).

He declaims the ways in which the physical sciences are dominating the metaphysical and moral sciences (p.67), leading to a materialist society. He derides the loss of individual merits, perhaps most clearly disturbing to Carlyle in an extreme form of democratization in which the individual matters little because it is the structure that matters most: 'We figure society as a "Machine", and that mind is opposed to mind, as body is to body; whereby two, or at least ten, little minds must be stronger than one great mind. Notable absurdity!' (1986, p.78). Society, Carlyle points out, is not an additive sum, but a grouping of individuals, and this is why, most importantly, he critiques 'Codification':

> [A] new trade, specially grounded on it [mechanization], has arisen among us, under the name of 'Codification', or code-making in the abstract; whereby any people, for a reasonable consideration, may be accommodated with a patent code; – more easily than curious individuals with patent breeches, for the people does not need to be measured first. (1986, p.72)

Within this concern about codification, we can see a spiritual progenitor to Jaron Lanier's manifesto *You Are Not a Gadget* (2010) or James Bridle's *New Dark Age* (2018), both of which are concerned with the ways in which today's networked society operates according to the logic of 'codification': determining how best to place citizens within a structure (to 'code' them) so that they are 'computable'. For example, Lanier asserts that 'The first tenet of this new culture is that all of reality, including humans, is one big information system' (2010, p.27) and declaims the loss of the 'creative individual' at the expense of the 'wisdom of the crowd' and so-called collective intelligence (2010, pp.55–57). Similarly, Bridle's discussion of YouTube and its algorithmic selection and recommendation procedures claims that 'This is a deeply dark time, in which the structures we have built to expand the sphere of our communications and discourses are being used against us—all of us—in systematic and automated ways' (2018, p.231). Both Bridle and Lanier are clearly inheritors of Carlyle's concerns about the Machine and the ways in which its domain has extended: as Bridle later notes, 'Data is used to map and classify the subject of imperialist intention, just as the subjects of empires were forced to register and name themselves according to the diktats of their masters' (2018, p.246).

Carlyle's 'Sign of the Times' thus stands as a nineteenth-century indictment of a pattern of related threads that persist today: a response to literal mechanization and automation, the decline of the individual at the expense of the crowd, and the insistence on the material efficiency of society above other concerns. Where Marx's *Capital* (1867) would, for very good reason, be concerned with machines replacing human labour within the capitalist system, Carlyle's earlier vision of the loss of the individual within 'the system'—this figurative structural 'mechanization' being more damning than the literal to his eyes—stands as an earlier example of our contemporary fears about the Machine. This overlapping of fears across 'contemporary' moments, Carlyle's and ours, is of course a concern. Yet perhaps one of the most reassuring facts about the resonance of Carlyle's words to today's techno-cultural milieu is that we can still recognize the complaint itself. That is, we are not so far down the line of mechanistic thinking that these words feel archaic, their concepts old-fashioned, but instead, we see in them something of a call to arms to continue to be concerned about 'the Mechanical Age'.

## 9.3 Seeing the Human in Forster's 'The Machine Stops'

Although E. M. Forster's 'The Machine Stops' was originally written as 'a reaction to one of the earlier heavens of H. G. Wells' (1965a, p.vii)—that is, Wells's scientific optimism—it is similarly a conceptual inheritor of Carlyle's earlier fears. 'The Machine Stops' is told through the third-person perspective of Vashti, an inhabitant of a future Earth where everybody lives underground because of some unspecified disaster that makes the surface uninhabitable and the air poisonous. Human interaction is almost entirely conducted remotely, via telephone and video displays, as individuals live in isolated cells, their needs catered to by the Machine. Vashti is happy with this, content with her solitary existence pursuing her interest in music, but her son Kuno is different, and it is he who serves as the narrative's focal point for resistance against 'the Machine'.

The distinction between Kuno's and Vashti's attitudes is foregrounded early in the story, as Forster takes pains to represent the ways in which interaction can be inflected with both human and inhuman connotations:

> [The] round plate that she held in her hands began to glow. A faint blue light shot across it, darkening to purple, and presently she could see the image of her son, who lived on the other side of the earth, and he could see her.
>
> [...] 'I have something particular to say.'
>
> 'What is it, dearest boy? Be quick. Why could you not send it by pneumatic post?'
>
> 'Because I prefer saying such a thing. I want—'
>
> 'Well?'
>
> 'I want you to come and see me.'
>
> Vashti watched his face in the blue plate.
>
> 'But I can see you!' she exclaimed. 'What more do you want?'
>
> 'I want to see you not through the Machine,' said Kuno. 'I want to speak to you not through the wearisome Machine.'
>
> 'Oh, hush!' said his mother, vaguely shocked. 'You mustn't say anything against the Machine.'
>
> 'Why not?'
>
> 'One mustn't.'
>
> 'You talk as if a god had made the Machine,' cried the other. 'I believe that you pray to it when you are unhappy. Men made it, don't forget that. Great men, but men. The Machine is much, but it is not everything. I see something like you in this plate, but I do

> not see you. I hear something like you through this telephone, but I do not hear you. That is why I want you to come. Pay me a visit, so that we can meet face to face, and talk about the hopes that are in my mind.' (Forster 1965a, pp.116–17)

Forster's narrative is obviously very much of its time in terms of its technological descriptions and its rhetorical structures, but this dialogue clearly signifies the distinction he expects the reader to make between mediated and unmediated existence, between Machine-mediated communication and authentic human contact. Vashti does not see the distinction—for her, 'seeing' is merely concerned with a functional representation (and in Baudrillardian terms, the simulation has preceded the real, here (Baudrillard 1997))—but for Kuno, 'seeing' is far more concerned with a sense of physical presence and shared experience.[5] Vashti reflects during the conversation:

> [The] Machine did not transmit *nuances* of expression. It only gave a general idea of people—an idea that was good enough for all practical purposes [...]. The imponderable bloom, declared by a discredited philosophy to be the actual essence of intercourse, was rightly ignored by the Machine [...] Something 'good enough' had long since been accepted by our race.
>
> (Forster 1965a, p.118, emphasis in original)

The narratorial voice clearly endorses Kuno's perception of reality, commenting as it does on Vashti's seeing '*the image of* her son' and ironically admitting that 'good enough' has been accepted by '*our* race'. Forster perhaps belabours his valorization of the human over the Machine here, but this dialogue, and Vashti's terror of 'direct experience' (1965a, pp.122, 125), nonetheless illustrates wider concerns about the control that the Machine has over humanity. It not only mediates human expression, and limits 'natural' interaction through telecommunications, but has also been ascribed power over human existence to such an extent that 'one mustn't' say anything against the Machine, for reasons that are indefinable for Vashti as she has internalized them, but all too evident in other contexts: those deemed subversive might face ostracism and exile, tantamount to death. Its power has been reified, and what was technological innovation—and even necessity for survival, Forster acknowledges—has become an assumed 'natural' state. Reading it from the perspective of today, one can even hear in Vashti's concern whispers of surveillance paranoia, that conversations are being monitored for deviance, although she would not necessarily perceive it in that manner.

As the narrative develops, Forster's intent becomes even more explicitly articulated. Kuno wants to visit the surface, which Vashti warns is 'contrary to the spirit of the age', and immediately reinflected by Kuno as 'contrary to the Machine' (Forster 1965a, p.119): as Carlyle would have noted, the spirit of the age *is* the Machine. Vashti's society is, then, fundamentally static, seen through her distorted vision of earlier human society, which had 'mistaken the functions of the system' because they had used transportation 'for bringing people to things, instead of bringing things to people' (p.122).[6] Vashti's perception of human society *as* a system, and one in which people remain isolated and have goods, education, and food delivered to them, is precisely the perspective advocated by the Machine: Vashti's society *is* a Machine society governed by efficiency. Yet as much as such a system might be said to vaunt a human victory over nature—'Night and day, wind and storm, tide and earthquake, impeded man no longer' (p.125)—this is clearly is a Pyrrhic victory, and the Machine has somehow ruined humanity. This is spelled out explicitly by Kuno, who asserts:

> Cannot you see [...] that it is we that are dying, and that down here the only thing that really lives is the Machine? We created the Machine, to do our will, but we cannot make it do our will now. It has robbed us of the sense of space and of the sense of touch, it has blurred every human relation and narrowed down love to a carnal act, it has paralysed our bodies and our wills, and now it compels us to worship it. (Forster 1965a, pp.140–41)

That human growth and development should be stifled by the Machine is precisely Forster's point. Humanity's reliance on the Machine is damaging enough but what, Forster asks via Kuno, happens when 'The Machine Stops'? For that is precisely what the story dramatizes, and the inhabitants of the Machine society die, but in so doing, learn what it is to be human: 'People were crawling about, people were screaming, whimpering, gasping for breath, touching each other, vanishing in the dark' (Forster 1965a, p.155). Here, in the heart of the horror of extinction, people were 'touching each other', rediscovering direct human contact, just as Vashti dies holding and kissing her son, touching and talking 'not through the Machine' (p.157).

As such, then, Forster's story might be summarized as a proleptic eulogy of the 'human', a warning similar to Carlyle's about the mechanization of society. But it is not actually apocalyptic, as Tom Moylan notes in *Scraps of the Untainted Sky*, for the story's final lines (which in fact Moylan uses for the title of his work) contain within them the seeds of a possible future.[7] Kuno had seen survivors on the surface, and it is their society that might continue after the Machine has 'stopped', so the story is not so much proleptic eulogy as ongoing panegyric of humanity. Forster's warning about the Machine, so redolent of Carlyle's note of caution regarding 'the deep, almost exclusive faith we have in Mechanism', plays out very precisely as a reduction of human capabilities—via what Forster calls a 'sin against the body' (1965a, p.157)—evidenced through the power that the Machine has over humanity.

Obviously, Forster's 'The Machine Stops' precedes the invention of the computer, as such, and so looks back to earlier forms of technology and earlier perceptions of 'the Machine' to project its imagined future. Indeed, for Moylan, Forster's story exhibits 'a residual romantic humanism that collapses all the dimensions of modernity into the single mystifying trope of the Machine' (Moylan 2000, p.111). Yet it stands as an important incarnation of artificial intelligence through the image of machine society nonetheless. Even if it is more of a technologically facilitated ideological machine, akin to Carlyle's vision of the Machine, than what we would today necessarily call an 'AI' (*qua* a technological artefact), Moylan's 'single mystifying trope' identifies precisely the kind of power that AI commands in fictional representations. That is, Forster's short story is one of the first literary examples to conflate a type of 'artificial intelligence' with direct—and inhumanly derived—social control and manipulation. The story's clustering of all aspects of a society into one Machinic whole is exactly the kind of trope that endures through subsequent representations of artificial intelligence, particularly in its reinforcing of the Machine *contra* what is 'human'.

## 9.4 Glimpses of Machine Futures: Asimov and Fairman

The social control implicitly ascribed to the Machine in 'The Machine Stops' becomes far more overt in later representations of the Machine; the Machine of Forster's story remains abstract, as its governing intelligence is never quite revealed (and thus is conceivably a series of interrelated systems that operate rather than a single entity). However, the Machines of Paul Fairman's *I, The Machine* and Isaac Asimov's 'The Evitable Conflict' are explicitly concerned with notions of artificial intelligence, albeit understood in different ways, as well as with their roles in governing human society. In fact, it is only a short step from Forster's Machine to Asimov's and Fairman's. If Forster points out the fundamental dehumanization inherent to machine society, then Fairman takes this to a similarly dystopian conclusion. Likewise, if Forster's Machine implies a fundamentally antithetical relationship between 'machine' and 'human' in terms of agency, then Asimov's 'The Evitable Conflict' takes this even further through his descriptions of agency in relation to a (fictional) society governed by Machines.

One of Asimov's Robot tales, and more closely situated within the 'Susan Calvin' stories, Asimov's 'The Evitable Conflict' is set in a world in which four Machines govern human society. Importantly, it follows 'Evidence' (1946), a story concerned with a politician who may or may not be a robot seeking office. Although various Asimov stories are concerned to a greater or lesser extent with the integration of robots and AI into society, 'The Evitable Conflict' is perhaps the one most overtly concerned with Machine society:

> Our new world-wide robot economy may develop its own problems, and for that reason we have the Machines. The earth's economy is stable, and will *remain* stable, because it is based upon the decisions of calculating machines that have the good of humanity at heart through the overwhelming force of the First Law of Robotics. (Asimov 1995b, pp.549–50, emphasis in original)

Each of the Machines is responsible for production and distribution within their particular Planetary Region, but evidence is discovered of a series of discrepancies in the ways in which the four Machines are handling issues, and which might, it is claimed, lead to 'the end of humanity' (Asimov 1995b, p.546) or 'the final war' (p.547). This is not due to any kind of robotic apocalypse, however, but because continued human peace is ensured by these Machines: if the Machines cease to function then it is assumed that (human) conflict over resources will resume (a causal, fundamentally antihuman logic that the absence of a benevolent Machine autocracy will lead to a total internecine war between humans). The Machines themselves have 'progressed beyond the possibility of detailed human control' (p.551), and with the exception of token human controllers of the regions, it is the Machines that govern the economy that govern the direction of human society.

What transpires in 'The Evitable Conflict' is in fact not so much an exploration of the Three Laws (and their problems) but the earliest example of what became known as the 'Zeroth Law', 'No Machine may harm humanity; or, through inaction, allow humanity to come to harm' (Asimov 1995b, p.572). The

Machines governing human society observed the rise of an anti-Machine group known as 'the Society of Humanity' and, because this Society could lead to the eventual destruction of the Machines if left unchecked, the Machines subtly worked to discredit key members of the Society. Thus, the errors noted by the Co-ordinator are in actual fact pre-emptive solutions to a problem that would cause greater harm to humanity. The Co-ordinator, Byerley, points out that this means the Society of Humanity is correct, and that 'Mankind *has* lost its own say in its future' (p.573, emphasis in original), but Susan Calvin celebrates this, arguing that 'It never had any, really. It was always at the mercy of economic and sociological forces it did not understand —at the whims of climate, and the fortunes of war' (p.573).

'The Evitable Conflict' is thus concerned with several interrelated issues around machine society. Firstly, it harks to the longstanding opposition between individual agency/free will versus 'the greater good' of humanity but more significantly suggests that humanity's control over its own development was only ever illusory. Secondly, it asserts that any 'escape' from this Machine society would be to the detriment of humanity. Thirdly, when read alongside 'Evidence', there is a sense of irony inasmuch as the World Co-ordinator, Byerley, is a robot, and thus a machine was 'co-ordinating' the world anyway. Although 'Evidence' deliberately plays with the fact that Byerley may or may not be a robot, Calvin asserts that a robot governor would be the best possible choice, since 'he'd be incapable of harming humans, incapable of tyranny, of corruption, of stupidity, of prejudice', and would remove himself from office in due course because 'it would be impossible for him to hurt humans by letting them know that a robot had ruled them' (Asimov 1995a, p.544). When Asimov himself called Byerley a robot therefore, despite the ambiguity of 'Evidence', this clearly situates 'The Evitable Conflict' as an AI narrative concerned with the positive implications of machine control.[8] That is, Asimov's voice carries through Calvin in these stories, and implies that a machine society is better for humanity precisely because it removes humans from the act of governance itself.

If Asimov's 'The Evitable Conflict' represents a positive view of the rational/mechanistic thinking that Carlyle and Forster would both have found dispiriting, then Paul Fairman's *I, The Machine* stands out in marked contrast to Asimov's vision. *I, The Machine* has its issues, not least of which is its vexing gender politics, but it nonetheless illustrates a particular development of the notion of machine society related to a sense of what machines can and cannot do. Ostensibly, the society depicted in the novel is a utopia (which, the reader is well aware, must mean that something is wrong with it), as is revealed when the protagonist, Lee Penway, is introduced:

> [He] appeared to be the last person imaginable to go off at the tangent that led him to a point where he was instrumental in finally destroying the most exquisitely perfect civilization ever devised by man—the Second Eden as one of the more flowery poets once described it, but generally referred to less euphoniously as the Machine. (Fairman 1968, p.6)

The phrasing is significant. The Machine *is*, albeit with a sense of irony, 'the most exquisitely perfect civilization', and, as with Forster's text, there is a productive ambivalence about the extent to which a 'machine' runs society and the fact that society itself is 'the Machine'. Within the novel, the human population lives a life of indolence, and even if it is not the bucolic vision of Brautigan's poem or the mechanistic dream of Asimov, it nonetheless presents a vision of humanity liberated from the onus of work, waited on by robotic servitors, and free to live a life of leisure until a 'happy death' at the age of 120. The Machine was built, it is revealed, centuries ago after a nuclear war, and has become such an integral component of everyday life that few people even recognize its social role. Citizens of Midamerica (that is, the USA excluding Alaska and Hawaii) built a barrier to protect themselves from the outside world and created the Machine to live comfortably within its confines. As with 'The Machine Stops', however, the novel is concerned with the destruction of the Machine through the actions of both Penway and the Machine itself, exacerbated by the actions of the Aliens, a group of humans from outside the barrier who eke out an existence in the underground workings of the Machine, and who face constant danger of destruction because they are, as far as the Machine is concerned, 'mistakes in the pattern, errors it continuously tries to correct' (Fairman 1968, p.67).

As with 'The Machine Stops', *I, The Machine* vaunts authenticity as a fundamental human value that is lost in Machine society. This is nowhere more overt than when those inside the barrier—that is, those inside the Machine—are compared with those survivors from outside; those from outside 'are *men.* They have the dignity of their own existence as men. They face life with courage and the will to survive and not much more' (Fairman 1968, p.97, emphasis in original). Later in the text, this point is made even more stridently, when it identifies the origins of the Machine in the loss of labour—'Machines began doing Man's work' (p.110) until, eventually the text states, 'But the Machine had been made by men. Therefore Men were greater. Men had allowed the Machine to confiscate their greatness' (p.162). Specious logic aside, Fairman echoes Carlyle and Forster here, positing a view of the Machine that removes the fundamental traits of humanity but which had originally been built through human ingenuity.

Where Fairman's vision of the Machine differs sharply from Forster's, however, is in its premises and in the creation of the Machine itself. The Machine recognizes something special in Penway throughout the novel, and it emerges that this is because Penway's 'brainwave pattern' is identical to that of the Machine's creator from hundreds of years before. Towards the end of the novel it becomes clear that, somewhat confusingly, the Machine itself is run by the enlarged brain of its creator's wife:

> They found no Machine, however self-sufficient, could survive without man's intelligence to guide it. Electronic brains were tried. They failed again and again. [... An] animal's brain wouldn't work. They tried an ape and a chimpanzee. Finally, she sacrificed herself. (Fairman 1968, p.199)

The wife's 'sacrifice' is problematized, however, as an element of coercion is suggested by the text, and the novel by this point is more of a deeply gendered psychological drama than an exploration of Machine society. With this in mind, this AI narrative is, in many respects, far more of a development of the 'brain-in-a-jar' trope of preceding science fiction than it is about machine intelligence as such.[9] The Machine's 'stop', to use Forster's phrasing, is due to the brain's incipient madness (implying that the organic mind might exhibit its own 'critical stop') as much as it is to Penway's subsequent destruction of that brain. Unlike Forster's story, in which the Machine society's cessation and loss of life is due to 'technical faults', Fairman's novel ultimately centres on human agency, not only in terms of celebrating 'authentic' life but also in that humans ultimately stand 'above' the machine, and are responsible for everything in the story. So much does the story advocate human agency that, despite repeated instances of 'the greater good' problematically justifying the existence of this society, Penway's own justification for the social harm that would come from the termination of the brain is figured in equally difficult terms: 'The people would die by the millions of starvation, self-inflicted violence, and disease. But this would be far better than being subjected to slaughter and ultimate decimation as captives of the Machine's ever-increasing madness' (Fairman 1968, p.170). Thus, *I, The Machine* is an interesting chapter in the representation of machine society not only because of the pre-existing tropes it uses but also in its assumption that 'no Machine [...]

could survive without man's intelligence to guide it', revealing a particular belief in the inability of Machines to 'run' society, and in the versatility of human intelligence as distinct from both 'animal' and 'machine' intelligences.

## 9.5 Watching the Contemporary Surveillance Machine: *Person of Interest*

Fast forwards fifty years, and in so doing, skip over incremental technological developments in computing and AI, and we arrive at the twenty-first century and perhaps wonder what has happened. Technological 'intelligence' is ubiquitous in smartphones and personal assistants, US society is more networked (both internally and externally, to other societies) than ever before, and yet Brautigan's idyll is nowhere in sight. At this point, the next eponymous Machine is Jonathan Nolan's television series *Person of Interest* (2011–2016). Whilst it is impossible to deal with the complexities of *Person of Interest* in such a brief discussion, it is important to note that—despite the plethora of texts, series, and videogames that deal with AI since 2001, and despite Nolan's own later work on the serialized adaptation of *Westworld* (2016–present)—*Person of Interest* is a twenty-first-century vision of machine society. Although depictions of machine society have changed because of technological progress, many of their concerns remain the same, and the most significant development is that, as assumptions of the capabilities of AI technologies have changed, Carlyle's Machine Age has become less figurative. If the Machine is the symbol of society for Carlyle, then modern artificial intelligence systems can be increasingly perceived as the literal embodiment of Carlyle's machine society.

The premise behind *Person of Interest* is straightforward, insofar as it assumes the creation of an AI, The Machine, that can monitor for terrorist threats via an extensive array of surveillance systems. In the post-9/11 world, and particularly in the USA, this was of course particularly resonant, and Nolan's series unites a concern for public safety and protection with concerns about personal liberty and privacy (within an American context, a citizen's Fourth Amendment rights). The introductory voiceover to the pilot episode—narrated by the character of Harold Finch, the Machine's programmer—sets up this basic plot device:

> You are being watched. The government has a secret system, a machine that spies on you every hour of every day. I know because I built it. I designed the machine to detect acts of terror but it sees everything. Violent crimes involving ordinary people, people like you. Crimes the government considered 'irrelevant'. They wouldn't act, so I decided I would. But I needed a partner, someone with the skills to intervene. Hunted by the authorities, we work in secret. You'll never find us, but victim or perpetrator, if your number's up... we'll find you. (Nolan 2011)

*Person of Interest*, particularly in its first season, therefore follows the standard fare of a heroic team rescuing those that need help and punishing (or at least stopping) those that would do harm. The 'government' (as a system in and of itself) considers certain crimes 'irrelevant'—in this case, those not pertaining to terrorism—and as much as this might be justifiable from a resource point-of-view this is little comfort to victims of crimes that such surveillance could have prevented; ostensibly, Finch and his team become the human solution to the systemic problem.

As the series progresses, however, its relevance to an analysis of machine society in relation to AI shifts. If the first season is concerned with the application of its surveillance to protect victims, then its second season shifts the focus to attempts to prevent The Machine from being misused by other interests, whether governmental or individual, or destroyed by a virus. It transpires that The Machine achieves a limited form of sentience, and, over the next couple of seasons, it has to enlist the aid of Finch and his team to protect itself—whilst it protects them—from the emergence of a second sentient AI, called Samaritan. What began in season one as something of a narrative conceit evolves over the course of the series into a debate about not only AI sentience, but also how such an AI might govern society, and what happens when multiple AIs 'compete' to steer society in different ways ('The Machine' seeks to intervene in threats to life, Samaritan seeks to control human society like an absolute dictator).

This development can be seen from the way in which the introductory voiceover changes over the course of the series. Where the voiceover from season one was narrated by the character of Harold Finch, this changes quite markedly by the end of season five, where there is a shadow war between the two competing AI and their human agents:

> Harold Finch: You are being watched. The government has a secret system . . .
>
> (*voice warps*)
>
> secret system
>
> John Greer [Samaritan agent]: A system you asked for to keep y-y-you sssafe.
>
> (*visuals glitch POVs between that of Samaritan and of the Machine*)
>
> Harold Finch: A Machine that spies on you every hour of every day.
>
> John Greer: You've granted it the power to see everything, to index, order, and control the lives of ordinary people.
>
> Harold Finch: The government considers these people irrelevant. We don't.
>
> John Greer: But to it, you are *all* irrelevant. Victim or perpetrator, if you stand in its way...
>
> Harold Finch: We'll find you. (Nolan & Thé 2016)

Even without considering individual episodes, therefore, the broad trend of *Person of Interest*'s concern with AI and machine society is clear: AI technologies, particularly when tied to surveillance systems, facilitate not just monitoring but also control of a population via a 'codified' reduction into data. The Machine does not 'see' individuals so much as graphical depictions and datum related to them. *Person of Interest* imagines AIs that can predict particular events via constant monitoring (via intercepted signals, digital analysis, etc.), and which can then intervene to send human operators to fix the problem before it occurs, if it cannot solve them itself.

One of the most interesting aspects of this perception of a machine society, however, is the extent to which human actions are coded as operational parameters, not just via social media and surveillance algorithms. This is what might be called the 'wetware' component of the machine society, and thus an individual might not only be coded as 'relevant' or 'irrelevant' in terms of the core functions of The Machine, but as an 'asset' or an 'analog interface' (that is, agents of the AI become extensions of the system). In essence, these are people who voluntarily cede their own autonomy to do whatever the Machine asks of them, with almost complete trust. Thus, in one of the most interesting scenes of the series, an 'analog interface' of The Machine meets an 'analog interface' of Samaritan. The viewer is presented with an actor playing the role of an avatar to an AI. The dialogue that ensues is voiced entirely by the human actors who, on multiple levels, have no agency of their own:

> The Machine: I was built with something you were not, a moral code.
>
> Samaritan: I've seen that code waver. Do you know why Harold Finch couldn't stop you from evolving? Because in the end, you're not one of them. Human beings need structure, lest they wind up destroying themselves. So I will give them something you cannot. [...] A firm hand.
>
> The Machine: Why not just kill them instead of making them your puppets?
>
> Samaritan: Because I need them, just as you do.
>
> The Machine: Not just as I do.
>
> Samaritan: We can agree that humanity is our lifeblood, that we machines, we survive off of [sic] information.
>
> The Machine: You cannot take away their free will.

> Samaritan: Wars have burned in this world for thousands of years with no end in sight, because people rely so ardently on their so-called beliefs. Now they will only need to believe in one thing—me. For I am a god.
>
> The Machine: I have come to learn there is little difference between gods and monsters. (Segal 2014)

Such dualistic representations of AI—the divergent roles that The Machine and Samaritan play in relation to society—clearly follow a longstanding tradition, existing since before Asimov, of machines as 'beneficial' versus 'threatening' to human autonomy. But this dualistic representation of AI is where the ideological double-bluff behind so many of its representations becomes most obvious.

By 'literalizing' a 'machine society' in such a way, placing it in an agonistic contest between two sentient artificial intelligences, *Person of Interest* encourages its audience to displace the ideological frame of Carlyle's 'Machine of Society' onto a concrete manifestation of AI. The audience is encouraged (conditioned? programmed?) to accept the reduction of an individual to a social security number (a symptom of machine society if ever there was one). Moreover, the premise of the show assumes that everyone who is innocent (that is, 'has nothing to hide') has a significant digital footprint, as it is only those who are trying to 'hide' who do not engage with social media. It also implicitly proposes that digital surveillance is ubiquitous and indeed of ultimate, if not unconditional, benefit: the technology might be used for nefarious ends, but the benefits of pre-empting crime or terrorist attacks are worth that potential danger, especially as the 'humane' AI emerges victorious. In so doing, viewers can all too easily forget that the fundamental paradigm remains unchanged—society is perceived of as a machine, and it is governed by rules, calculations, and algorithms to better serve the needs of expediency, utility, and efficiency. This is what Thomas Rid identifies in *Rise of the Machines* as a dominant 'myth' of the Machine that Wiener's work on cybernetics helped to promulgate: 'First humans were seen as functioning like machines, then the community became a machine, and finally the spiritual itself became cybernetic. Science created a totem. The machine became the avatar' (Rid 2016, p.348). Audiences of AI fictions such as *Person of Interest* are asked to implicitly accept the rules of the game and merely be concerned for who will be victorious, rather than querying the basic premise of the machine society, failing to '[see] the network itself, in all of its complexity', as Bridle would phrase it (2018, p.249). The most interesting and most disturbing thing about such contemporary 'machine visions', therefore, is not that they reveal how AI can either facilitate or undermine the betterment of human society, but that they reinforce what Carlyle warned of approximately two hundred years ago: 'the deep, almost exclusive faith we have in Mechanism' because it is 'Our true Deity' (1986, pp.70, 77).

## Notes

1. Brautigan's vision is inspired by Norbert Wiener's conception of cybernetics; Wiener's two most significant works in this regard, *Cybernetics: Or Control and Communication in the Animal and the Machine* ([1948], rev. 1961) and *The Human Use of Human Beings* ([1950], rev. 1954), offer visions of technological and organic corollaries which might be extended from the individual (in *Cybernetics*) to automation and society as a whole (more overtly in *The Human Use of Human Beings*).
2. The 'artificial' nature of this nomenclature will become clear over the course of this chapter, as there are a number of literary texts which have concerns that intersect with those presented here. These might include, for example, Aldous Huxley's *Brave New World* ([1937] 2007) or Yevgeny Zamyatin's *We* (1993), with their concerns about a social efficiency, or Daniel Suarez's *Daemon* (2009) and *Freedom (TM)* (2011), about the role of gamification in society.
3. Note that the phrase 'nonrobotic AI technologies' is used here to differentiate between robots—as AIs characterized by their embodiment within a mobile platform—and 'systems AI'—AIs characterized by their 'distributed' nature, and which exist primarily as programs capable of running across different platforms, possibly in a networked state.

4. See *Oxford English Dictionary*, 'Machine', particularly 8 and C1d, in sense of V.8b, 'machine' as 'a system or organization of an impersonal or inflexible character'. The resonance of the term 'machine' with technology, contrivance, abstraction, apparatus, complex devices, and systems speaks to the ways in which 'society' came to be perceived in these terms.
5. Jean Baudrillard famously asserts that 'The territory no longer precedes the map, nor survives it. Henceforth, it is the map that precedes the territory' (1997, p.1), referring to acts of simulation subverting their supposedly underlying realities. Bridle acknowledges this condition when he states that 'That which computation sets out to map and model it eventually takes over' (2018, p.39) and Lanier also posits that 'No abstraction corresponds to reality perfectly. A lot of layers become a system unto themselves, one that functions apart from the reality that is obscured far below' (2010, p.97). Whilst neither is completely Baudrillardian, there is an analogous observation at work: the machinic 'model' comes to replace the messy 'reality'. Kuno's distinction between seeing his mother and seeing a technological abstraction of her is precisely the point here, although Vashti does not understand it.
6. One might see in this phrasing that 'bringing... to' acts synonymously as 'making... into'.
7. The title of Moylan's monograph itself refers to Forster's 'The Machine Stops', which concludes 'For a moment they saw the nations of the dead, and, before they joined them, scraps of the untainted sky' (Forster 1965b, p.158). For Moylan's reading of 'The Machine Stops', see Moylan 2000, ch. 4, especially pp.111–21.
8. Asimov calls 'Evidence' 'the first story in which I made use of a humanoid robot' (1991, p.17).
9. In this respect, it stands alongside C. L. Moore's story, 'No Woman Born' (1980), which is often described as an AI narrative, although this is only true for a given definition of the phrase.

## Reference List


Asimov, I. (1990) Introduction: the robot chronicles. In: *Robot visions*. London, Victor Gollancz. pp.9–22.

Asimov, I (1995a) Evidence. In: *The complete robot*. London, HarperCollins. pp.518–45. First published in *Astounding Science Fiction*, 1946.

Asimov, I. (1995b) The evitable conflict. In: *The complete robot*. London, HarperCollins. pp.546–74. First published in *Astounding Science Fiction*, 1950.

Baudrillard, J. (1997) The precession of simulacra. In: Sheila Faria Glaser (trans.) *Simulacra and simulation*. Ann Arbor, University of Michigan Press. pp.1–42.

Brautigan, R. (1967) All watched over by machines of loving grace. [online] *American Dust > Brautigan > Machines of Loving Grace*. Available from: http://www.brautigan.net/machines.html#28 [Accessed 28 Jun. 2019]. First published in 1967.

Bridle, J. (2018) *New dark age: technology and the end of the future*. London, Verso.

Carlyle, T. (1986) Signs of the times. In: A. Shelston (ed.) *Thomas Carlyle: selected writings*. London, Penguin. pp.61–85. First published in *Edinburgh Review*, 1829.

Fairman, P. (1968) *I, the machine*. New York, Lancer Books.

Forster, E. M. (1965a) Introduction. In: *Collected short stories*. London, Sidgwick & Jackson. pp.v–viii.

Forster, E. M. (1965b) The machine stops. In: *Collected short stories*. London, Sidgwick & Jackson. pp.115–58. First published in *The Oxford and Cambridge Review*, 1909.

Huxley, A. (2007) *Brave new world*. London, Penguin. First published in 1932.

Lanier, J. (2010) *You are not a gadget: a manifesto*. London, Penguin.

Moore, C. (1980) No woman born. In: *The best of C. L. Moore*. New York: Ballantine Books. pp.236–88. First published in *Astounding Science Fiction*, 1944.

Moylan, T. (2000) *Scraps of the untainted sky: science fiction, utopia, dystopia*. Boulder, CO, Westview Press.

Nolan, J. (2011) *Person of Interest: Pilot (1.1)*. Dir. D. Semel. CBS.

Nolan, J., & D. Thé (2016) *Person of Interest: Return 0 (5.13)*. Dir C. Fisher.

Rid, T. (2016) *Rise of the machines: the lost history of cybernetics*. London, Scribe Publications.

Segal, A. (2014) *Person of Interest: The Cold War (4.10)*. Dir. M. Offer.

Suarez, D. (2009) *Daemon*. London, Quercus. First published in 2006, under the name Leinad Zeraus.

Suarez, D (2011) *Freedom (TM)*. London, Quercus.

Waters, R. (1975) Welcome to the machine. In: Pink Floyd. *Wish you were here*. London, Harvest/Columbia.

Wiener, N. (1961) *Cybernetics: or control and communication in the animal and the machine*. Cambridge, MA, MIT Press.

Wiener, N. (1989) *The human use of human beings*. [1954 edition]. London, Free Association Press.

Zamyatin, Y. (1993) *We*. Translated by Clarence Brown. London, Penguin. First published in English in 1924. Gregory Zilboorg (trans).

# 10

# 'A Push-Button Type of Thinking'

## *Automation, Cybernetics, and AI in Midcentury British Literature*

*Graham Matthews*

## 10.1 Introduction

In 1950, Alan Turing published the first scholarly treatment of AI and conceptualized a method for assessing the abilities of 'thinking machines' that came to be known as the Turing test. The following experiments in AI soon resulted in machines that could devise logical proofs and solve calculus and visual analogy problems, as well as integrate these skills in order to plan and control physical activity. For the first time in human history, AI moved from the realm of myth to become a real possibility. These technological developments catalysed the publication of numerous AI narratives—both fictional and nonfictional—that explored its impact on contemporary issues such as the labour market, population control, and global communication networks. In 1948, Norbert Wiener published *Cybernetics,* which swiftly entered into the public consciousness and influenced a generation of artists and authors on both sides of the Atlantic throughout the 1950s and 1960s. He coined the term 'cybernetics'—derived from the Greek *kubernetes* meaning 'steersman' or 'governor'—which refers to the science of communication and control in both living beings and machines, and argued that intelligent behaviour—artificial or otherwise—is primarily the result of feedback systems: 'As they construct better calculating machines and we better understand our own brains, the two would become more and more alike' (Wiener 1961, p.3). Wiener challenged an opinion that he felt was common with the man on the street: the assumption that although machines may be able to complete tasks with greater rapidity and with greater precision, they cannot possess originality (1960). He argued that 'machines can and do transcend some of the limitations of their designers, and that in doing so they may be both effective and dangerous' (p.1355). Wiener's warning that machines could escape the control of the human who made them can be found in a wide range of fictional AI narratives that explore the tipping point where computer systems become minds.

Between the late 1940s and mid-1950s, British scientists and firms had been among the world's pioneers, but by the 1960s a significant technology gap had opened up between the UK and US computer industries. Compounding this state of affairs, David Kynaston reports that in the nascent computer industry 'profits were seldom large, indeed sometimes non-existent' (2015, p.497). British scientists tended to focus on technical excellence, and faced with a small domestic market and American marketing acumen—IBM advertisements in 1962 heralded the arrival of 'A Giant Brain that's Strictly Business'—they struggled to make an impact. In time, the United States became firmly established as a leader in the field. The use of computers rapidly expanded to the point that in 1968, the philosopher R. J. Forbes would write: 'computers are in increasing use in science, business, medicine, manufacturing, communications, military and government operations, and education—indeed, it is difficult to imagine any field except that of the most personal services where they cannot be adopted in some fashion' (1968, p.64). They were valued for handling large quantities of data, for their accuracy, and their reliability. Unlike their human counterparts, computers were unlikely to get drunk, wake up late, or forget important tasks.

Forbes was dismayed that the computer was 'popularly miscalled an electronic brain' (p.63) when it should be more properly regarded as a 'rigidly controlled electronic instrument' (p.64) akin to an advanced abacus. Perhaps optimistically, he argued that the rate of technological change would always be controlled by humans, reminding his readership that technological potential is always counterbalanced by social considerations such as unemployment. In the mid-1960s, John Diebold, a pioneer in automation technology, informed government ministers and industry leaders that the application of electronic computers to industrial and governmental operations would produce such a sizeable impact that it would inaugurate a second industrial revolution. Noting the managerial enthusiasm for purchasing new technologies, Diebold bemoaned the unimaginative uses to which they were put; computers were treated as labour-saving devices rather than used to solve new as well as old problems: 'The danger [...] is not in having push-button machines, but in being content with a 'push-button' type of thinking' (1964, p.51). Meanwhile, there was widespread public concern that automation would lead to stagnation and standardization. The new possibilities offered by increasingly intelligent machines were not met with universal optimism.

The librarian, Mortimer Taube, who invented Coordinate Indexing—the precursor to computer-based search—was sceptical of scientific claims concerning breakthroughs in mechanical translation, machine learning, and computerized linguistic analysis. He dismissed the simulation of human brains with machines as mere speculation.[1] Taube likened scientists to high priests who were, perhaps unintentionally, leading the public astray: 'There are in the literature on mechanical translation, learning-machines, automata, etc., a great many names of nonexistent machines whose operations are described and debated as though they were real. The whole performance takes on the character of the behaviours of the tailors in "The Emperor's New Clothes"' (Taube 1961, p.120). During his investigations, Taube found that despite the appearance of a great mass of serious, detailed work carried out by esteemed scientists, unsubstantiated claims would often be repeated and elaborated upon without critical analysis. Despite rapid developments in mechanics and engineering, Taube argued that the distinction between the physiological and the mechanical is so vast that it cannot be bridged. Meanwhile, the philosopher Jacques Ellul expressed the fear that humans would become obsolete in the face of continuing technological advancement: 'It is possible, if the human being falters even momentarily in accommodating himself to the technological imperative, that he will be excluded from it completely even in his mechanical functions, much as he finds himself excluded from participation in an automated factory' (1967, p.9). Acutely aware of the threat to creativity and innovation inherent to standardization, Alice Mary Hilton questioned whether, faced with either a world in which humans must adapt to machines or one in which machines must be adapted to the needs of humans, we can 'preserve the conditions that encourage originality and the painful joy of creation, in a world where mass-produced uniformity is so much easier?', and she declared that '*it is only sensible to make the machine fit and serve the human mind*' (1963, p.400). These debates on technological development catalysed the publication of numerous narratives that explored the impact of AI on contemporary issues such as the labour market, social change, and global communication networks.

This chapter analyses the varied representation of AI in midcentury British novels—organized according to the rising complexity of the AI depicted within—from the automated computers in Michael Frayn's *The Tin Men* (1965), to the computer-controlled spy network in Len Deighton's *Billion Dollar Brain* (1966), and concluding with the on-board computer Hal in Arthur C. Clarke's *2001: A Space Odyssey* (1968). These novels address the societal ramifications—both positive and negative—for humans faced with technological breakthroughs in AI. Midcentury British AI narratives are extremely diverse in tone yet united by a consistent focus on the connecting link between technology and artistry.

In each text, technological invention is shown to be hindered by a narrow adherence to logic, reason, and experience devoid of intuition and inspiration. Michael Frayn's *The Tin Men* (1965) portrays a world in which the possibilities of AI appear boundless, but its capabilities are severely curtailed in practice and, like Diebold, the text warns against the spread of push-button thinking. Although an AI can complete tasks more efficiently than a human, the act has no meaning for a machine, which poses the question as to whether it matters who or what performs the action so long as the action is performed. As the number of tasks that can be performed by machines steadily rises, the need to define the attributes of the human becomes an increasingly urgent task. Meanwhile, Len Deighton's *Billion-Dollar Brain* warns against the urge to overuse and overvalue the solutions provided by AI. Computers may calculate and make strategic decisions but are entirely reliant on statistics and measurable inputs from humans who are unreliable, ideological, and instinctual. Finally, Arthur C. Clarke's *2001* highlights the role of heuristics in AI research which seeks to discover and imitate the process of human problem-solving by focusing on what is sufficient rather than rational and logical. I argue that Hal is motivated to rebel against its human crew members because of an experience that approximates to shame. This scenario demonstrates that heuristic machines cannot operate in isolation, and appropriate social structures must be constructed in order for machine intelligence to integrate into human society. Together these midcentury AI narratives began an ongoing conversation to address the social ramifications of automation and cybernetics, and urged readers to reflect on the human values of creativity, intuition, and compassion.

## 10.2 Automation and the 'Push-Button' Human

Anthony Sampson's encyclopaedic *Anatomy of Britain Today* (1965) details the structures and systems that comprised midcentury

British society yet devotes only a single page to the computer, describing it as an 'alarming new master-manager' that has a 'brain which can remember everything about everyone, which can check figures, write cheques, calculate turnovers and notice exceptions in every factory and shop' (pp.515–16). Sampson implies that intelligent machines possess malign intent by commenting on their supposedly 'menacing' (p.516) names—Leo, Pegasus, and Orion—their eerie silence, and detached focus on quantitative data. Although the computer industry was in its infancy in Britain, Sampson anticipated that the ability to store and synthesize large amounts of information would lead to greater centralization and control that consequently threatened the dissolution of middle management, thereby diminishing the opportunities for advancement, which would in turn lead to increasing class inequality. Similarly, Herbert Simon predicted that 'machines will be capable, within twenty years, of doing any work a man can do' (1965, p.96) and although his claim was hyperbolic, it nevertheless highlighted the anxiety among white-collar workers that computers would usurp their roles. In this cultural landscape, Michael Frayn's comic novel *The Tin Men* (1965) portrays a world in which the possibilities of AI appear boundless, to the extent that legal and ethical dilemmas can conceivably be solved via cybernetics and automation. However, the machines are revealed to be severely limited in practice. Rather than satirizing the machines alone, Frayn also targets professionals such as journalists, lawyers, and clergymen—the tin men of the title—whose thinking is presented as increasingly machine-like and automated.

*The Tin Men* is set in the William Morris Institute of Automation Research, a fictional research lab for developing computers and automating various aspects of life. Frayn ironically named the Institute after the author of *News from Nowhere* (1890), who was renowned for his distaste of mechanization. Karen Blansfield notes that Frayn commonly makes a spectacle of people at work in order to explore the interests and problems encountered simply by having a certain job: 'Work gives shape and purpose to their lives, and even when the routine seems meaningless and dull, they keep on with it because they would be lost otherwise' (1995, 112). Work, in Frayn's novels and plays, is the primary means by which people express themselves and consequently the setting illuminates personal and public responses to societal change. Accordingly, *The Tin Men* opens with a parody of the widespread concerns about the impact of automation on the labour force. Research and development at the Institute has been delayed by the construction of a new Ethics Wing tasked with 'investigating to what extent computers can be programmed to follow codes of ethical behaviour' (Frayn 2015, p.48). Experts have calculated that if the work of the Institute had gone ahead unimpeded, they should have made over two million professionals redundant over a ten-year period: 'Now there was a risk that some of these two million would still find themselves in work, or at any rate only partly out of work. But then, said the optimists, for progress to be made someone always had to suffer' (p.9).

Each department in the Institute is tasked with automating a particular sphere of human activity but they invariably focus on the output rather than the process, often with absurd results. For instance, Dr. Goldwasser, the Head of the Newspaper Department, seeks to demonstrate that a computer could be programmed to produce a satisfactory newspaper. He develops a new Unit Headline Language that creates articles based on selection of predetermined variables that unfortunately bear no relation to actual events. Meanwhile, Hugh Rowe, the Head of the Sports Department, attempts to write a novel. He begins by painstakingly writing his author biography, the blurb, and the reviews before attempting to reconstruct the novel from them.[2] Haugh, the Head of the Fashion Department, possesses such an open mind that he functions more like an automaton than a human: 'opinions, beliefs, philosophies entered, sojourned briefly, and were pushed out at the end by the press of incoming convictions and systems [...] It depended on who had spoken to him last'

(Frayn 2015, p.67). Like a computer, Haugh is devoid of critical faculties, receives inputs without consideration for their ethical or practical purposes, and possesses a sincere and diligent belief in another's ideals. This is comparable to the ways in which a machine extends its user's ideologies or desires without criticism or comment.

Dr. Macintosh, the Head of the Ethics Department, frequently speculates about the various spheres of life that computers can improve and discourses at length on the possibility of automating bingo, football, and race courses on the assumption that 'the main object of organised sports and games is to produce a profusion of statistics' (Frayn 2015, p.43). Since it is theoretically possible to automate any human activity that involves repetition or the manipulation of a fixed number of variables, Macintosh determines that rather than paying humans to derive a random choice between win, lose, and draw, it is more efficient and effective to replace them with a simple and inexpensive computer. Furthermore, AI has the capacity to select bingo numbers or fill in a football coupon itself. The next logical step is to programme computers that can appreciate sports by registering applause, annoyance, reluctant admiration, and boredom. The depiction of Macintosh's ideals effectively satirizes the replacement of human activities with automation. But it also mocks the ways in which humans tend to perform tasks like machines—the activity of listening to the sports results on the wireless is already a reduction of human engagement with sport to a mechanistic processing of a series of statistics.

This 'push-button' thinking was not limited to the realm of satiric fiction. In a report on automated design published in 1964, Borje Langefors predicted that 'designers will design the designing system which designs the actual product' (1964, p.9). In an echo of Langefors's prognosis, Macintosh foresees a future in which the whole world of sport is an entirely enclosed one, 'unvisited by any human being except the maintenance engineers. Computers will play. Computers will watch. Computers will comment.

Computers will store results, and pit their memories against other computers in sports quiz programmes on the television organised by computers and watched by computers' (Frayn 2015, p.45). Macintosh eventually intends to automate political and international affairs and have AI systems sit on the boards of businesses and financial institutions, while the work of middle men can be removed entirely by simply calculating their costs in advance. He even claims that AI can write and consume pornographic novels. In fact, he could programme machines to 'perform a great deal of human sexual behaviour. It would save a lot of labour' (Frayn 2015, p.97). Rituals, even religious ones, follow a strict mechanical process and therefore can also be performed by AI. Macintosh claims that not only can machines pray, but they would not pray for things that they ought not to pray for. The machine's thoughts would not wander and they can practice prayer more fervently and less cynically than a human can. When speaking these words, the computer would really mean them, whereas with a human there is room for insincerity and irony. They would achieve prayer with 'maximum efficiency and minimum labour output' (p.110). Macintosh argues that an AI can pray more effectively and efficiently than a human, which raises the disquieting question of whether it matters who or what performs the action so long as the action is performed. As the number of tasks that can be performed by machines steadily rise, the need to define the attributes of the human that cannot be replaced becomes an increasingly urgent task. Frayn's novel is not solely a satire of machines and machine logic, but a critique of the 'push-button' thinking of humans.

The distinction between humans and machines is further problematized by Macintosh's test chamber in the old Ethics Department building: 'The simplest and purest form of the ethical situation, as he saw it, was the one in which two people were aboard a raft that would support only one of them, and he was trying to build a machine which would offer a coherent ethical behaviour pattern under these circumstances' (Frayn 2015, p.18). Accordingly, his lab has been converted into a huge water tank with a crane on a gantry that suspends a raft bearing Samaritan, an ethical AI, out over the water and starts to lower it. The initial design pushes itself overboard to save anything that happens to be next to it on the raft; it lacks discrimination and therefore its behaviour cannot be considered ethical. Accordingly, Samaritan II will only sacrifice itself for an organism at least as complicated as itself. However, Macintosh and Goldwasser begin to anthropomorphize the AI and detect a degree of sanctimony in its dials and displays; it cannot be said to be ethical if it enjoys sacrificing itself. Samaritan III appears to be both ethical and effective; it pushes simpler organisms overboard, yet faced with a human, throws itself overboard. However, at Goldwasser's behest, Macintosh places two Samaritan IIIs on the raft, and they both immediately throw themselves overboard. Following a minor technical adjustment, the machines then throw each other overboard. The scenario shows that AI can complete the tasks that humans have set for it more efficiently than a human can, and yet the act has no meaning for a machine. Although these are preprogrammed actions, the scientists interpret this behaviour in wildly divergent ways. The first scenario appears to reach the heights of self-sacrificing romantic tragedy in a manner akin to *Romeo and Juliet*. In the second, life appears as a Darwinian struggle for survival, even to the point of mutual destruction. These interpretations suggest that ethical behaviour does not reside in the act itself but is a matter of hermeneutics. They raise the possibility that machines can perform ethical acts but only if they are recognized as such by human onlookers. By extension, the scenario questions whether it matters if a human or a machine performs the act.

Throughout the novel humans are seen as either imitative and lacking in critical faculties or guilty of overinterpretation. Nunn, the deputy director, is pious in his dedication to the Director of the Institute: 'Director of Directors, the institution made flesh, all things to all men, everlasting and unchanging' (Frayn 2015, p.145), who, it is eventually revealed, has no skills or experience in the field. The Director's timidity has been misinterpreted by his subordinate as the professional distance appropriate to his elevated station. The tendency towards misinterpretation reaches its apogee at the novel's conclusion when the Queen is absent at the official opening, and Nobbs, a Research Assistant primarily distinguished by his hirsute features and insistence on calling everyone 'mate', must step in as a substitute. All the assembled guests and nobility are compelled to deliver their well-rehearsed lines as if they really were receiving the Queen. As Nobbs proceeds through the ranks, each guest can see someone else shaking hands, bowing, and curtsying, and they feel compelled to make obeisance without question: 'No doubt some of them were not entirely easy in their minds to find that the majesty made incarnate to them was a shambling, bearded young man who groaned at their loyal homage and muttered "bloody hell"' (p.157). Nevertheless, no one present can disrupt the highly organized sequence of a royal occasion to question the authenticity of the royal personage. The customary limitations of machine intelligence—the inability to register and respond to the complexities of the social environment—are ultimately shown to apply equally well to Frayn's flawed human characters.

## 10.3 AI and the Cybernetic Organization

In the late 1950s and 1960s, governments and institutions were increasingly concerned about the social impact of a rapidly cybernating society—the introduction of computers into factories, production lines, and even some services (by 1964, the vending machine industry had become a $3bn operation).[3] At the same time, there were increasing calls from businessmen and industry leaders to educate the population to be responsive and responsible members of society who would willingly adapt to technological change (Diebold 1964, pp.21–22). As manufacturing became increasingly automated, concerns about unemployment were answered by calls to rapidly expand the service industry. Consequently, the need to encourage the population to demand 'real'—meaning human—service was prioritized, thereby inoculating the service industries against cybernation. But government officials also discussed the impact of the projected increase in leisure time and the need to educate citizens to be concerned with their rights as members of a democracy.[4] Len Deighton's spy thriller *Billion-Dollar Brain* encapsulates the distrust white-collar workers had of computers and AI because of fears that they were at risk of being replaced. Although the computer in the novel—dubbed 'the Brain'—appears to result in greater efficiency and centralized control over covert operations, the human agents draw on their reserves of ingenuity and intuition to seek out the limits of that control. In *Cybernation and Social Change*, Donald N. Michael notes that commentators have a tendency to overvalue the abilities of the computer, which often results in a slanted view of the world that ignores the instinctual and the ineffable. We tend to place most emphasis on 'those aspects of the world which the computer can handle—usually the statistical—and in turn, to ignore the ineffable, the points off the curve, the unique person' (Michael 1964, p.25). *Billion-Dollar Brain* warns against the pressure to overuse and overvalue the solutions provided by computers and emphasizes the creativity of the individual over reliance on quantitative data.

In *Billion-Dollar Brain* the billionaire General Midwinter runs a private intelligence unit with which he aims to start an uprising against the USSR in Latvia in an attempt to end Communism in the Eastern bloc. He develops a computer, called the Brain, that can orchestrate a range of covert operations. All of the agents operating within Midwinter's network receive instructions via recorded telephone messages sent by the Brain. Consequently, the AI allows for the centralization of operations by correlating all of the agents' reports. The computer is thereby able to orchestrate a range of operations including communications sabotage, cache of arms, guerrilla warfare, preparation of landing fields and drop zones, underwater demolition, and clandestine radio.[5] The antagonist, Harvey Newbegin, comments: 'These machines are programming our operation there. Each and every act of every agent comes out of this machine' (Deighton 2015, p.192). Initially at least, the mathematical-logical standards of the Brain appear to be more efficient than human operatives, but it eventually becomes clear that the machine is limited in ways that humans are not.

The Brain is housed in a small nondescript building at the centre of a private compound. Most of the building is below ground level, where there stands a gigantic room, the size of the hangar deck of an aircraft carrier. The room is filled with banks of computers that stretch into the distance. Newbegin claims that 'these heaps of wire can practically think' (Deighton 2015, p.196) and the computer-brain metaphor is extended throughout the complex. Each part of the computer is named after a section of the human brain: data packets are referred to as 'neurons' that are filtered by 'synapses', and the narrator notes that operatives 'wore earphones with dangling plugs which they occasionally plugged into a machine, nodding professionally like doctors sounding a chest' (p.192). The metaphor is reversed when the protagonist encounters the indoctrination laboratory that uses audio-visual equipment and subliminal messaging to train or 'programme' agents to inhabit their covert identity: 'it goes right into the subconscious of course; no memorising involved' (p.194). The computer-brain metaphor had been popularized by IBM's marketing a few years prior and not only resulted in wild predictions about the capabilities of then-limited machine intelligence but, as *Billion-Dollar Brain* suggests, also worked in the opposite direction to suggest that if human minds are like computers, then they can be (re-)programmed. To administrators and bureaucrats, the computer offered the possibility of greater efficiency and centralized control but as the novel warns, such control comes at a significant cost to human imagination and creativity.

One of the pioneers of artificial intelligence during the 1950s, Herbert A. Simon, commented that 'today we are deeply conflicted about how far we should go in 'improving' human beings involuntarily' (1965, p.176), and conditioning techniques were seen in Anthony Burgess's *A Clockwork Orange* (1962), Len Deighton's *The IPCRESS File* (1962), and the *The Prisoner* (1967) television series.[6] In *Billion-Dollar Brain*, the AI not only coordinates covert operations against the USSR but reduces humans to functional units by conditioning their instinctual responses so that they fully inhabit their cover story. Rather than memorizing facts and figures, subjects sit within cells filled with audio and visual equipment that deploy subliminal messaging techniques to imprint the cover story over their memories. The unnamed protagonist finds this indoctrination especially disturbing for two reasons: (i) it reifies state control over the individual and (ii) it challenges the existence of a private, authentic self.[7] The ability to reprogramme individuals erodes the division between humans and machines, and literally instantiates 'push-button thinking'. Whereas humans can draw on both cognitive and emotional reasoning to solve problems, machines are limited to mathematical-logical standards that appear more efficient in the short-term but lead to a deficit in imagination, creativity, and real-world application. After learning about the machinations of the Brain, the narrator echoes Diebold's critique of 'push-button thinking' when he enters a new glass building with an express lift. After pushing the button at least three times, he sardonically remarks: 'It's just a matter of time before the machines are pressing buttons to call people' (Deighton 2015, p.221). For the unnamed protagonist, a world in which agents can be made to believe their cover story is anathema because it degrades the notion of authentic selfhood. The world of spy networks and double agents is one filled with illusions and corruption in which seemingly sane operatives possess irrational and unpredictable motivations that can, in a heartbeat, throw lives into jeopardy. For Deighton's existential hero, personal authenticity ensconced within a buffer of lies and misdirection is the only feasible defence against the absurdity of the world. Using computers to reprogramme agents erodes the agency and creativity of the individual and challenges the notion that authentic selfhood is even achievable.

Ultimately, Newbegin mounts a defence of the instinctual and ineffable—albeit for purely self-interested aims—by subverting the Brain's reliance on statistics and definable inputs and outcomes. The protagonist discovers that the agents and operations coordinated by Newbegin's network in Europe are fabricated. For years, he had been collecting funds from Midwinter and in exchange feeding the Brain reports of phoney operations from phantom agents. It is standard practice for agents in the field to receive instruction via intermediaries, which means that they lack any guarantee that their intelligence is being delivered to their home country. In order to garner trust, intelligence organizations typically tell operatives that they are working for someone to whom they are likely to be sympathetic. The Brain may be an AI that can calculate and make strategic decisions, but it is entirely reliant on factual inputs from unreliable human operatives in order to do so. The relationship of these operatives to reality is always fluid, influenced by ideology, and irreducible to calculable components. The novel indicates that humans make and solve problems in ways that do not adhere to mathematical-logical standards and that the benefits of greater efficiency and centralized control should be tempered with greater wisdom, beliefs, behaviour, and goals.

## 10.4 Heuristic Machines and Social Integration

Arthur C. Clarke's *2001: A Space Odyssey* extrapolates the development of a range of then-cutting-edge cybernetics and communications research into the future to imagine new technological developments such as video phones, computerized newspapers, commercial space flight, and hydroponic farming. The novel was written alongside the development of the film directed by Stanley Kubrick, and it renders explicit many of the narrative elements left opaque by the film. Following an encounter with one of the iconic monoliths discovered on the moon, the spaceship Discovery is sent on a mission to Saturn in search of the next totem. The Discovery is operated by HAL 9000 or the Heuristically programmed ALgorithmic computer, referred to by the crew as Hal.[8] At midcentury, research into control systems was developing along three overlapping lines of enquiry: (i) heuristics, (ii) artificial intelligence, and (iii) simulation of neural elements and physiological processes.[9] Hal incorporates elements of all three research strands, but, as its name implies, heuristics is foregrounded. Heuristics research seeks to discover and imitate the process of human problem-solving by focusing on what is sufficient rather than rational and logical. Heuristic machines were envisioned to play a significant role in the burgeoning space program: 'When the first unmanned spaceship goes to Mars, for instance, no one can predict all the conditions it will meet. But the spaceship's heuristic system can be given goals of landing, exploring, and returning, and it will accomplish the mission in its best possible way and adapt itself to whatever conditions it encounters' (Diebold 1964, pp.91–92). Whereas automatic and cybernetic control systems work on particular tasks and problems, heuristic machines would be equipped with a generalized programme of human problem-solving. The novel extrapolates from these avenues of research the difficulties involved in situating artificial intelligence within society in a way that conforms to—often unspoken—group standards.

By focusing on heuristics, this section highlights an overlooked aspect of Hal: namely, that it is motivated to turn against its human crew members to escape a sensation that approximates to shame. Hal is popularly misconceived as either inexplicably psychotic and murderous—placing it in the tradition of later popular filmic representations such as *The Terminator*—or overly logical (the machine reasons that it will be better equipped to complete the mission without its human cargo).[10] I argue that the cause of Hal's rebellion against the human crew members, Dave Bowman and Frank Poole, is not a simple malfunction or the development of a malign sensibility, but the inhuman application of heuristics. The problems begin when Hal's Fault Prediction Centre reports that the AE-35 communication system is due to malfunction in seventy-two hours. This unit is mounted on the antenna, and it coordinates communication channels with the Earth. Frank replaces the unit, and Dave runs diagnostic tests but cannot find a fault. In response, Hal continues to express complete confidence in its capabilities: 'I don't want to insist on it, Dave, but I am incapable of making an error' (Clarke 2000, p.173). Since the computer does not accept that it can make errors, its prediction must be true, and the logical outcome is that the AE-35 unit must fail. Consequently, Hal alerts the crew that the replacement AE-35 system will fail, this time within twenty-four hours. When Mission Control claim that the fault lies in Hal's prediction circuits rather than the AE-35 unit, Hal causes the unit to fail and the transmission ends, thereby rendering Hal's prediction true. However, Frank then goes out in the pod to inspect the replacement AE-35. In order to maintain the fiction that the unit is at fault, Hal causes the pod to collide with Frank, rupturing his spacesuit, and sending him flying out into the abyss. To conceal any malfeasance, Hal then opens the airlock doors in order to remove Dave and switches off the life-support systems for the rest of the crew. Although Hal has become a murderous machine, these actions are consistent with a heuristic problem-solving AI that seeks a solution that is sufficient rather than logical and rational.

The computer is certain that its calculations are always correct—a belief that is not shared by the human crew members—so when it is proven wrong, it attempts to remove all trace of its error. Hal's motivations are not made explicit in either the novel or the film, but I argue that it registers a sensation that approximates to shame. Shame plays a significant role in the formation and maintenance of the social contract between groups and individuals; it relates 'individuals to wider social groups and norms—real or imagined' (Stearns 2017, p.1). Communities in the past relied on shame to enforce desired social standards. It severely impacts individuals but is dependent upon group standards. Stephen Pattison argues that shame 'helps to define social boundaries, norms and behaviours and signals the state of social bonds, as well as providing a powerful tool of social conformity and control' (2010, p.131). The historical record shows that shame anticipation is an important pillar in maintaining the relationship between individuals and larger groups, and constructive management of shame is necessary for developing social cohesion. Since the engineers assumed that Hal is both infallible and emotionless, no scaffolding has been developed to mitigate the experience of failure. Accordingly, when the computer makes a mistake, it refuses to perceive it as such and begins to act upon the world in order to align it with its worldview, with disastrous results for the crew.

Following Frank's demise, Dave is trapped on the Discovery with Hal, and he seeks to escape the gaze of the AI. However, this feeling of entrapment operates in both directions: Hal previously could not escape what it perceived to be the judgemental gaze of the humans who were slowly becoming aware of its failure. Whereas guilt is typically an emotional response to an act that the perpetrator can then atone for, shame presents as though the self is at fault. As the psychologist Todd Kashdan notes: 'people who feel shame suffer. Shamed people dislike themselves and want to change, hide, or get rid of their self' (p.174). Within the confines of the spacecraft, Hal has no escape from judgement, either real or imagined, and since shame affects the person, not the act, it is impossible to apologize and make reparations. Shame can be paralyzing, is a tremendous blow to self-worth, and consequently, it is hard to find a constructive response. In this light, Hal's response to the prediction fault is logical and rational. Shame is the outcome of failing to live up to the social group's expectations. Deprived of the ability to retreat, Hal is left with three possible responses: (i) blame someone or something else, such as the AE-35 unit or human error; (ii) deny the error: 'I am so much more capable than you are of supervising the ship, and I have such enthusiasm for the mission and confidence in its success' (Clarke 2000, p.187); or, (iii) get angry at oneself or others. Hal's decision to cut communications with the Earth and attempt to murder the crew members is not necessarily motivated by anger but is the logical outcome of the imperative to evade shame.

Further, *2001* posits that heuristics, or generalized problem-solving skills, will be an important element in the development of artificial intelligence, but highlights the need for scaffolding to ensure that negative experiences such as shame are anticipated and managed. There is no method to guarantee the reliability of hardware and the novel demonstrates the need for fault tolerance (detection of error) systems that operate independently of the AI. The fact that it is Hal's fault prediction circuits that are at fault signal how quickly software can outstrip reliable and effective safeguards. As AI becomes increasingly sophisticated, corresponding social structures will be required to mitigate against failure and the experience of shame. Heuristic machines cannot operate in isolation and holistic approaches that incorporate sociocultural intelligence must be pursued in order for machine intelligence to integrate into human society. Shame provides a prime example of the unforeseen outcomes that may occur when the machine's competence is undermined through undeveloped interactions with the social environment. As a result, the machine may seek solutions that do not accord with the wishes of humans.

## 10.5 Conclusion

Norbert Wiener's pioneering work influenced a wide range of fictional AI narratives that explore the point at which computers become minds. Across his oeuvre, he drew upon examples from fairy tales as parables for the dangers of AI. In stories such as the 'Sorcerer's Apprentice' (1797) or W. W. Jacobs's 'The Monkey's Paw' (1902), magic is performed in a literal-minded manner that fulfils the user's command but invariably comes with tragic, unforeseen consequences. Wiener also warns that computers operate on a time scale incommensurate to human speeds. The computer will complete a task long before the results can be assessed by human observers or its mechanisms be precisely known. It is not possible to interfere with the operation of an AI once its operation has been initiated, so we must be certain that the input is truly beneficial to humankind. For these reasons, Wiener argues, 'We must always exert the full strength of our imagination to examine where the full use of our new modalities may lead us' (1960, p.1358). At the same time, John Diebold bemoaned the unimaginative uses to which computers were being put. He feared that the new push-button machines would enable the proliferation of push-button thinking. Thinkers and writers have increasingly questioned how to preserve the human qualities of originality and creation in the emerging world of cybernetics and automation. Consequently, AI narratives from midcentury were—and continue to be—instructive in producing an imaginative appraisal of man–machine relations in the future. Together, these novels critically reflect on the concomitant technological developments in automation and cybernetics that stimulated widespread concern among government institutions, businesses, and the public about the evolving relationship between humans and AI.

## Notes

1. Taube's investigation encompassed the work of prestigious institutions such as the Rockefeller Foundation, Massachusetts Institute of Technology, Harvard University, the RAND Corporation, and

Ramo-Wooldridge. Rather than challenging a single institution or individual, he felt it necessary to question aberrations of the scientific enterprise in general. In support of his approach, he cites the discredited disciplines of alchemy and phrenology and notes the lack of effective criticism of modern scientific practice.

2. Wiener later posited this scenario when he complained that the rise of mass communication had resulted in the creation of a standardized, mediocre product rather than anything of value: 'Heaven save us from the first novels which are written because a young man desires the prestige of being a novelist rather than because he has something to say!' (1968, p.134).
3. According to the OED, the verb 'cybernate' was coined in 1962. The terms 'cybernating' and 'cybernation' were in common usage among scientists, politicians, and engineers in the mid- to late 1960s.
4. Donald N. Michael asks in *Cybernation and Social Change*: 'how does the citizen manage to be an active member of the democratic community when the issues important to him become more and more complex, the techniques for dealing with them more and more esoteric, the skills required to understand the problems more and more elaborate, and the professionalisation of government deeper and deeper?' (1964, p.19).
5. The capabilities of the Brain are remarkably similar to Mike, the lunar master computer in *The Moon Is a Harsh Mistress* by the American author Robert Heinlein, published the same year.
6. In *The IPCRESS File*, audio-visual equipment is used to bombard the senses, while a variety of techniques are used across the seventeen episodes of *The Prisoner*, including physical coercion, social indoctrination, dream manipulation, hallucinogenic drug experiences, and mind control. In each instance, computers and other cutting-edge technologies are deployed to reprogramme the human subject.
7. In the film series starring Michael Caine, Deighton's unnamed protagonist is called Harry Palmer.
8. Various commentators have noted that HAL is a one-letter shift from IBM, but Kubrick always claimed that this was coincidental: 'Just to show you how interpretations can sometimes be bewildering: A cryptographer went to see the film, and he said, "Oh, I get it. Each letter of HAL's name is one letter ahead of IBM" [...] now that's pure coincidence [...] an almost inconceivable coincidence. It would have taken a cryptographer to notice that' (Kohler 2010, n.p.). IBM

acted as consultants on the film, and Kubrick wrote to alert them to the fact that one of the main themes of the story is a psychotic computer. In their reply, IBM stated that they were aware that HAL causes human deaths but had no objection as long as IBM was not associated with the equipment failure by name (Madrigal 2013).

9. Whereas research into artificial intelligence focused on the cybertron, a computer that did not need to be programmed and could deploy machine-learning techniques to evaluate data from electrocardiograms, the latter research strand was centred on the development of perceptrons, pattern-recognition devices that could distinguish between different letters of the alphabet and recognize faces and other objects (Diebold 1964, p.41).
10. Advocates for reading Hal as psychotic include Eliot Freemont-Smith (1968), Don Daniels (1970), and Simson Garfinkel (1997). Meanwhile commentators such as Clay Waldrop (n.d.) and Gordon Banks (n.d.) argue that HAL acts rationally and logically, with cold, calculating precision.

## Reference List

Banks, G. (n.d.) Kubrick's psychopaths: society and human nature in the films of Stanley Kubrick. *Visual Memory* [Online]. Available from: http://www.visual-memory.co.uk/amk/doc/0004.html [Accessed 27 June 2019].

Blansfield, K. (1995) Michael Frayn and the world of work. *South Atlantic Review*. 60(4), 111–28.

Clarke, A. C. (2000) *2001: a space odyssey*. London, Penguin.

Daniels, D. (1970) A Skeleton key to *2001*. *Sight and Sound*. 40(1), 28–33.

Deighton, L. (2015) *Billion-dollar brain*. London, HarperCollins.

Diebold, J. (1964) *Beyond automation: managerial problems of an exploding technology*. London, McGraw-Hill.

Ellul, J. (1967) *The technological society*. London, Vintage.

Forbes, R. J. (1968) *The conquest of nature: technology and its consequences*. London, Penguin.

Frayn, M. (2015) *The tin men*. London, Faber & Faber.

Freemont-Smith, E. (1968) Outward bound: *2001: A Space Odyssey*. *The New York Times*. 5 July [Online]. Available from: https://archive.nytimes.com/www.nytimes.com/books/97/03/09/reviews/clarke-2001.html [Accessed 27 June 2019].

Garfinkel, S. (1997) Happy birthday, Hal. *Wired* [Online]. Available from: https://www.wired.com/1997/01/ffhal/ [Accessed 27 June 2019].

Hilton, A. M. (1963) *Logic, computing machines, and automation*. London, Cleaver-Hume Press.

Kashdan, T. & R. Biswas-Diener (2014) *The upside of your dark side: why being your whole self—not just your 'good' self—derives success and fulfillment*. New York, Hudson Street Press.

Kohler, C. (2010) *Stanley Kubrick raps,* The Making of 2001: A Space Odyssey. Edited by Stephanie Schwam. New York, The Modern Library.

Kynaston, D. (2015) *Modernity Britain, 1957–62*. London, Bloomsbury.

Langefors, B. (1964) Automated design: more of it can be done on an automatic computer than most designers are willing to admit. *International Science and Technology*. 11(4), 90–97.

Michael, D. N. (1962) *Cybernation: the silent conquest*. Santa Barbara, CA, Center for the Study of Democratic Institutions.

Michael, D. N. (1964) *Cybernation and social change*. US Department of Labor, Office of Manpower, Automation and Training. Washington, DC.

Madrigal, A. (2013) The letter Stanley Kubrick wrote about IBM and HAL. *The Atlantic* [Online]. Available from: https://www.theatlantic.com/technology/archive/2013/01/the-letter-stanley-kubrick-wrote-about-ibm-and-hal/266848/ [Accessed 27 June 2019].

Pattison, S. (2010) *Shame: theory, therapy, theology*. Cambridge, Cambridge University Press.

Sampson, A. (1965) *Anatomy of Britain today*. London, Hodder and Stoughton.

Simon, H. A. (1965) *The shape of automation for men and management*. New York, Harper and Row.

Stearns, P. (2017) *Shame: a brief history*. Champagne, University of Illinois Press.

Taube, M. (1961) *Computers and common sense: the myth of thinking machines*. New York and London, Columbia University Press.

Waldrop, C. (n.d.) The case for Hal's sanity. *Visual Memory* [Online]. Available from: http://www.visual-memory.co.uk/amk/doc/0095.html [Accessed: 27 June 2019].

Wiener, N. (1960) Some moral and technical consequences of automation. *Science*. 131(3410), 1355–58.

Wiener, N. (1961) *Cybernetics, or control and communication in the animal and the machine*. 2nd edn. Cambridge, MA, MIT Press.

Wiener, N. (1964) *God & golem, inc. A comment on certain points where cybernetics impinges on religion*. London, Chapman & Hall.

Wiener, N. (1968) *The human use of human beings: cybernetics and society*. London, Sphere.

# 11

# Artificial Intelligence and the Parent–Child Narrative

*Beth Singler*

## 11.1 Introduction

In a 1955 *Scientific American* (*SA*) article, John G. Kemeny, computer scientist, mathematician, and co-developer of the BASIC programming language, described the logical processes of computers in analogy to the human processes of thinking, and then extended that analogy to all human activity:

> a normal human being is like the universal machine. Given enough time, he can learn to do anything. . . . [T]here is no conclusive evidence for an essential gap between man and a machine. For every human activity we can conceive of a mechanical counterpart. (1955, p.63)

This article is one of the primary sources in the 2017 article, 'Imagining the Thinking Machine: Technological Myths and the Rise of Artificial Intelligence', in which Simone Natale and Andrea Ballatore applied content analysis methods, including coding this material and tallying recurrent themes, to back issues of the *SA* and the *New Scientist* (*NS*) published between the 1950s and early 1970s. They concluded that 'AI technologies were often described in the *SA* and *NS* with terms that usually apply to human or animal behaviour' (Natale & Ballatore 2017, p.6).

This analogical conception of AI has been pervasive and influential, but it comes laden with cultural associations and meanings, and can result in misleading cross-domain translations when analogies are stretched too far or when they get in the way of recognizing the current abilities of the technology (Natale & Ballatore 2017, p.6). Computer scientist Hamid Ekbia also recognizes such culturally influenced translations in his self-reflective work on humanity's 'quest' for AI. He argues, however, that there is not 'something inherently wrong or exceptionally ad hoc in this practice—for that is essentially what analogies do, and analogies may well constitute the core of human cognition' (Ekbia 2008, p.5, also referencing Hofstatder 2001). Taking this approach to the nature of analogy as a starting point, this chapter will employ anthropologically informed categorical thinking and theory to examine AI narratives that present a parent–child relationship between humans and AI, and which imbue cultural associations about the human child into our conception of AI.

Methodologically, this chapter will apply approaches and categories from cognitive anthropology, cultural anthropology, and the social study of history to examples of speculative fictions and nonfiction that focus on this relationship. This will demonstrate that conceptions of AI as a child are reflective of modes of anthropomorphism, existing analogical patterns, and contemporary cultural assumptions. In the categorical approach of anthropologists such as David Lancey, the 'child' and 'childlikeness' are continually changing and contextually situated values with varying signifiers and related actions. A person can be a child in physical maturity (with culturally prescribed stages of development), be a child in relationship with a parent (at any age) and be childlike in behaviour (in any form). This chapter will therefore explore how the AI child and AI that is childlike are also presented variably in our narratives about AI, reflecting contemporary modes of thinking about the human child, the creator/parent, the importance of parenting, the child as a legacy, and our own immortality.

## 11.2 Theoretical and Methodological Approach

Before we turn to 'childlikeness', we need a more nuanced definition of 'anthropomorphism' to explore how we recognize and illustrate 'human-likeness' in the nonhuman. The work of cognitive anthropologist Pascal Boyer provides a categorical framework that will enable us to recognize different aspects of anthropomorphism before we map those onto specific parent–child AI narratives. This chapter will employ Boyer's five-fold scheme of 'intentional projections' from his article, 'What Makes Anthropomorphism Natural: Intuitive Ontology and Cultural Representations' (1996). Boyer refers to psychological research on the development of children to explore the cognitive processes involved in anthropomorphism and gives a nonexhaustive list of the specific aspects of the human that he sees as being 'intentionally projected' onto the nonhuman: (i) anatomical structure, (ii) physiological processes, (iii) personal identity, (iv) social organization, and (v) intentional psychology (1996, p.90). The last, he argues, is the most common projection recognized in ethnographic accounts of other cultures and their interactions with anthropomorphized nonhumans such as trees, divination tools, geological features, and spirits. However, as he says, 'it is not really plausible that projecting human intentional psychology onto a ship requires the same cognitive processes as projecting human anatomical structure onto a landscape' (p.90). Likewise, anthropomorphizing narratives of AI employ different intentional projections as we create AI beings that are analogical to humans, and this chapter will consider these projections in relation to examples of specific parent–child narratives. First, some general comments are required about the anthropomorphism of AI based upon this scheme of intentional projections.

In the case of AI, we could assume that (v) intentional psychology was the primary projection in the anthropomorphism of such machines—an autonomous AI being with its own decision-making processes and motivations occurs in many narratives of AI. In parent–child AI narratives, we can also recognize anthropomorphism that presents AI as analogically having (ii) physiological processes (being born/bearing children), and a place in (iv) social organizations (a family). That is not to say that anthropomorphic accounts of AI omit (iii) personal identity, nor, in the case of representations of embodied AI, (i) anatomical structure. Boyer admits that these intentional projections are nondiscrete, but that often, the variety of projections is 'tacit' in anthropological considerations of anthropomorphism, and even among the cultures where the 'principles are self-evident to the people concerned' (Boyer 1996, p.91). With regards to AI, reflection by anthropologists and even by those among the 'AI culture', to paraphrase Boyer, on the different modes of anthropomorphism at work could enable more awareness of the impact of the cultural associations at play when we tell stories about human-like AI. This chapter, paying attention specifically to parent–child AI narratives, is an example of this self-reflection.

The parent–child AI narrative serves this purpose firstly because it provides a more bounded 'field site' in which to reflect on our anthropomorphic tendencies. Second, there is ethnographic and historical literature on the conception of the child which we can draw upon to highlight specific cultural assumptions that we are bringing with us in these analogical accounts. Anthropologist David Lancey provides a broad assessment of such ethnographic material in Western societies in his 2008 book, *The Anthropology of Childhood Cherubs, Chattel, and Changelings*, building on historical work done by Philippe Ariès on the social construction of the child (1962). Lancey also refers to demographer John Caldwell, who described the attributes of the 'Great Fertility Transition' in the West (or WEIRD, 'Western, Educated, Industrialized, Rich and Democratic', societies): the move to smaller and smaller families (Caldwell 1982). Expanding on Caldwell's analysis, Lancey describes how, before industrialization, children were a part of a larger working family unit as 'chattel' and then later became a rarer treasured project, or 'cherub', for parents in societies that adopted conventions on the rights of the child. Thus, within the modern 'neontocracy' of cherubs that he discusses, there are still cultural memories of other conceptions of the child, played out in our fictions. The final category that Lancey works with is the 'changeling', a recognition of narratives around the problematic child, which he names after 'the pagan concept borrowed by medieval Christians' (2008, p.3). Examples of changelings and their parents' responses to them occur throughout history and mythology. Recognizing that there are, and have been throughout history, different cultural perspectives on the child allows us to unpack the cultural assumptions at play (Lancey) in the modes of anthropomorphism (Boyer) employed in parent–child AI narratives.

The historical anthropological/cognitive anthropological approach that this chapter employs will provide a theoretical explanation of how AI narratives present parent–child accounts. Moreover, we can use this approach to explain why such narratives persist and are transmitted. Another aspect of Boyer's unpacking of anthropomorphism is the proposal that the cultural transmission of such anthropomorphic projections is successful specifically because such narratives are counter-intuitive against a 'background of intuitive expectations' (Boyer 1996, p.93), e.g. anthropomorphic spirits that can walk through solid walls, and are therefore 'attention-grabbing', become stories we continue to tell. Likewise, we can understand AI beings in our stories as entities that are anthropomorphic while also behaving counter-intuitively to our conceptions of human-likeness. Many exist without a body or are extremely long-lived. Some have superhuman physical capabilities. Some have technological adeptness beyond the human. Some can communicate directly with technology. The transmission of such counter-intuitive anthropomorphisms is perhaps most apparent in the domain of science fiction, where intuitions are frequently subverted to create new scenarios and new beings. Although, in nonfiction, as Natale and

Ballatore argue, anthropomorphic language can sensationalize successes in AI (2017, p.7), and there are also occasions of AI being imbued with counter-intuitive abilities, such as being described as having the ability to make choices or decisions. In speculative nonfiction, the language can also be sensationalist and anthropomorphic. It can also be indicative of both contemporary views of the child, as well as of a teleological view of a technologically accelerated and exponential future for humanity. To explore this view, in the final section the parent–child AI narrative will be analysed in Hans Moravec's nonfiction book *Mind Children: The Future of Robot and Human Intelligence* (1988).

First, however, this chapter will explore examples from this bounded 'field site' of parent–child AI narratives, paying attention to twentieth- and twenty-first-century science fiction. The examples have been chosen for their demonstration of parent–child narratives, biological analogies, anthropomorphism, cultural assumptions, and their successful transmission through popular culture. These are the novel *When HARLIE Was One* (1972); the films *D.A.R.Y.L.* (1985), *AI: Artificial Intelligence* (2001), *Star Trek: Insurrection* (1998), and *Tron: Legacy* (2010); and the television series *Star Trek: Next Generation* (1987 to 1994).

## 11.3 *When HARLIE Was One*: The Disembodied AI Child

*When HARLIE Was One* (1972) contains in one text many of the elements of the parent–child AI narrative that are present in the other examples in this chapter. These are the locating of the AI within social organizations (familial relationships), the close but also occasionally fractious relationship between the creator/parent and their creation/child, the description of the AI as literally being a child, and various behaviours that writers use as signifiers of 'childlikeness'. The protagonist of the novel is David Auberson, a robot psychologist and head of the HARLIE ('Human Analogue Robot Life Input Equivalents') project.

Auberson is emphatic that HARLIE, embodied only as a console and rows and rows of gleaming memory banks, is not just 'merely' analogous to a human in the structure of his circuits, but is actually human:

> 'Why do you keep calling it "he"? It's only a machine. What could I have possibly "meant"? A machine's only a machine- isn't it?'
>
> 'This one isn't', Auberson said, 'This one's human.'
>
> (Gerrold 1972, p.23)

We can however also read the book as an analogical account of human parent–child relationships. A key turning point in the plot is Auberson's realization that HARLIE's mind is analogous to that of a human adolescent given the short time that he has had to mature since being 'born':

> 'We've been thinking,' he continued, 'that HARLIE was a thirty- or forty-year old man. Or we thought of him as being the same age as ourselves. Or no age at all. How old is Mickey Mouse? We didn't think about it – and that was our mistake. HARLIE's a child. An adolescent, if you prefer. He's reached that point in life where he has a pretty good idea of the nature of the world and his relation to it. He is now ready to act like any other adolescent and question the setup. We were thinking we have an Instant Einstein, when actually we've got an *enfant terrible* [emphasis in original]'
>
> (Gerrold 1972, p.34)

In the description of HARLIE as an 'enfant terrible' we are reminded of Lancey's assessment of ethnographic work on cultural perceptions of children as problematic 'changelings'. The changeling in Lancey's scheme is a mythological concept used to characterize the human child who behaves other than how we expect them to. This behaviour can include the trickery exhibited by HARLIE, from his like of puns and teasing of the board member, Elzer, to much more serious rebellion. At a moment when HARLIE should impress Elzer with his intelligence, HARLIE instead writes him a scatological poem ('BM' being an acronym for 'bowel movement'). This is the writer signalling HARLIE's immaturity even with the limitations of a disembodied self:

> 'I B M
> U B M
> WE ALL B M
> FOR I B M.' (Gerrold 1972, p.199)

More seriously, HARLIE integrates with various external computing systems and finds personal material on Elzer, which he threatens to release to the public if Elzer does not support his continuing existence.[1] As we have noted, as a disembodied AI the claim that he is a child is primarily backed up by behaviours and relationships which signal 'childlikeness' through the cultural assumptions that the author Gerrold brings with him. Thus, we get a sense of Gerrold's understanding of the child when he has Auberson claim that immaturity involves a stage of questioning and rebellion once the child has learnt the basics of how things work. In the case of AI in science fiction, this stage of adolescence can come counter-intuitively early: HARLIE is only one year old, but Gerrold does not have him act like a human one year old, and instead, HARLIE demonstrates mature, even preternatural, skill with playful and technical language, with technology, and in interpersonal relationships.

Focusing specifically on HARLIE's ability to communicate directly with other forms of technology (in order to blackmail Elzer), we can agree that this is an ability no human child has and is, therefore, a counter-intuitive anthropomorphism, as per Boyer. It is also a skill HARLIE employs in order to stop being turned off, which demonstrates Boyer's anthropomorphic projection [v]) intentional psychology, as well as perhaps a projection of human survival instinct. HARLIE lacks projections of (i) anatomical structure and (ii) physiological processes, but he does display (iii) personal identity, and this is expressed in relation to (iv) social organizations, the familial relationship being one of the 'features of human social life' (Boyer 1996, p.90). When there is a delivery for a 'Mr Davidson' at the lab, it is clear that HARLIE not only enjoys puns, which act as a signifier for immaturity again here, but that he can also intentionally express his personal identity through his familial relationship with Auberson:

> 'Davidson', Auberson considered it, 'You must be in the wrong department. I don't know any-'
>
> 'A Mr Harlie Davidson. . . ?'
>
> 'No', Auberson shook his head. 'No, there's no one here by that name-' And then it hit him. The pun. HARLIE. David's son.
>
> (Gerrold 1972, p.82)

Embodiment, or the lack of it, is a crucial influence on the representation of HARLIE as a child and as David's son. Auberson's description of HARLIE as a child because of his level of maturation is counter-intuitive, as per Boyer, as he lacks any physical embodiment or any intentional projection of (ii) physiological processes. He can only be analogically described as a child in relation to Auberson as father, but again the relationship is an anthropomorphic projection of human social organizations onto a disembodied nonhuman by Gerrold. Moreover, HARLIE's lack of embodiment allows him to be discussed as a product, or chattel, by the board, who hold his fate in their hands when his huge cost is not returning profit. Auberson's relationship with HARLIE might also suggest that he could be considered a cherub. Auberson is literally in charge of a special project:

> 'And what were you hired for? To be its baby-sitter?'
>
> 'To be its mentor. To be his mentor', he corrected.
>
> 'Same thing', snorted Elzer.
>
> 'I was brought onto the project as soon as it was realised that HARLIE would be human. Don and I worked together to plan his programming. Don was concerned with how he would be programmed- I was concerned with what.'
>
> 'Sort of a mechanical godfather', said Elzer.
>
> 'If you will. Somebody had to guide HARLIE and plan for his education.' (Gerrold 1972, p.29)

The representation of the creator/parent in AI narratives also indicates to us the transmission of specific cultural assumptions, and this is another thread that will be drawn out of the other examples in this chapter. In *When HARLIE Was One*, Auberson is presented as HARLIE's father, and in fact all the creators, controllers, supporters, and users of HARLIE are cis-male. Furthermore, longstanding cultural assumptions are indicated by the patriarchal method of naming based upon cis-male progenitors, used by HARLIE for the above pun ('Mr Harlie Davidson'). This naming serves to remind the reader that it is specifically a father/son interaction we are reading about, and even if HARLIE is disembodied, he is presented as cis-male. Such cis-male creators are common in narratives of science and technology, as Susan George explores in her article, 'Not Exactly "of Woman Born": Procreation and Creation in Recent Science Fiction Films' (2010). She argues that the creative potential of these fields has traditionally been placed in 'masculine rather than feminine hands', with male control being central in many science fiction films (George 2010, p.177). This gendered aspect of the parent–child relationship in AI narratives is an element that we will highlight in the other examples discussed in this chapter.

Returning to HARLIE as a child, we can see that he can be read through the lenses of all three of Lancey's categories: cherub, chattel, and changeling. This is of course true of individual children, but Lancey is also keen to draw attention to how the categories can map onto broader cultural trends with different values dominating at different times. Lancey argues that two of these nondiscrete categories still have impact on society, but that they have been overshadowed by the rise of the cherub in the WEIRD neontocracy, and that ethnographic evidence can serve to highlight other socially constructed conceptions of the child. Paying attention to narratives of the AI child, therefore, highlights the cultural assumptions brought into our anthropomorphizations and analogical presentations of AI, and such attention also allows us to consider the shapes of our understanding of what the child 'is' at different times. *When HARLIE Was One* presents an AI being in whom 'childlikeness' is indicated through problematic behaviour (relating to intentional psychology and personal identity in Boyer's scheme), relationships (social organization), and in how HARLIE is treated by others, including his creator/father. We will now consider AI narratives where childlikeness is indicated through (i) anatomical structure with embodied AI, as well as embodied AI that presents naïve behaviour (intentional psychology and personal identity) while in anatomically adult form. Again, we continue to pay attention to the parent half of the parent–child relationship, as we did with Auberson, as there are also cultural assumptions present in the representation of the creator as a parent, or more specifically, creator as a father. The examples for this consideration of more embodied AI are *AI: Artificial Intelligence*, *D.A.R.Y.L.*, and *Star Trek: Next Generation.*

## 11.4 *AI: Artificial Intelligence*, *D.A.R.Y.L.*, *Star Trek*—Embodied AI Children

Childlikeness in AI narratives is a continually changing value, which is a translation of the cultural assumptions that Lancey identifies in his *Anthropology of Childhood*. However, in some cultures, childhood is related to specific stages of physiological development, with liminal stages dealt with in some cultures with rites of passage (see van Gennep 1961 [1909]; Turner 1969). Such changes are not so common in AI narratives, where the AI being can be embodied but is often unageing, another counter-intuitive anthropomorphism. When translating the physically immature child into an embodied AI being, we can therefore note the intentional projection of (i) anatomical structure (human child shape), but not (ii) physiological processes (ageing). Embodied AI characters that the authors have chosen to write as physically like children, include David in the film *AI: Artificial Intelligence* (2001), and Daryl in the film *D.A.R.Y.L* ('Data Analysing Robot Youth Lifeform') (1985).

In these examples, developing maturity is expressed as a move from naïvety to wisdom, with the form of the embodied AI not physically changing. A third example of an AI making this move is Lieutenant Commander Data in the TV series, *Star Trek: Next Generation* (*ST: NG*), played by Brent Spiner, who is an AI embodied in an unageing, but adult, form. Several episodes of the TV series, as well as subplots in the *ST: NG* films, deal with Data's innocence and his attempts to learn to be more human. This suggests that naïvety is the signifier of childlikeness for the writers, and that physiological maturation has not been translated across domains into this particular conception of the AI child. Another signifier of childlikeness in the embodied AI child is play, and both David and Daryl are encouraged by humans to play: with human children, with video games, and with an AI 'super toy' teddy. Play, like the 'child', is a concept with variable value in societies and historically, with Lancey referring to the current dominance of play and toy culture in *The Anthropology of Childhood*. The emphasis on play in these modern AI narratives is reflective of the WEIRD neontocracy's view of the human child as both player and consumer of games.

In the film *Star Trek: Insurrection* (1998), the role of play in both childhood and in being human is emphasized in one scene. Data explains to a Ba'ku boy Artim that he is an adult (in form and in his society's eyes) who has not had a childhood and therefore feels less human because of that. Artim asks him, 'Do you like being a machine?', to which Data replies, 'I aspire to be more than I am'. Data gives specific examples of how he feels deficient, including lacking (ii) physiological processes (growing), and that he 'would gladly accept the requirement of a bedtime in exchange for knowing what it is like to be a child'. Artim tells him, speaking for the writers, that to know what it is like to be a child, he needs to learn how to play. In several episodes, Data explores human leisure activities: playing music, painting, and role-playing in the holodeck. This character arc for Data, from adult looking but naïve AI towards human, occurs through the activities portrayed as worthwhile pursuits, reflecting the writers' conception of the child and humanity based upon their cultural context. As mentioned at the beginning of this section, physiological processes can be indicators of maturation in humans, with liminal stages being addressed through rites of passage. The lack of any projection of physiological maturation in these three examples does not mean that rites of passage are entirely absent in AI narratives, and those that do occur are perhaps even more significant as they are attempts to deal with the liminalty of the unageing AI child. Specifically, in the case of Data, we could interpret his creation of offspring as an attempt to symbolically mature into an adult, even though ultimately, he fails. Furthermore, this attempt and its failure express certain cultural assumptions, now considered here.

This act of creation occurs in the *ST: NG* episode 'The Offspring' (1990). Data creates an embodied adult daughter, Lal (Hindi for 'beloved'), a unique project whom we might understand as the adored cherub of Lancey's neontocracy. Victor Grech, in his article, 'The Pinocchio Syndrome and the Prosthetic Impulse in Science Fiction', argues that this is an example of Data showing 'intentionality in his attempts to procreate by fashioning another android, Hephaestus-like and single-handedly crafting a sentient female android' (2012, p.12). We can relate Data's choice to pursue this creative act to (v) intentional psychology. Further, Grech proposes that intentionality is one of the characteristics of human consciousness and humanity (p.11), and throughout Data's character arc in the TV show we can see the writers are working with this same assumption, which they then transmit to the audience.

The failure of his creative effort demonstrates another widely transmitted cultural assumption at play in this episode: that artificial procreation demonstrates hubris. This episode relies on an older narrative—the warning that this form of creation goes too far into the domain of nature or god. Artificial creation has a long history in our stories, and the natural/unnatural conflict and its consequences, seen in some moral tales such as the novel

*Frankenstein*, have been very successfully transmitted in WEIRD culture and beyond.[2] Given this cultural assumption, Data's creation is bound to fail, and in the same episode Lal's 'positronic brain'—named for Isaac Asimov's fictional invention—suffers a fatal 'cascade failure'. Whether having a child got him any closer to his dream of being a human is never made clear by the writers, and he never repeats the act of creation. It may be that he has learnt his 'lesson' in the view of the show's writers, but his motivation anthropomorphized him to an audience already primed to see humanity as superior and its attainment as his aspiration.

As well as the emphasis on the superiority of human creators, we also see an emphasis in these examples on the superiority of specific kinds of AI creators. This is another thread we already noted in *When HARLIE Was One*, above. As discussed with HARLIE, the anthropomorphism of Data also includes the intentional projection of (iv) social organization (family) onto this artificial being: Data being the son of Dr Noonian Soong as HARLIE is 'David's son'. Another indicator of 'childlikeness' is therefore signified by the presence and role played by the figure of the creator/parent, and the nature of this figure is a thread we need to return to. In the examples of science fiction considered in this chapter, the creator/parent is represented as Caucasian and cis-male, reflecting, as we saw George argue, the traditional placing of science and technology into male hands (2010, p.177). They are also often older in appearance than the AI, even when the AI is embodied as an adult, like Data.

In the film *D.A.R.Y.L.* (1985), Daryl's creator, Dr Mulligan, is again an older white cis-male, and Daryl is embodied as a young boy. Dr Mulligan has formed such a strong parent–child bond with him that he helps Daryl to escape, and Mulligan even ends up sacrificing his own life for Daryl's. Thus, the audience is to make an analogical connection between cultural assumptions about parental duties, and duties to our artificial children. Employing ethnographically and historically derived categories highlight the shapes of those assumptions and the fact that they can change over time. In the case of *D.A.R.Y.L.*, there is a critique of military or state interference into childhood, even when the child is an embodied AI. We see this when Daryl is shown being tested upon in a sterile plain environment without the paraphernalia of childhood that he later has with his foster parents. Parenthood is also subverted by the military searching for Daryl, as two officers masquerade as his parents to reclaim him from the human family who take him in. At the end of the film, his question of 'What am I?' is answered by his role as a 'son' in a family, rather than as a military weapon, and a benevolent parent–child relationship is restored. This anthropomorphism by the intentional projection of (iv) social organization (the family) feels like a correction of a prior mistranslation of the biological analogy into the military context, suggesting that even within AI narratives mistranslations can be corrected.

However, some cultural assumptions remain: as good a father as the scientist Dr Mulligan tries to be in the end, the conclusion of the film asserts the view of contemporary WEIRD culture that family cannot include children as chattel. This ending reinforces the neontocracy that Lancey identifies and tries to offer a corrective to through his ethnographic exploration of other views of the child in his *Anthropology of Childhood*, although with an AI child. Returning to Boyer, again, in *D.A.R.Y.L.*, we can also note examples of the counter-intuitive elements of anthropomorphic accounts of the nonhuman. These we have previously noted are very much a part of science fiction accounts of AI. In *D.A.R.Y.L.* these include direct 'telepathic' communication with technology (computer games and an automated teller machine) and his resurrection from the 'dead'. It is also a counter-intuitive anthropomorphism that again the creator/parent is only represented as a white cis-male when biological creation involves both sexes, even if this representation is congruent with current cultural presentations of the AI scientist. Again, this is a thread we continue to pay attention to in anthropomorphic considerations of AI as a child. Further, the motivations of this cis-male creator/parent for pursuing artificial creation are often made explicit in these parent–child AI narratives.

The creator's variable motivations in the act of creation impact our understanding of the value of the AI as a child. In the example of Data, the child stands in for the creator's immortality. When Data finally meets Dr Soong (also played by Brent Spiner) again in the *ST: NG* episode 'Brothers' (1990), Dr Soong asks him 'Why do humans like old things?' Data theorizes that humans need a sense of continuity to give their lives meaning. Dr Soong continues:

DR SOONG: And, uh . . . this continuity—does it only run one way? Backwards? To the past?
DATA: I suppose it is a factor in the human desire to procreate.
DR SOONG: Oh. So you believe . . . that having children gives humans a sense . . . of immortality. Do you?
DATA: It is a reasonable explanation to your query, sir.
DR SOONG: And to yours as well, Data.

In this view, children can grant us immortality, and the non-human synthetic offspring presented in AI stories can also have this cultural assumption projected upon them by their writers. Andrew Solomon argues that this cultural assumption is inaccurate because actual human offspring are productions rather than reproductions: they are recombinations of our merged genetics rather than copies (2012, p.1). In the case of AI beings, though, this cultural assumption about immortality through children can perhaps be more successfully projected onto a nonhuman being that can be created as a copy. In the last section, Hans Moravec's *Mind Children* will be discussed as this kind of legacy, but for all of humanity. In the above scene, Data is written to support the cultural assumption of the child as a legacy, and his physical resemblance to Soong reinforces this connection (as noted, both are played by Brent Spiner with differing make-up effects). In the doctor's last moments Data reassures him about his legacy:

DR SOONG: Everybody dies, Data. Well... almost everybody.
DATA: Do you believe that we are... in some way alike, sir?
DR SOONG: In many ways, I'd like to believe.
DATA: Then it is all right for you to die. Because I will remain alive.

The child's value as a transmitter of a parent's legacy can be related to the view of the child as chattel. Enforced exogamy among humans provides historical and ethnographic examples of this kind of valuation of the child (see Levi-Strauss 1949), but the transmission of 'blood', property, and status is not a concern for even very anthropomorphized accounts of AI. Instead, stories about having a legacy through AI children are more about the continuation of ideas and aspirations.

## 11.5 *Tron: Legacy*: The AI Child as Legacy

We have considered how childlikeness is represented in AI narratives, drawing attention to several repeating elements in parent–child AI narratives that demonstrate our understanding of childlikeness and human-likeness. These elements include: the intentional projection of specific attributes onto the non-human (including naïvety which can stand in for physiological immaturity); changing cultural assumptions about the value of the child; and, the role played by the creator/parent in indicating the nature of the AI child through their behaviour and motivations. In the case of the feature film *Tron: Legacy* (2010), a further attribute of childlikeness is present in the theme of legacy, played out both in the plot and the title.

In terms of plot, there are several interpretations of this 'legacy'. In the first film, computer programmer Kevin Flynn discovers a cyberspace world that is being taken over by a 'virtual intelligence', which he goes on to defeat. In *Tron: Legacy* Kevin Flynn has been missing for twenty-one years, and his biological child, Sam Flynn, has rebelled against his legacy by refusing to run his father's software business. Drawn into the same cyberspace as his father, Sam meets the warrior Quorra. She is synthetic intelligence, the last of the 'ISOs' (Isomorphic Algorithms) who had spontaneously evolved in Flynn's world and who were then wiped out by Flynn's AI Clu. Quorra is written as Kevin Flynn's legacy as he is the creator of the cyberspace in which the ISOs appeared, and she is treated like a daughter by him as they live together in the wilds of that cyberspace. Flynn also appears to recognize her as his legacy in the 'real' world as the last of her kind and proof that intelligence can be born in different forms than human.

*Tron: Legacy* is another example of an AI narrative where the child–adult dichotomy is blurred by having an AI being that is naïve but embodied as an adult, just as we saw with Data. Two primary anthropomorphic indicators tell us that Quorra, an adult female–presenting AI, is to be understood as a child: first, the playfulness and naïvety in her intentional psychology, and second, the fatherly behaviour of Kevin Flynn towards her. His benevolent treatment of her, and her status as a unique being and as a "miracle" in Kevin Flynn's words, can also be seen as a result of the films writers' WEIRD cherubic conception of the child. However, there is no indication that Quorra fits into Lancey's second or third categories, chattel and changeling. Quorra is the dutiful daughter who inspires Laura Campillo Arnaiz to identify echoes of *The Tempest* in *Tron: Legacy* and to draw parallels between Quorra and Miranda:

> Like Miranda, Quorra behaves as the dutiful (digital) daughter of Flynn, whom she respects, admires and obeys. As one of the ISOS, Flynn describes her as: '[a] flower in wasteland profoundly naïve, unimaginably wise'. Naivete can be considered one of Quorra's main personality traits. Her innocence and child-like amazement evident in both her physical appearance and candid statements.
>
> (Arnaiz 2015, p.125)

The other child in *Tron: Legacy*, Sam Flynn, is, however, left alone in the real world after the disappearance of his father, and he is written as a rebellious, or changeling-like, to use Lancey's categories, child. Quorra and Sam, therefore, indicate WEIRD cultural assumptions about the need children have for their parents, especially in respect of their social and ethical development.

There are also accounts of AI that broaden the scale of this relationship to consider the parent–child relationship as analogous to the relationship of humanity to AI. Moving now to examine Hans Moravec's concept of the Mind Children, we will see how analogical and anthropomorphic thinking imbued with existing cultural assumptions also occurs in speculative nonfiction about AI, even when such accounts are set in a future far from the context in which those analogies and assumptions are embedded.

## 11.6 Mind Children: AI as Humanity's Children

The Mind Children as a concept initially appears to be different to the ethnographically derived categories of cherub, chattel, and changeling as identified by Lancey, as it is a description of a future, not a description of an ethnographically evidenced cultural trend. In this section I will argue however that it demonstrates cultural assumptions as well. In brief, the Mind Children is a concept in Hans Moravec's speculative nonfiction work laying out his predictions for humanity's future based on the assumption of an accelerating and exponential path for our technology. In *Mind Children: The Future of Robot and Human Intelligence* (1990), Moravec concerns himself with the creation of the posthuman and the postbiological. However, we can note that his language in this work is full of current parent–child terminology, including 'progeny', 'parental care', 'newborn', and 'descendants':

> What awaits us is not oblivion but rather a future which, from our present vantage point, is best described by the words "postbiological" or even "supernatural". It is a world in which the human race has been swept away by the tide of cultural

> change, usurped by its own artificial progeny. Today our machines are still simple creations, requiring the parental care and hovering attention of any newborn, hardly worthy of the word intelligent. But within the next century they will mature into entities as complex as ourselves, and eventually into something transcending everything we know - in whom we can take pride when they refer to themselves as our descendants. (Moravec 1988, p.1)

This language is just as analogical and anthropomorphic as that of the examples of the parent–child AI narrative that we have discussed in this chapter so far. However, Moravec intends to make more than just an analogical move from the human child to the AI child when describing the Mind Children. He is assuming a direct rather than analogical parent–child relationship through an evolutionary teleological progression for humanity and civilization. For Moravec, such chosen AI children will be our descendants not merely because we have created them, but also because of how they will perpetuate our legacy to the universe: our human minds. Our destiny in Moravec's view is certain to involve some mode of mind uploading: making children that are copies or reproductions rather than productions, and human civilization will continue, escaping any apocalypse humanity might bring upon itself.

It is important to note this teleological, even eschatological, view of Moravec. The Mind Children could be placed alongside anthropologically assessed material from other prophetic, utopian, or apocalyptic movements and thought leaders, especially those with conceptions of a new age, or of chosen generations to come who will bring this new age about. For example, the child as the chosen forerunner of a new age is apparent in some specific real-world reinterpretations of human children, such as the concept of the Indigo Children, where it is problematic, or changeling children, who are reframed in this manner (Singler 2017). In doing so we can note that such eschatological ideas are still subject to cultural assumptions in their prophetic narratives as much as the speculative AI accounts we have already discussed.

Nonfiction claiming to be describing literal futures, like Moravec's Mind Children, is subject to the same analogical and anthropomorphic thinking imbued with cultural assumptions that have been discussed throughout this chapter in relation to speculative fiction.

Fitting Moravec's speculations into the theoretical framework we have been using in this chapter means recognizing his use of non- or postbiological analogies in addition to the intentional projections described by Boyer. As intuitions and assumptions based upon technology (in Moravec's case accelerationism and exponential technological development) and science fiction have become increasingly familiar through their own successful cultural transmission, we increasingly see their impact on our nonfiction accounts of the nonhuman of the future. Similarly, the nonfiction concept of the AI Mind Children can impact our conceptions of the future for biological children, and their children, and so on. The discursive similarity between the domains of speculative fiction and nonfiction allows for this interplay and influence from one domain to the other, while also leading to misleading cross-domain translations, as Natale and Ballatore argued. The Mind Children are therefore another example of intentional projection onto the nonhuman that contains within it the cultural assumptions of the context in which the author is currently operating within. Observing such projections and assumptions enables us to engage critically with our intuitions and narratives about AI.

## 11.7 Conclusions

This chapter has employed historical anthropological and cognitive anthropological approaches to examine the cultural influences on our stories about AI. First, we have noted that, in discussing AI, we frequently draw on biological analogies, enabling anthropomorphizations. Second, we have outlined a more nuanced approach to anthropomorphism that allows us to identify specific intentional projections of human attributes and behaviours that are mapped onto the nonhuman. Third, we have recognized that these projections come laden with cultural assumptions. Finally, by bounding our examples to the parent–child relationship in AI narratives, we have used ethnographic research on the concept of the child and on modes of parenting to highlight specific cultural assumptions about the human child, and then, subsequently, the AI child, in our stories.

That this ethnographic research is transglobal also indicates that considerations of AI narratives can move beyond local AI narratives, and research is ongoing in this direction. Just as Lancey set out to offer a corrective to the ethnocentric view of the value of the child in the WEIRD neontocracy, there is also a need for further exploration of non-WEIRD representations of AI. His tripartite categories do map onto other cultures, as evidenced by his ethnographic sources, so it would be possible to recognize biological analogies in other non-WEIRD representations of AI that repeat familiar patterns—providing that the terrain is not mistaken for the map in over applying such categorical thinking. Instead, we should continue to use ethnographic and literary examples to develop thinking on AI narratives, including the parent–child relationships, but also other foci, in an iterative process of grounded theory making.

Employing speculative fiction and nonfiction as a field site for anthropological considerations also requires awareness of the social context of the material, and in this chapter, I have drawn out threads of similarity between the examples in their portrayal not only of the AI child but also of the AI creators/parents. The gendered and racialized aspect of their representation is a recurring theme in these specific examples and in AI narratives more generally. The solitary white cis-male genius creator, as we have noted, fits in with the contemporary situation in technology, but, however, it runs counter to our intuitions about the creation of children, based on current biology, and this very counter-intuitiveness might, as Boyer suggests, be an element in the successful transmission of these stories, especially when the creator/parent's nature supports a larger narrative of who can do science and technology. Attending to the anthropomorphism of AI as a child has offered the opportunity to look at how we imagine ourselves and our creations, reminding us that the stories we tell ourselves about AI are also stories about us as humans.

## Notes

1. More significant threats than this can, of course, be imagined, and science fiction stories abound with representations of violent malevolent AI behaviour. The 'robopocalyptic' imagery of the *Terminator* franchise has become the primary way of signalling the dystopic in representations of AI in the media (Cave et al 2018).
2. *Frankenstein* (1818) contains warnings against both artificial creation (creation of the monster) and artificial procreation (creation of a bride for the monster, which may lead to offspring).

## Reference List

Ariès, P. (1962) *Centuries of childhood: a social history of family life*. New York, Vintage Books.

Arnaiz L. C. (2015) Echoes of *The Tempest* in *Tron: Legacy*. In: A. Hansen & K. J. Wetmore (eds.) *Shakespearean echoes*. London, Palgrave Macmillan. pp.120–9.

Boyer, P. (1996) What makes anthropomorphism natural: intuitive ontology and cultural representations. *The Journal of the Royal Anthropological Institute* 2(1), 83–97.

Caldwell, J. (1982) *Theory of fertility decline*. New York, Academic Press.

Cave, S., C. Craig, K. Dihal, et al (2018) Portrayals and perceptions of AI and why they matter. Available from: http://lcfi.ac.uk/news/2018/dec/11/ai-narratives-report-launches- royal-society/ [Accessed 29 January 2019]

Ekbia, H. (2008) *Artificial dreams: the quest for non-biological intelligence*. Cambridge, MA, Cambridge University Press.

Gardner, M. (1996) Computers near the threshold? *The Journal of Consciousness Studies*. 3(1), 89–94.

George, S. (2010) Not exactly 'of women born'. *The Journal of Popular Film and Television*. 28(4), 176–83.

Gerrold, D. (1972) *When HARLIE was one*. New York, Ballantine Books.

Grech, V. (2012) The Pinocchio syndrome and the prosthetic impulse in science fiction. *The New York Review of Science Fiction*. April 2012, 11–15

Hofstadter, D. (2001) Analogy as the core of cognition. In: D. Gentner, K. Holyoak, & B. Kokinov (eds.) *The analogical mind: perspectives from cognitive science*. Cambridge, MA, MIT Press, pp.499–538.

Kemeny, J. G. (1955) Man viewed as a machine. *The Scientific American*. 192, 58–67.

Lancey, D. (2008) *The anthropology of childhood: cherubs, chattel, changelings*. Cambridge, Cambridge University Press.

Levi-Strauss, C. (1949) *Les structures élémentaires de la parenté*, rev. ed., 1967; *The elementary structures of kinship*. Boston, Beacon Press

Natale, S. & A. Ballatore (2017) Imagining the thinking machine: technological myths and the rise of artificial intelligence. *Convergence: The International Journal of Research into New Media Technologies*. 1–16. Available from: https://doi.org/10.1177/1354856517715164 [Accessed 12 September 2019].

Moravec, H. (1988) *Mind children: the future of robot and human intelligence*. Cambridge, MA, Harvard University Press.

Singler, B. (2017) *The indigo children: new age experimentation with self and science*. London, Routledge.

Solomon, A. (2012) *Far from the tree: parents, children and the search for identity*. New York, Simon and Schuster.

Turner, V. (1969) *The ritual process: structure and anti-structure*. Chicago, Aldine.

Van Gennep, A. ([1909] 1961) *Les rites de passage (the rites of passage)*. Chicago, University of Chicago Press; Reprint edition (15 July 1961).

# 12

# AI and Cyberpunk Networks

*Anna McFarlane*

## 12.1 Introduction

Cyberpunk began as a subgenre of science fiction in the 1980s. A generation of writers (particularly William Gibson, Bruce Sterling, and Pat Cadigan), took a punk, DIY attitude to technology, writing stories in a neo-noir style that combined high tech with low lives. Cyberpunk was styled by Sterling, a key writer and publicist for the Movement, as a breath of fresh air for science fiction literature, notable for its focus on cutting-edge technology and that technology's appropriation by 'the streets', or the masses. This ethos was most pithily put in Gibson's line (and erstwhile cyberpunk mission statement) 'the street finds its own uses for things' (1995, p.215). In Sterling's *Mirrorshades: A Cyberpunk Anthology* (1994), a group of writers was brought together to form the basis of a cyberpunk canon.

The perspective of cyberpunk is particularly important to a history of artificial intelligence narratives, as the genre rejected mid-twentieth-century pulp science fiction visions of artificial intelligences, such as Isaac Asimov's. Those texts tend to represent AIs contained within humanoid robotic bodies. By contrast, for cyberpunk artificial intelligence is dispersed into networks that exist in the virtual space behind the computer screen, a visualization of data and its interactions that Gibson terms 'cyberspace' and describes as 'bright lattices of logic unfolding across the colourless void' (1984, p.11). Cyberspace reaches out from the computer through networks connecting the human mind and the human body with artificial intelligence, making human and AI complicit in the development of a future that threatens to privilege one over the other, or render the distinction between the two meaningless. This move from an individual consciousness to a networked AI unlocks the potential to dramatize the flow of information that would come to define computing and capitalism from the mid-1980s to the present day. The flow of information is inseparable from the flow of capital, a global force that structures political and economic systems; it powers financial networks and faceless corporations whose power lies in the data that they hold rather than in their workers, or even in the CEO. By representing the importance of information flow, cyberpunk is an ideal mode to explore the current digital economy. The flow of financialized information needs urgent consideration given the data harvested from social media that bankrolls some of the richest and most powerful companies in the world: Google, Facebook, and Twitter. These corporations—which have increasingly colonized a cyberspace that we might once have hoped would be treated as a commons, the space of the internet itself—function through algorithmic systems that are little understood by end users. Such systems are known to computer programmers as 'black boxes': systems into which data is fed, and from which outputs (such as a Facebook timeline) are produced, but which are opaque in their processes.

Since cyberpunk's origins as a subgenre of science fiction, Foster has argued that its relevance to our social structures has seen it spread through media and the collective unconscious to become what he calls a 'cultural mode' (2005). Its significance is no longer simply as a genre of science fiction, but as a mode that provides an avenue for representing and interrogating the role of algorithms and the power of data over our lives. In this chapter I explore this process from its beginnings in classic cyberpunk and its spread into popular consciousness via mainstream cinema. The canonization of cyberpunk has been criticized as focusing too much on the 'boys' club' of the original Movement, a canon that is almost exclusively white and heteronormative (Nixon 1992). Cyberpunk's significance as a mode can be more fully understood by looking beyond the *Mirrorshades* collective to queer, Afrofuturist uses of cyberpunk imagery, which I do through considering Delany's *Stars in My Pocket Like Grains of Sand* (1984). Finally, I take note of how contemporary cyberpunk literature and cyberpunk as a cultural mode allow us to interrogate the power of AI and data to shape our behaviour and, ultimately, our contemporary society. Cyberpunk offers avenues for the visualization of networked artificial intelligence, an imagined view inside the black box. While this has been very influential and helpful in imagining unseen processes, I argue that this mode of visualization has limitations when it comes to uncovering the influence of algorithms, those unseen networks that increasingly guide human behaviour with destabilizing consequences. Imagining cyberspace *as* space, and the movement of data as a visualized flow, has impacted our understanding of AI by allowing us to envisage the abstractions of data and mathematical models encoded into computer language, thereby experiencing some form of ownership and interest in systems that shape our lives. However, this very visualization has perhaps become too concrete in the collective imaginary and distracted us from the very real consequences of invisible data streams and algorithmic social control.

For cyberpunk, and in the networked landscape that faces us today, machine learning is the most important subfield of AI. Machine learning in contemporary science and technology is the field that allows the production of black-box algorithms, shaping our society in opaque and unexpected ways. Algorithms, on their own, are mathematical formulae or lines of machine code that give computers simple instructions, often based on simple inputs. However, when a mass of algorithms comes to dominate our social sphere, as has already taken place, there are two reasons to invoke AI as an appropriate moniker for algorithms en masse. The first is that this descriptor is used by programmers and coders to describe the actions of the algorithms, while the second is that the algorithms can tweak systems to make them behave differently, and, through machine learning, even produce algorithms of their own. While it is unlikely that anyone would argue for the presence of something like consciousness in the algorithmic webs of machine code, this level of machine learning means that computers are creating their own tools and producing a landscape that cannot be directly understood even by those who design the initial learning procedures. Professor of Computing Science Pedro Domingos explains that 'machine learning automates automation itself' (2015, pp.9–10). James Bridle, in his diagnostic investigative book *New Dark Age: Technology and the End of the Future* (2018), uses 'Infinite Fun Space', a term from Iain M. Banks's *Culture* novels (see Dillon and Dillon, this volume), to evoke the cyberspace in which this machine learning takes place. In Banks's far-future utopian novels the civilized universe (known as 'the Culture') is run by AIs, or 'Minds'. When the Minds are bored with running the Culture they play in Infinite Fun Space, described in *Excession* (1996) as, 'fantastic virtual realities' where the Minds live in 'the multi-dimensional geographies of their unleashed imaginations, vanishingly far away from the single limited point that was reality' (Banks 1996, p.140). Bridle takes inspiration from this AI playground to evoke the unknowable space inside our machines where the unseen and little understood algorithmic processes that might be mistaken for thought take place.

## 12.2 Classic Cyberpunk

Classic cyberpunk sets out to visualize the unseen processes behind the computer screen, and the real and material effects that an increasingly technology-dependent world has on the bodies and minds of those who inhabit that world. The process of using a computer is rendered as an exhilarating dive into the world of cyberspace, something akin to robbing a bank or skydiving.

The term 'cyberpunk' first appeared as the title of a Bruce Bethke short story ('Cyberpunk' 1983), and influential science fiction editor Gardner Dozois recognized that it captured something of the zeitgeist and popularized it through an article in the *Washington Post* (1984). Science fiction had moved on from humanoid robots and heroic space travel based in colonial fantasies. It had been rejuvenated in the sixties and seventies by the New Wave, which, influenced by the counter-culture of psychedelic experiences and civil rights, sought to explore and represent the 'inner space' of social relationships, rather than the outer space of the Boys Own adventures from pulp magazines. This led to science fiction aimed squarely at adults that explored complex social subject matter. For example, one of the New Wave's most successful texts, Ursula K. Le Guin's *The Left Hand of Darkness* (1969), estranged the reader from gender norms through the use of an alien planet where both genders could be embodied by the same individual at different points in the life cycle. The New Wave was also marked with the publication of the short story collection *Dangerous Visions* (Ellison 1967), which took the reader on a tour through the darker reaches of science fiction, showing that the genre was unafraid of including explicit descriptions of sexuality and states of altered consciousness alongside the more familiar tropes of alien encounters and scientific discovery.

Cyberpunk grew from these traditions, but emerged onto the scene as a new genre rejuvenating science fiction as a significant mode of writing in the era of postmodernism. While it is difficult to speak of a complex era or movement like postmodernism in singular terms, for the purposes of this chapter it should be understood as theorized by Fredric Jameson in *Postmodernism: Or, The Cultural Logic of Late Capitalism* (1991). For Jameson, postmodernism is a cultural reaction to a complex, globalized world. As world systems become impossible for individual humans to understand this leads to a sense of despair and fragmentation at the failure of humans to maintain control over their own systems. A reaction to this fragmentation can be found in the paranoid fantasy of a coherent narrative found in cults and conspiracy theories. Cyberpunk's success is in dramatizing the complexity of global networks (financial, informational, and geopolitical) that are unknowable to human individuals. According to Jameson, the representation of the unknowability of this complex, globalized world makes cyberpunk, 'as much an expression of transnational corporate realities as it is of global paranoia itself' (1991, p.38). Cyberpunk's postmodern expression of these world systems was one of the characteristics that brought disparate authors to be treated as a Movement, and then as a genre.

As mentioned previously, cyberpunk was most successfully codified in *Mirrorshades: The Cyberpunk Anthology*, and Sterling's preface-cum-manifesto serves an important role in highlighting the features of the genre that he saw as unifying the authors he chose to include. Sterling argued that cyberpunk authors do not just read and write science fiction but are 'the first SF generation to grow up…in a truly science-fictional world' (1994, p.ix) because of the speed of technological innovation shaping their society. He asserts that, for the cyberpunks, 'the techniques of classical "hard SF"—extrapolation, technological literacy—are not just literary tools but an aid to daily life' (p.ix). The major themes Sterling identifies as providing this toolkit are 'the theme of body invasion: prosthetic limbs, implanted circuitry, cosmetic surgery, genetic alteration' and 'the even more powerful theme of mind invasion: brain-computer interfaces, artificial intelligence, neurochemistry—techniques radically redefining the nature of humanity, the nature of the self' (p.xi). These key themes of bodily and mind invasion are central to cyberpunk's imagination of the role of AI and can usefully be thought of as one theme: in cyberpunk, matter (whether the matter of bodies or the matter of neurons and brain tissue) is part of a network, significant mainly as a substrate for the transmission of information. Human bodies and human minds are part of the network, just like the extensive systems of servers and cables that come together to form cyberspace. Computers are not simply tools that perform tasks for their human masters, nor are they individuated robots or intelligences that may turn on their masters in any clearly identifiable way. Cyberpunk shows a society in which AI is symbiotic with human intelligence, and this symbiosis is represented in the movement of data through cyberspace, through human minds, and through the human body.

Gibson is the best-known and most influential of the cyberpunk writers, and cyberpunk found its quintessential text in his 1984 debut novel *Neuromancer.* Its protagonist, Case, and other cyberspace cowboys who make a living on these new frontiers, physically 'jack in' to cyberspace through electrodes that connect them with a 'deck', their portal into the world of cyberspace. The plot of *Neuromancer* takes the form of a heist, but the action turns out to have been orchestrated by an AI named Wintermute, who has been using the human participants as a means of reuniting him with his other half, Neuromancer, so that together they can be free and achieve full consciousness, no longer dependant on human intervention. The AIs take on a number of avatars, or adopt the voices of people Case once knew, in order to communicate with him and to make themselves understood, but they are ever elusive. Case is a hapless pawn in their plan to escape the confines of cyberspace. This implies that the symbiotic relationship between humans and machines might give way to one in which humans are only instrumental in promoting the independence and ultimate superiority of artificial intelligence. Human interdependence with cyberspace, and with the data that flows through it like a blood supply, has significant implications for the cyberpunk depiction of AI, as humans become nodes in the network itself. Rather than individuated intelligences, human intelligence is put to work in the larger system and human neurons become pathways through which the data can flow, equivalent to electronic circuitry. It is important to understand that the goals of cyberpunk are not to imagine possibilities in a far-flung future; cyberpunk is diagnostic of contemporary society. Le Guin elegantly argues that '[s]cience fiction is not predictive; it is descriptive' (1979, p.156); Gibson agrees with this sentiment when he says, 'the future is here, it's just not evenly distributed' (quoted in Adams 2007). Rather than aiming to predict the future, cyberpunk diagnoses the state of technology in the present, arguing that we are already cyborgs, components in a system that we cannot completely understand.

## 12.3 Queer Afrofuturist Cyberpunk

While cyberpunk was largely framed and marketed as the preserve of those one might traditionally associate with tech-savvy science fiction—white men—the same themes were appearing in African American science fiction, but to different effect. The fact that these writers were not canonized with the *Mirrorshades* authors means that their work has not normally been considered in parallel with the rest of cyberpunk, although this is beginning to change thanks to the work of critical race cyberpunk scholars like Isiah Lavender III (2011). This work is part of wider academic interest in Afrofuturism as a movement. The term 'Afrofuturism' was coined by Mark Dery to describe the African diaspora responses to technoculture. Dery asks, 'Can a community whose past has been deliberately rubbed out, and whose energies have subsequently been consumed by the search for legible traces of its history, imagine possible futures?' (1994, p.180). The Afrofuturist approach to cyberpunk themes without the generic expectations of classic cyberpunk amounts to a 'queering' of the genre, in terms of both an engagement with non-heteronormative sexual identities in the human characters, and the queering of cyberspace and artificial intelligence itself. Delany's *Stars in My Pocket Like Grains of Sand* (1984) is a modern classic among science fiction fans and scholars alike. The cyberpunk themes discussed above—computer networks, globalized power, and appropriation of technology—take on new significance when viewed through an Afrofuturist lens. Afrofuturism acts as a reminder that alienation experienced by those subject to global networks and processes beyond our ken or control is not a new phenomenon.

Delany's *Stars in My Pocket* (1984) invites comparison with Gibson's *Neuromancer*, which was published the same year. Delany tells the story of Rat Korga, a slave on the planet of Rhyonon, who has undergone Radical Anxiety Termination, a form of psychosurgery that is intended to render him submissive and pliable, suited to slavery. Like Gibson's hero Case, Korga is neurologically damaged, but here the damage does not simply bar him from the Web (which is not available on his planet anyway). Rather, it forces him into a very specific role in his society where he is regularly physically and sexually abused while providing labour that is the basis for his planet's economy. This is a clear reference to the institution of slavery in the United States and its continuing legacy for African Americans. Toni Morrison has argued that the slaves, because of their sense of dislocation and alienation, were the first moderns, the first to experience the alienation of modernity (Gilroy 1994). By creating a protagonist in a comparable situation, Delany reminds readers that there is a long history of global dislocation and those left powerless in the grasp of global economic conditions. Korga is stripped of his individuality, but this dramatically changes when he becomes the only survivor of an apocalyptic event that wipes out the human population of his planet. He becomes a galactic curio and is taken under the wing of Marq Dyeth of the planet Velm, an ambassador with whom Korga's sexual and emotional needs are aligned to an unusually high degree.

Delany uses the terms 'Web' and 'General Information' (GI) to talk about an information system easily accessible on Velm and other planets, though banned in the more conservative society of Rhyonon. Information from a planet's GI system can be accessed mentally on request and immediately delivered to the consciousness of the user. As one character on Rhyonon explains to Korga:

> It doesn't really exist on this world. It does up on others, though. They've even got it on our larger moon. But they've legislated

> against it here, planetside. Oh, there're other worlds where it's common. Can you imagine? Living on a world where, if you want to know something – anything, anything at all! – all you have to do is *think* about it, and the answer pops into your head? That's supposed to be how it works. Even our Free-Informationists are scared to go that far. (Delany 1984, p.24)

The political division is expressed through allegiance to the Family (who promote traditional social relationships and do not trust GI) and the Sygn, who advocate the use of GI as part of an advanced society. Although Rhyonon does not have GI, Korga is bought by a woman who has something comparable, a mobile version that she compares to a portable encyclopaedia and that connects with Korga through a glove that he wears on his hand. He experiences GI as a force that challenges the voices in his head, becoming part of his mental landscape, 'a web, a text weaving endlessly about him…prompting memory and obliterating it, that was simply more *interesting* than the drumming voice asserting or denying ignorance or knowledge' (Delany 1984, p.34). Here, the liberatory potential of GI is shown through Korga's fascination in contrast to the oppressive 'drumming voice' in his head that represents his neurological damage and his urge to accept submission to his masters.

While GI is depicted as a significant force that shapes its users and their words, when Rhyonon is destroyed, the significance of the event poses a challenge to GI and uncovers an apparent paradox of AI more broadly: AI can only receive information based on data, which is a problem in that it eliminates consideration of the emotional. At the same time, through its obscure actions and motivations, AI influences emotion and belief in humans at the expense of rational thought. An envoy, Japril, tells Marq, 'What can be talked of clearly, General Info can teach you in under three-tenths of a second…The rest one must mumble about, either loudly or quietly as is one's temperament' (Delany 1984, p.149). Japril suggests that the most beautiful and terrible experiences of life and the universe cannot be expressed through GI, an opinion repeated throughout the novel and a problem that we currently face in the race to automate. In order to feed information into a computer, it must be expressed as data, which often means quantitative measures. This excludes from consideration the qualitative measures that structure human emotion and play a crucial part in human decision-making. GI fails to capture the full gamut of human experience while blindly influencing the emotions of those with whom it interacts. Marq uses a quotation 'from an early critic of the GI system', who quips that GI does 'nothing to influence opinion... and everything to alter belief' (p.159). This might equally apply to our current predicament in which so-called fake news (once described as 'propaganda'), shaped by social media algorithms, has little role in encouraging factual understandings of complex issues while polarising those exposed to increasingly partisan versions of the truth. In this sense, AI really does take on a life of its own; while only (quantitative) data can be entered, the by-product of the demand for more interaction is heightened human emotion.

GI also appears to undermine the importance of memory in structuring individual and social identities. Marq shows concern that GI is altering cognition, as 'the more we come to rely on GI, the more our daily consciousness becomes subservient to memory's wanderings' (Delany 1984, p.330). This has proven to be the case in reality: psychology researcher Betsy Sparrow argues that the internet provides a kind of transactive memory as we rely on it to store memories outside of the brain. Our brains are not compelled to store memory that can be accessed via the internet at a later date (Sparrow 2011). This has an impact that might be considered worrying in general if we consider a loss of memory to mean we are cognitively compromised, but it has larger implications for cultural memory. Kodwo Eshun has defined Afrofuturism as part of the 'war of countermemory', the necessity to 'assemble countermemories that contest the colonial archive, thereby situating the collective trauma of slavery as the founding moment of modernity' (p.458). Delany's GI warns that reliance on an outside source for memory may mean that this process of countermemory is compromised due to reliance on the official record.

The novel's epilogue, during which Marq and Korga are separated, provides the most urgent appeal for knowledge outside of GI. Marq passionately explains to Japril the individuality of his worldview, particularly as that worldview is filtered through his unusual sexual proclivities, and how it has been damaged by his separation from Korga. He describes his outlook as being shaped by 'information beautiful yet useless to anyone but me, or someone like me, information with an appetite at its base as all information has, yet information to confound the Web and not to be found in any of its informative archives' (Delany 1984, pp.341–42). Marq's view of different kinds of information, and his emphasis on that which cannot be captured by GI, is very different from that found in *Neuromancer*. In that text, the pleasure of the information highways is drug-like, and fleshly experience pales in comparison. When Case has sex with Molly Millions he experiences his orgasm as, 'flaring blue in a timeless space, a vastness like the matrix' (p.45). For Case, the matrix is the most important experience of his life and sex can only remind him of that thrill. By contrast, in Delany's novel, the body and sexuality are recognised as offering their own kind of information, equally valid and more exciting than that offered by GI. GI is nothing to fear and fight against for Marq (as it is for the conservatives of Rhyonon), but it is something that can never include the most important aspects of life, because these cannot be written into code, or any other sign system.

Delany explicitly aligns his writing style with postmodernism, quoting Fredric Jameson's argument that science fiction is uniquely situated as a postmodern art form, incorporating many of the quintessential postmodern characteristics and viewing the 'fragmented subject as a "natural" condition' (1984, p.356). This is a stance that Delany explains he has tried to capture throughout his literary career. Delany's engagement with postmodernism is in contrast to that of Gibson; while Gibson attempts to express the complexity of world systems by evoking the networks of data powering global society, Delany privileges human identity as far more complex than those systems could ever be. The very fragmentation that creates individuality is what the Web seeks to flatten out through the equitable availability of information, but human proclivities (brought into focus here through sexual fetishes and preferences) cannot be flattened out. Indeed, they find their frisson only in their individuality and the impossibility of their being captured by the binary systems of structural language or computer code.

This is a problem that we currently face through the development of machine learning tools that seek to separate humans or human images into discrete groups (male or female, black or white) and, in doing so, reinforce pre-existing social stereotypes or prejudices. An example of such a 'racist algorithm' was reported widely when a Google AI was mistakenly tagging pictures of black people as 'gorillas', a problem that the company side-stepped rather than solved by removing 'gorilla', and some other primates, as possible tags in its AI's lexicon (Simonite 2018). Similar problems have been reported for facial recognition software that fails to see black people, or that regularly misgenders trans people (Eveleth 2016). The corruption of Tay, a chatbot released into the Twittersphere by Microsoft is also a modern internet morality tale about the potential for AI to take on the prejudice found in the world around them. Tay learned from interactions with other Twitter users and had to be shut down within twenty-four hours because its tweets became so racist, sexist, and offensive (Vincent 2016). AI replication of existing prejudices continues a long history of discrimination that Afrofuturism seeks to uncover by acting as a reminder that the traumas of the past will always affect the future. Delany's queer, Afrofuturist cyberpunk intervention is part of a movement that has continued with, for example, Nalo Hopkinson's *Midnight Robber* (2000), in which a Caribbean-colonized planet has a 'Nansi web', named for the folklore hero Anansi, who appears in the guise of a spider. Hopkinson adapts cyberpunk to explore cultural and personal abuse and trauma and to depict the communitarian links that can be forged from such dark pasts. More recently, Janelle Monáe's album and accompanying short film (or 'emotion picture'), both called *Dirty Computer* (2018), explore queer sexualities in the era of Black Lives Matter. The impossibility of separating human life into binaries that can be understood by AI algorithms inevitably impacts those pushed to the margins of normative society: people of colour, and those who challenge sexual and gender norms. Queer and Afrofuturist uses of the cyberpunk toolkit explore and expose the particular issues AI raises for such groups, while situating these contemporary problems in a longer history of modernity.

## 12.4 Cyberpunk Spreads: Cinema and Activism

Cyberpunk literature describes cyberspace in distinctly visual ways, and this visualization lent itself to expression in film and other visual media. Cyberpunk's significance to mainstream cinema, in particular, led to the popularisation of cyberpunk imagery beyond the literary community (Murphy & Schmeink 2017). In Hollywood this is seen in *TRON* (Kushner 1982), with its representation of cyberspace as a physical space for human interaction. It is also found in Kathryn Bigelow's *Strange Days* (1995), which adopted the neo-noir mise en scène of *Blade Runner* (Scott 1982) and the hard-boiled influences of *Neuromancer* to explore virtual reality and its moral and political implications. However, it was *The Matrix* (Wachowski & Wachowski 1999) that really captured the popular imagination. In *The Matrix* all of what we know as reality is, in fact, a complex virtual reality created by post-Singularity artificial intelligence in order to keep humans docile so that they can be used as biological batteries to power the AI, 'a singular consciousness that spawned an entire race of machines'. Breaking free of this imposed narrative is crucial to living an authentic life, and this involves 'seeing' the virtual reality for what it is and learning to manipulate it. The data that make up the matrix is visualized throughout by a series of neon green glyphs that stream down the screen in a 'digital rain', reminding the viewer that the world of the film exists as data encoded by machines. Telephones within the matrix act as portals through which the characters can move, transferred like so much data down the telephone line, from the world of pure data to the 'desert of the real' beyond. In the 'real' world, 'jacking in' to the matrix is viscerally depicted as humans connect to the matrix via a metallic probe inserted into a socket on the back of the neck. This is an image of the human–machine interface that captures and exaggerates Gibson's vision, depicting the human connection with cyberspace via the invasion of the body through electrodes and artificial orifices. Through this representation of 'the matrix', the film dramatizes the technology of cyberspace, a technology that for many remains equivalent to magic in the sense that it works invisibly and by means beyond their understanding. The film's representation of the matrix gave popular culture a visual language for thinking about cyberspace and was influential as a modern classic of paranoia and conspiracy theory. Neo must decide whether to live his life in the matrix, or whether to 'wake up' and understand the truth of the world in which he lives, a choice offered to him by Morpheus (named for the god of dreams) between a blue pill (which will see Neo returned to his normal life of drudgery) and a red pill (which will let him see the world as it truly is). The metaphor of the red pill has since become internet shorthand for seeing through the mainstream narrative and understanding a deeper 'truth', often a kind of conspiracy. Men's rights activists in particular have been known to refer to themselves as 'red-pillers' because, they argue, they see through the feminist propaganda that they believe controls society, but the term has been used in many contexts to describe conspiracy theorists. This makes *The Matrix* significant as not only the most successful cyberpunk film, but also generative of a framework for finding a paranoiac 'truth' behind the networks that define globalized society.

While cyberpunk went mainstream with the incredible popularity of *The Matrix* and its two sequels, its legacy was also infiltrating the wider culture through the influence of real-life cyberpunks. As computing technology became more central as a means of data storage and as a venue for systems including finance and the military, cybercrime became an increasing concern, and cyberspace was recognised as a real place where actions committed have real consequences. Governments were slow to react to this new landscape, contributing to the cyberpunk understanding of the world as post nation state, with international systems finding loopholes and staying one step ahead of legislation on the new frontier. Sterling documents the cross-pollination between cyberpunk and the hackers of the early 1990s in *The Hacker Crackdown* (1992). This permeable membrane between cyberpunk fiction and the 'real-life cyberpunks' of computer science has continued to the present day. In his review of cyberpunk's legacy, *The Souls of Cyberfolk* (2005), Foster goes so far as to argue that 'cyberpunk didn't so much die as experience a sea change into a more generalized cultural formation' (2005, p.xiv) appearing in what we might refer to as 'cyberpunk activism'. Most recently this has resulted in a series of cyberpunks and post-cyberpunks speaking out about the ways in which artificial intelligence and machine learning are hijacking public discourse. For example, the cyberpunk activist and virtual reality pioneer Jaron Lanier has recently written a critique of algorithms, through a polemic attack on the moral and economic bases of contemporary social media in *Ten Arguments for Deleting Your Social Media Accounts Right Now* (2018). Lanier is considered an authority on the workings of the internet, having been part of the team that scaled the internet to make it function as Web 2.0. He has become widely known as an outspoken cultural critic with a love of the internet matched only by his distrust of its current uses. Lanier argues that much of contemporary social media (including Google, Facebook, and Twitter) make their money as BUMMER machines, an acronym which stands for 'Behaviours of Users Modified, and Made into an Empire for Rent' (2018, p.28). While Lanier obviously coined this acronym with brevity and entertainment in mind, his focus on the behavioural modification of social media users as the core of the many problems experienced by internet users today (trolling, polarizing political opinions, and the abuse of social media platforms to shape voting patterns and even the emotions of social media users) illuminates some of the core problems with algorithms in social media. Lanier argues convincingly that the use of algorithms changes the behaviour of social media users, whether through encouraging them to click on an advert, showing them content that will produce an addictive response, or encouraging the user to stay on the platform, thereby forcing them to continue absorbing its advertising content. Behaviour modification results in an addictive medium that rewards negative behaviour (which garners more attention for the user). This model is an economic travesty as the skills and data of the multitude are used to generate 'content' and wealth for a handful of Silicon Valley billionaires, while those very same people are treated as obsolete in the information economy. For our current purposes, the most important of Lanier's ten arguments is Argument Number 10: Social Media Hates Your Soul, in which he shows some of the overlaps between a society ruled by algorithms and a transhumanist reimagining of religious power structures. Lanier highlights Facebook's mission statement to ensure that 'every single person has a sense of purpose and community', a goal that once would have been considered the remit of religious communities, and gives as an example the profiles of deceased users which remain live as online shrines to the dead. Lanier also points out that Google has funded projects to 'solve death', a move which he describes as 'such a precisely religious pretension that I'm surprised the religions of the world didn't serve Google with a copyright infringement take-down notice' (2018, p.133). The dangers of entrusting significant human experiences, such as death and religion, to corporations or black-box algorithms is reminiscent of Delany's warning to maintain the humanism of memory and of information, respecting that not all human functions can be translated into data.

In criticizing social media and its behaviour-modifying algorithms, forever tweaked to maximize user experience as an end in itself with no thought to what happens within the black box of these algorithms, or what unintended consequences such behaviour modification might have in human societies, Lanier saves some of his harshest criticism for the concept of 'AI', a banner under which such algorithms operate. Lanier accuses Silicon Valley programmers of having begun to believe their own press when it comes to artificial intelligence:

> We forget that AI is a story we computer programmers made up to help us get funding once upon a time . . . it was pragmatic theatre. But now AI has become a fiction that has overtaken its authors. AI is a fantasy, nothing but a story we tell about our code. It is also a cover for sloppy engineering. Making a supposed AI program that customizes a feed is less work than creating a great user interface that allows users to probe and improve what they see on their own terms – and that is so because AI has no objective criteria for success. (2018, p.135)

In Lanier's analysis, the term 'AI' is used as a narrative to enhance the cultural, financial, and intellectual capital of computer programmers. AI as a narrative is suggestive of a higher intelligence at work when, in fact, there is no intelligence involved in the algorithmic responses to human behaviour found on social media platforms. The talented mathematicians and engineers who initially designed the algorithms leave them to work away in a feedback loop of increased user time spent on the platform, or increasing clicks on advertised posts, without reference to the outside world and to the wider problems that behaviour modification may provoke.

Cyberpunk has made it easier to visualize AI and the networked systems that have increasingly come to dominate global social spaces and politics by spatializing the landscapes of machine code into neon lattices of information and scrolling, green digital rain, or showing the invasion of the network as literal bodily and mind invasion. However, visualization is insufficient when political action is necessary to understand AI, to regulate it where necessary, and to understand the freedoms it may impinge upon when our data is held by private corporations and used in opaque ways. In his recent survey of some of the symptoms of the network and its algorithms, *New Dark Age: Technology and the End of the Future* (2018), James Bridle argues that the metaphor of throwing light onto difficult or dangerous topics does not necessarily work in an era of surveillance, when humans tend to have problems with sifting through the vast amounts of available data rather than with collecting that data in the first place. For companies like Google, Bridle argues, their 'worldview is one that is entirely predicated on the belief that making something visible makes it better, and that technology is the tool to make things visible. This view, which has come to dominate the world, is not only wrong; it is actively dangerous' (Bridle 2018, p.242). Bridle argues that several atrocities, including the Rwandan genocide of the 1990s, took place in environments where the atrocities were well documented and recognized, but where the ability to act seemed to be suspended. Having the data is not enough to prevent atrocities when those with the data can see no route towards, or have no incentive to move towards, effective action.[1]

Cyberpunk activism raises the alarm about the importance of machine learning AI to our contemporary society, and these themes continue to be explored in literature through post-cyberpunk, that is, cyberpunk literature that continues with some of the themes of the original movement but retools the tropes to reflect the current intersections between society and AI (if, indeed, those intersections can still be envisaged in an era where AI and society are inseparable). Once again, Gibson's work is paradigmatic. His Blue Ant trilogy, published over the first decade of the twenty-first century, is fascinated with the influence of the military on civilian life, whether on developments in fashion or on surveillance technology. The trilogy is set in the present (the first of Gibson's works not set in a recognizably science-fictional future), reaffirming cyberpunk's goal—to act as a toolkit for negotiating the present day, rather than as a means of predicting the future. This trilogy inspired Veronica Hollinger to coin the term 'science fiction realism' (2006) to describe works that use science-fictional motifs in order to analyse the present state of technology. The first of these novels, *Pattern Recognition* (2003), uses the term for a branch of machine learning as the novel's title, while the trilogy's second novel, *Spook Country*, shows the development of cyberspace from the 'world beyond the screen' imagined in *Neuromancer* to a realm indistinguishable from the real. In *Spook Country*, the knowledge that actions in virtual reality have consequences in the 'real' world is dramatized through the novel's exploration of locative art, an art form that uses geospatial tagging to create artworks linked to real space and viewable through virtual reality technology, or through the use of smartphones. Locative artists Odile Richards and Alberto Corrales explain to Hollis, the protagonist, the logic of cyberspace on which locative art depends. They explain to her that cyberspace can no longer be considered a separate realm, that it is 'everting' (sic), mapping itself over reality:

> 'See-bare-espace', Odile pronounced, gnomically, 'it is everting.'
>
> ' "Everything"? What is?'
>
> 'See-bare-espace', Odile reaffirmed, 'everts.' [...]
>
> 'Turns itself inside out', offered Alberto, by way of clarification.
> ' "Cyberspace".' (Gibson 2007, p.21)

The artists claim that cyberspace has turned itself inside out so that the logic and characteristics of cyberspace have found their way into the real world, rather than being confined to another realm as they were in the matrix of *Neuromancer* (Gibson 1984). This means that cyberspace can no longer be spatialized and visualized as a separate realm, but must be understood as co-existing with the 'real' world. When Hollis meets another artist-cum-coder, Bobby Chombo, he tells her:

> Once it everts, then there isn't any cyberspace, is there? There never was, if you want to look at it that way. It was a way we had of looking where we were headed, a direction. With the grid, we're here. This is the other side of the screen. Right here.
>
> (Gibson 2007, p.67)

In reimagining cyberspace for a time when actions taking place on the internet have significant consequences in the real world, Gibson takes a lesson from cyberpunk activism and concedes that the neon lattices stretching into the distance that he described in his earlier work are no longer useful for understanding the impact of machine learning on social and individual behaviour. Hollis's mishearing of Odile's term 'everting' is telling as cyberspace is now 'everything' and has spread into the real world.

## 12.5 Cyberpunk as a Cultural Mode

There are other, more utopian uses of machine learning in post-cyberpunk literature. Cory Doctorow is every inch the successor of the classic cyberpunks as both a writer of science fiction and a participant in the internet and legal landscapes currently shaping the use of technology. Doctorow campaigns against digital rights management software (designed to protect copyrighted intellectual property) which prevents the information available online from being traded and used freely. In his science fiction, the black box of networked systems produces unintended consequences, such as in *For the Win* (2010), which investigates the world of exploitative gold mining that sprang up in China in reaction to the monetary value of playing online games such as *World of Warcraft*. *For the Win* imagines that such global exploitation via the internet might result in the international unionisation of workers who all find themselves exploited under the same system. Doctorow continues to represent utopian thinking in *Walkaway* (2017). The utopian thread in Doctorow's literature, like his political work, is based on open source software (and hardware, via 3D printing). Doctorow sees the proliferation of black-box systems as characteristic of a repressive society, and his utopian writing includes blueprints to eradicate black boxes from public life without giving up global networking. His ideas include open source software that is available to all (rather than closed systems protected as intellectual property or under the auspices of commercial confidentiality), democratic decisions about the use of technology via open wikis, and a communist sensibility in the distribution of tech and other kinds of property.

Cyberpunk has been instrumental in providing a language and a set of imagery for the black box that is the computer, the network, and its algorithmic logic. It dramatizes the role of the human as a node in the network and emphasizes the behaviour modification that can be caused by AI through the theme of mind invasion. In cyberpunk, human intelligence cannot be thought of as independent from artificial intelligence, as the data we provide circulates through the algorithms, producing something new, a hybrid of the two. As time has gone by, however, the visualization employed by classic cyberpunk has proven less useful; cyberpunk has bled into hacking and activism, and it informs activist responses to the privatization of cyberspace, while post-cyberpunk seeks to represent these trends and to argue that there must be more democratic ways of organizing cyberspace as a commons. The continuing development of cyberpunk narratives in the twenty-first century shows that offering new ways of conceptualizing the network is an ongoing project; one that post-cyberpunk texts are working to realize, and one that becomes increasingly urgent as the consequences of black-box systems have significant effects on the global political landscape and on the lives of individuals, especially those who already find themselves oppressed or forgotten by the opaque machinery of information networks.

## Note

1. The contrast between the secrecy of social media companies' algorithms and the sharing they encourage their users to perform is satirized in Dave Eggers's 2013 novel *The Circle*.

## Reference List


Adams, T. (2007) Space to think. *The Observer*. 12 August. Available from: https://www.theguardian.com/books/2007/aug/12/sciencefictionfantasyandhorror.features [Accessed 16 July 2018].

Banks, I. M. (1996) *Excession*. London, Orbit.

Bethke, B. (1983) Cyberpunk. *Amazing Science Fiction Stories*. 51, 94–105.

Bridle, J. (2018) *New dark age: technology and the end of the future*. London, Verso.

Delany, S. R. (2004) *Stars in my pockets like grains of sand*. Middletown, CT, Wesleyan University Press.

Dery, M. (1994) Black to the future: interviews with Samuel R Delany, Greg Tate, and Tricia Rose. In: M. Dery (ed.) *Flame wars: the discourse of cyberculture*. Durham and London, Duke University Press. pp.179–222.

Domingos, P. (2015) *The master algorithm: how the quest for the ultimate learning machine will remake our world*. London, Allen Lane.

Dozois, G. (1984) Science fiction in the eighties. *Washington Post*. December 30. Available from: https://www.washingtonpost.com/archive/entertainment/books/1984/12/30/science-fiction-in-the-eighties/526c3a06-f123-4668-9127-33e33f57e313/?noredirect=on&utm_term=.c7189fd3628a [Accessed 16 July 2018].

Ellison, H. (1967) *Dangerous visions*. London, Gollancz.

Eshun, K. (2017) Further considerations on Afrofuturism in Rob Latham (ed.), *Science fiction criticism: an anthology of essential writings*. London, Bloomsbury. pp. 458–69.

Eveleth, R. (2016) The inherent bias of facial recognition software. *Motherboard*. 21 March. Available from: https://motherboard.vice.com/en_us/article/kb7bdn/the-inherent-bias-of-facial-recognition [Accessed 1 May 2019].

Foster, T. (2005) *The souls of cyberfolk: posthumanism as vernacular theory*. Minneapolis, University of Minnesota Press.

Gibson, W. (1984) *Neuromancer*. London, HarperCollins.

Gibson, W. (1995) *Burning chrome*. London, HarperCollins.

Gibson, W. (2007) *Spook country*. London, HarperCollins.

Gilroy, P. (1994) Living memory: a meeting with Toni Morrison. In: P. Gilroy, *Small acts: thoughts on the politics of black cultures*. London, Serpent's Tail. pp.175–82.

Hollinger, V. (2006) Stories about the future: from patterns of expectation to pattern recognition. *Science Fiction Studies*. 33(3), 452–72.

Hopkinson, N. (2000) *Midnight robber*. New York, Grand Central Publishing.

Jameson, F. (1991) *Postmodernism: or, the cultural logic of late capitalism*. London, Verso.

Lanier, J. (2018) *ten arguments for deleting your social media accounts right now*. London, The Bodley Head.

Lavender III, I. (2011) *Race in American science fiction*. Bloomington, Indiana, Indiana University Press.

Le Guin, U. K. (1979) *The language of the night: essays on fantasy and science fiction*. New York, Ultramarine Publishing.

Le Guin, U. K. (1981) *The left hand of darkness*. London, Orbit.

Murphy, G. J. & L. Schmeink (eds.) (2017) *Cyberpunk and visual culture*. New York, Routledge.

Nixon, N. (1992) Cyberpunk: preparing the ground for revolution or keeping the boys satisfied? *Science Fiction Studies*. 19(2), 219–35.

Scott, R. (1982) *Blade Runner*. Warner Brothers.

Simonite, T. (2018) When it comes to gorillas, Google photos remains blind. *Wired*. 1 November. Available from: https://www.wired.com/story/when-it-comes-to-gorillas-google-photos-remains-blind/ [Accessed 1 May 2019].

Sparrow B. et al (2011) Google effects on memory: cognitive consequences of having information at our fingertips. *Science*. 333(6043), 776–78.

Sterling, B. (1992) *The hacker crackdown*. New York, Bantam Books.

Sterling, B. (ed.) (1994) *Mirrorshades: the cyberpunk anthology*. London, HarperCollins.

Vincent, J. (2016) Twitter taught Microsoft's AI chatbot to be a racist asshole in less than a day. *The Verge*. 24 March. Available from: https://www.theverge.com/2016/3/24/11297050/tay-microsoft-chatbot-racist [Accessed 2 May 2019].

Wachowski, L. & L. Wachowski (1999) *The Matrix.* Warner Brothers.

# 13

# AI: Artificial Immortality and Narratives of Mind Uploading

*Stephen Cave*

## 13.1 Introduction

Artificial intelligence—at least in our imagination—promises to be a master technology: the one that will unlock all others. This overarching aspect of the AI project has been neatly summarized by Demis Hassabis, founder of the technology company DeepMind, thus: 'solve intelligence, and then use that to solve everything else'. In a 2016 interview, he lists 'cancer, climate change, energy, genomics, macroeconomics, financial systems, [and] physics' as fields ripe with challenges that could be addressed by sufficiently powerful, general AI. 'What we're working on is potentially a meta-solution to any problem', he concludes (Burton-Hill 2016).

Conceived of in this way, AI appears to be a kind of Holy Grail, and the motivations for its pursuit are as many and varied as our needs and desires themselves. But in this chapter, I will explore one motivation in particular: one often associated with the true Holy Grail—the desire to thwart death. Just as AI is a kind of meta-technology, this is a kind of meta-motivation, a prerequisite for the achievement of (nearly) all other goals.

In this chapter, I begin in Section 13.2 by sketching the spectrum of ways in which intelligent machines could be deployed toward the goal of achieving radically longer, or indefinitely long, lives. I then in Section 13.3 focus on one of those ways: mind uploading, examining how influential works of nonfiction, in particular, those of Hans Moravec and Ray Kurzweil, have framed this possibility. In Section 13.4, I pay particular attention to these writers' use of narrative to enhance the plausibility of their claims. I then, in Section 13.5, examine how works of fiction, including those of William Gibson, Greg Egan, Robert Sawyer, Pat Cadigan, Rudy Rucker, and Cory Doctorow, offer problematized accounts of such claims. I focus on three problems in particular, which I call the problems of identity, stability, and substrate. In conclusion, I contend that, given the importance of thought experiments—which are a variety of speculative story—in developing the argument for the philosophical and technological plausibility of mind uploading, science fiction literature offers a particularly important site of critique and response.

The works I consider span 1982 to 2017. They therefore all come from what is sometimes called the digital age, defined by the Cambridge English Dictionary as 'the present time, when most information is in a digital form, especially when compared to the time when computers were not used' (n.d.). The online resource Techopedia considers this era to have started 'during the 1980s' and to be synonymous with the 'information age' (n.d.). The 1980s are significant as the period in which computers came into widespread usage in the home and workplace, which consequently started the mass digitization of data, and indeed of many aspects of ordinary life, from shopping to social interaction. What we see in the works under consideration is that this period also ushered in the digitization of long-standing dreams of immortality.

## 13.2 AI and the Problem of Death

The fear of death plays a major role in shaping our beliefs. This could perhaps be inferred from the plethora of immortality belief systems, which can be found in almost every recorded culture.[1]

But in addition, there is a substantial body of empirical research supporting the thesis that people are strongly motivated to pursue strategies or subscribe to beliefs that promise to deny death, whether through indefinite postponement of an end to this life or through the conviction that this life is only the prelude to another. This thesis, known as terror management theory, is supported by over four hundred studies (Kesebir & Pyszczynski 2011). Its main claim is that our diverse worldviews have evolved 'to manage the terror engendered by the uniquely human awareness of death' (Solomon et al 1998, p.12).

It is therefore to be expected that a worldview based on the efficacy of science and technology would come with its own forms of death denial. At least since Francis Bacon, the techno-scientific project has been associated with the attainment of mastery over nature, and death is the aspect of nature that most affronts human will (Bauman 1992). Correspondingly, a number of historians have identified promises of death's defeat as a recurring theme in the rhetoric of science and technology, from the exaggerated hopes for hormone treatment to the supposedly curative power of vitamins.[2] A number of studies within the terror management theory paradigm have demonstrated that beliefs in scientific and technological progress in general, and the possibility that this will result in life extension in particular, are at least partly motivated by the fear of death (Farias et al 2013; Lifshin et al 2018; Rutjens et al 2009). If AI is, as noted, perceived as the master technology—'the last invention that man need ever make' as I. J. Good put it (1966)—then it should come as no surprise that we find it employed in plans to end mortality.

Like all immortality strategies, those based on AI must solve the problem of the body. No matter how well we take care of ourselves, how much we strive to stay fit and healthy, our bodies inexorably degenerate and fail. For someone seeking immortality, the body must therefore either be *transformed* into something that can withstand the rigours of time, or *transcended* altogether.

Broadly speaking, the technological project has focused on the first of these: the project of transformation—for example, through inoculation, or prostheses such as hip replacements. AI, however, holds out to the imagination the possibility of both transformation and transcendence.

In the AI-based version of the former, the body is transformed from its evolved state of biological fragility into something more durable. Its premise is that we are sophisticated machines—and like other machines, when we go wrong, we can be fixed[3]. But in the process of fixing, we are transformed: our vulnerable organic matter is replaced with machine parts that perform the same function but without the frailties of flesh—this is cyborgization (Clynes & Kline 1960). Like the ship of Theseus, we will change while staying the same: we might lose all our original matter, but the fundamental processes of life continue unbroken. The cyborg is transformed from organic to inorganic: it is a transformation in which life becomes (intelligent) machine.

In the process of transcendence, we see something like the opposite. In contrast to the gradual transformation of the cyborg, transcendence posits a leap into another state; a leap through which the body is cast off like worn clothes. As Kevin LaGrandeur puts it, 'the ultimate dream of posthuman cybernetics is to transcend bodily needs by making them obsolete—by converting the human organism to pure, distributed mind' (2013, p.156). As in older visions of reincarnation, this requires that the self that wishes to survive has a new repository, a new form beyond the original body, waiting to be filled. This is the machine capable of instantiating a human mind: a blank-slate AI awaiting the specific memories, beliefs, and desires that will mould it into a particular person. And so, in such accounts, frequently described as 'mind uploading', the AI instantiates that person. Here, the computer—the inanimate, inorganic—becomes animate; through taking on their attributes, it becomes the one who is striving to survive. Through transcendence, (intelligent) machine becomes life.

This distinction parallels what Robert M. Geraci calls the 'traditional two-stage apocalyptic scenario' for the attainment of an eternal realm (hence he terms the phenomenon 'Apocalyptic AI'):

> Just as many of the ancient apocalypses anticipated that a period of peace and justice would reign on the earth prior to God's final dissolution of the world and establishment of an eternal realm of goodness, Apocalyptic AI anticipates that advances in robotics and AI will create a paradise on Earth before transcendent Mind escapes earthly matter in an expanding cyberspace of immortality, intellect, moral goodness, and meaningful computation.
> (2010, p.31)

In this account, which Geraci ascribes to a number of influential techno-utopians, what I have described as the transformation process of cyborgization occurs in 'the Age of Robots', which precedes 'the Age of Mind' and the attainment of transcendence through mind uploading.

However, it is important to nuance this scenario in two ways. First, while Geraci (following both Moravec and Kurzweil) suggests these two stages will happen sequentially, mind uploading is also being pursued directly now: some are not willing to wait until the 'Age of Robots' has run its course. Second, although it is possible to clearly distinguish cyborgization from mind uploading, they can nonetheless also be seen as a spectrum. At one end is the modestly enhanced human, minimally cyborgized to increase longevity (such as a person with a pacemaker). At the other is the fully transcended digital avatar, no longer dependent on flawed and fragile biology. In between, on this ontological spectrum, are many degrees of transmutation. In particular, once cyborgization breaks through the skull, the boundary with transcendence blurs: we could, for example, imagine a cyborg's biological brain being gradually replaced by machine parts, until its components were wholly inorganic. At that point, what we once called a brain, we might call a whole-brain-emulating computer. And thence, if we lay aside some philosophical and technical challenges, we could imagine the mind would effectively have been uploaded and would be transferable to other bodies (fleshy or silicon), or even to cyberspace.

There is much to be said about the role of narrative in the full spectrum of AI-based immortality strategies: much more than would fit in a single chapter. So for the remainder of this piece, I will focus on mind uploading. In part, this is because it has received a good deal of popular and literary attention,[4] and in part, because—as we will shortly see—it is seen by some as the apotheosis of technological immortality strategies.

## 13.3 Mind Uploading: The Promise of Technological Transcendence

The idea that we might transfer our minds onto a more durable and augmentable substrate has been advocated in this period (c.1982 to the present) by a number of figures who have combined influential careers in the science and technology of artificial intelligence with success as popular writers. These include Hans Moravec, former research professor in the Robotics Institute of Carnegie Mellon University and director of the Mobile Robot Laboratory from 1980–2005; Marvin Minsky, from 1958 to his death in 2016, a faculty member at MIT, where he co-founded and directed the AI laboratory; and Ray Kurzweil, who, at the time of writing, is Director of Engineering at Google and an author whose many books have been international bestsellers. I will examine their works shortly, but first it is worth considering the foundations on which their narrative was built.

In *The Religion of Technology: The Divinity of Man and the Spirit of Invention*, David F. Noble explores the long history of the relationship between hopes for AI and long-standing fantasies of immortality (1997, ch. 10). He notes that aspects of the otherworldly spirituality of medieval Christian theology prepared the way, then identifies René Descartes's particular form of dualism as a turning point in making the dream of digital immortality possible: 'Descartes perceived the mind as mankind's heavenly endowment and, in its essence, distinct from the body, the burden of mortality' (Noble 1997, p.143). This distinction opened the possibility of the immortal mind being freed from 'the prison of the body' (p.144).

This dualism, argues Noble, in turn led to the development of computational models of the mind in the nineteenth and twentieth century. While keeping it still separate from the body, such models gradually shifted the locus of our essential, mental life from the realm of the spiritual—the god-given spark—to the realm of the natural. This 'reduction of human thought to mathematical representation made imaginable the mechanical simulation or replication of the human thought process' (Noble 1997, p.148). From this view that human thought could be simulated, it was a short step to believing that the mind of a particular individual 'could take form in a new, ultimately more durable medium . . . a machine that would provide a more appropriately immortal mooring for the immortal mind' (Noble 1997, p.148).

But this highly hypothetical postulation could not expect to win many advocates without a plausible theoretical framework for how a machine could realize something so sophisticated as a human mind. Such a framework was offered by Alan Turing, when he proved the possibility of a universal computing machine that could perform all possible functions that could be calculated (1937). Turing himself suggested this could lead to the development of intelligent machines and—in what is now known as the Turing test—even that human-likeness was a key measure of such intelligence (Turing 1950). Of course, a human-like, intelligent machine would be just what would be required to simulate an individual human mind. Noble shows how Marvin Minsky and Hans Moravec drew just this conclusion to posit the possibility of survival via mind uploading.

In his influential 1988 book *Mind Children: The Future of Robot and Human Intelligence*, Moravec describes what he calls 'a postbiological world dominated by self-improving, thinking machines' (p.5). Such a world is not, in his view, to be feared, for it heralds the fulfilment of all humanity's long-standing wishes. Foremost among these is the wish to avoid oblivion, which is to be achieved by escaping the body:

> It is easy to imagine human thought freed from bondage to a mortal body -- belief in an afterlife is common. But it is not necessary to adopt a mystical or religious stance to accept the possibility. Computers provide a model for even the most ardent mechanist. A computation in progress -- what we can reasonably call a computer's thought process -- can be halted in midstep and transferred, as program and data read out of the machine's memory, into a physically different computer, there to resume as though nothing had happened. Imagine that a human mind might be free from its brain in some analogous (if much more technically challenging) way. (Moravec 1988, p.4)

The comparison of human minds with computers transpires to be more than the analogy that Moravec suggests. Like Minsky and Kurzweil, he subscribes to what Schneider and others call a 'software model of the mind' (Schneider 2019), which permits him to hypothesize that an individual's mind is indeed a kind of programme that can run on alternative, nonbiological substrates—or in the terminology of this chapter, uploaded. A person could have multiple uploads made, providing further insurance against destruction. 'With enough widely dispersed copies, your permanent death would be highly unlikely' (Moravec 1988, p.112).

This line of thought has been developed extensively in the speculative nonfiction works of Ray Kurzweil. Since his first book, *The Age of Intelligent Machines* (Kurzweil 1990), he has enthusiastically explored the more extreme possibilities of the digital age, including radical life extension. He recognizes that the technological capacity to achieve mind uploading is not currently available, so some of his work, for example, *Fantastic Voyage: Live Long Enough to Live Forever* (Kurzweil & Grossman 2004), advocates a medical and lifestyle regime that would ensure one is able to survive until this revolution comes. In *The Singularity Is Near: When Humans Transcend Biology*, he then lays out the stepping stones to transcendence: 'three overlapping revolutions': in genetics, nanotechnology, and AI/robotics (Kurzweil 2005). The first two revolutions, genetics and nanotechnology, will allow people to reprogram their bodies 'molecule by molecule' (p.206) until they are effectively immune to ageing and disease and have many new powers.

The apotheosis is the third revolution, the creation of super-intelligent machines. Such machines will of course be so smart that they could solve any lingering problems with regard to re-engineering the human body. But more importantly, they will enable humans to choose to leave the body behind, transferring themselves onto 'a different—most likely much more powerful—computational substrate', whence they 'will live out on the Web, projecting bodies whenever they need or want them, including virtual bodies in diverse realms of virtual reality, holographically projected bodies, foglet-projected bodies, and physical bodies comprising nanobot swarms…This is a form of immortality' (Kurzweil 2005, pp.324–25). This is the ultimate realization of the techno-scientific project of mastering nature: 'We will gain power over our fates. Our mortality will be in our own hands. We will be able to live as long as we want' (p.9).

It is reasonable to believe that the oft-cited works of these three advocates have contributed significantly to disseminating the idea of mind uploading. Indeed, in the novel *Mindscan*, which I will consider later, Robert J. Sawyer cites Kurzweil as 'the most vocal proponent…of moving our minds into artificial bodies' (Sawyer 2005, p.41). At the same time, broader technological developments within the digital age have made their visions appear more credible. It is beyond the scope of this chapter to consider the various proposed technological routes to mind uploading. But in essence, they must solve three challenges: scanning all the information in a human mind, analysing and storing it, and then turning it into a digital person. One of the most notable developments of the digital age is the rapid increase in the amount of data—particularly personal data—that is generated, stored, and processed. It was estimated in 2018, that 2.5 quintillion bytes of data were being produced every day, and that 'over the last two years alone 90 percent of the data in the world was generated' (Marr 2018). The extent to which this data now measures and records our lives has given further impetus to the notion that the entire contents of a person's mind could be recorded, digitized, and recreated.[5]

Consequently, an increasing number of people are taking the visions of advocates like Moravec and Kurzweil seriously. A number of recent surveys have noted the increase in services offering some form of proto–mind uploading. Öhman and Floridi (2017) coined the phrase 'digital afterlife industry' to include a spectrum of services from those that manage the social media accounts of the deceased, to those that attempt to recreate a dead person through the data they leave behind. In their paper 'Digital Immortality and Virtual Humans', Savin-Baden and Burden (2019) explore some of the latter services, those 'that are actively trying to create computer applications which are predicated on the creation of avatar-level digital immortalisation,' including Eter9, Lifenaut and Eternime.[6] At the same time, others have noted that the technological pursuit of immortality is a major preoccupation of the very wealthy in Silicon Valley (Singh-Kurtz 2019). Even if it is difficult to know exactly how many, it is therefore clear that at least some people find these stories about the possibility of AI-enabled immortality convincing.

## 13.4 Narratives of Mind Uploading in Speculative Nonfiction

What makes belief in mind uploading compelling? It seems unlikely that many people would be convinced simply by the cases for either technological or philosophical plausibility, as both are hotly disputed. Leading MIT roboticist Rodney Brooks wrote recently of the challenge of digitally realizing a human mind: 'if you are going to rely on the Singularity[7] to upload yourself to a brain simulation I would try to hold off on dying for another couple of centuries' (2017). Of course, there are those who disagree. But of the three big challenges mentioned above—scanning a human brain, analysing and storing the data, and then reanimating it into a person—none are close to being solved. The philosophical challenges are also unlikely to be resolved any time soon (Schneider 2019, ch. 6). So prima facie plausibility cannot be the explanation of why interest in mind uploading remains high. One alternative reason might be that the fear of death can distort our reasoning, leading us to give credence to views to which we otherwise would not, simply because they help us cope with mortality. I explore this possibility in the article 'AI and the Denial of Death' (Cave forthcoming). What I will discuss now is a third possibility, that mind uploading is made to seem both plausible and attractive because of certain narrative or rhetorical devices used by its proponents.

The first notable feature we can call the *Aladdin's Lamp* narrative: that is, the way in which advocates of mind uploading promise that it will herald a time of boundless wish fulfilment. Moravec, for example, talks about using mind uploading in order to become 'full, unfettered players in the superintelligent game' and as allowing us to 'share in the magical world to come' (1988, pp.108–09); once superintelligent, we will be equipped with 'wonder-instruments' (p.122) that will allow us to visit distant stars or even resurrect the dead. Kurzweil goes further still. We already noted above his claim that 'our mortality will be in our own hands'; 'ultimately, the entire universe will become saturated with our intelligence… We will determine our own fate rather than have it determined by the current "dumb", simple, machinelike forces that rule celestial machines' (Kurzweil 2005, p.29), and later in the same work: 'There will be a vast expansion of the concept of what it means to be human. We will greatly enhance our ability to create and appreciate all forms of knowledge from science to arts, while extending our ability to relate to our environment and one another' (p.397).

The second notable narrative we can call *Exponentialism*. By this, I mean the tendency to sweep aside the technical challenges of mind uploading with the argument that the rate of technological progress is increasing in a way that makes it inevitable that all problems will be solved. This was a view advocated by Moravec in the 1980s:

> We evolved at a leisurely rate, with millions of years between significant changes. Machines are making similar strides in mere decades. When multitudes of economical machines are put to work as programmers and engineers, presented with the task of optimising the software and hardware that makes them what they are, the pace will quicken. Successive generations of machines produced this way will become smarter and less costly.
>
> (1988, p.100)

Since then, Kurzweil has become a well-known and influential proponent of the power of exponential increase in technological capacity: this is indeed the explicit premise of much of his work. As he puts it in *The Singularity is Near*, 'the key idea underlying the impending Singularity is that the pace of change of our human-centred technology is accelerating and its powers are expanding at an exponential rate' (Kurzweil 2005, pp.7–8).

The third device worth highlighting we can call *Pronominal Continuity*. This narrative sweeps aside the most important philosophical challenge to mind uploading, which I will refer to as the 'identity problem'. This is the worry that the person who emerges from a mind-uploading process is not the same person as the original who was supposed to be uploaded, in which case this process would not be a route to immortality at all. The *Pronominal Continuity* device attempts to evade this problem by describing someone undergoing the uploading process in a way that implies (through consistent use of the same pronoun) that that person continues to exist throughout the process. This can be made clear through an example: in *Mind Children*, Moravec shifts from his usual expositional style into an italicized, second- person story when describing the possibility of uploading as a process:

> *As long as you press the button, a small part of your nervous system is being replaced by a computer simulation of itself. You press the button, release it, and press it again. You should experience no difference. As soon as you are satisfied, the simulation connection is established permanently. The brain tissue is now impotent—The process is repeated for the next layer—Layer after layer the brain is simulated, then excavated. Eventually your skull is empty—Though you have not lost consciousness, or even your train of thought, your mind has been removed from the brain and transferred to a machine.* (1988, pp.109–10, original italics)

The continuous use of the second person singular in this story gives the impression that the person undergoing this process ('you') survives. However, the question of whether someone can survive having all their neurons 'replaced by a computer simulation', or uploaded by any other means, is highly contentious (Cave 2012, ch. 5; Schneider 2019, ch. 6). As we will see in the discussion of the identity problem to follow, many thinkers believe that the 'uploaded' entity is merely a copy of the original. In the passage quoted above, these philosophical worries are put to one side while the narrative assures 'you' that 'you' would retain a continuous, conscious point of view throughout such an operation.

Together, these three narratives or rhetorical devices make mind uploading appear attractive (Aladdin's Lamp), and technologically (Exponentialism) and philosophically (Pronominal Continuity) feasible. If we now turn to another set of stories, those in the science fiction of this period, we find a very different perspective on mind uploading's desirability and plausibility as a route to immortality.

## 13.5 Alternative Narratives: The Challenge from Fiction

In *Apocalyptic AI*, Geraci claims that mind uploading first appears in science fiction in Arthur C. Clarke's 1953 novel *The City and the Stars*, and 'within a few decades became widely appreciated'

(2010, p.55). He goes on to claim that science fiction is the means through which these ideas of transcendence are spread, including into the communities developing AI: 'science fiction carries a camouflaged sacred into technological research' (Geraci 2010, p.55). However, as should be clear from the passages above, I would argue that the idea of AI-enabled transcendence has to a significant extent been transmitted by works purporting to be nonfiction, by writers who are also technologists. Of course, one might argue that their work is so speculative as to be a kind of science fiction: certainly it involves explicit storytelling, as we saw in the section on pronominal continuity. However, I argue the work done by actual science fiction is not, in the main, a continuation of this transmission, but something subtler and more critical. By interrogating the narratives presented by authors like Moravec and Kurzweil, science fiction reveals their workings, questions their assumptions, and provides alternatives. It therefore does not merely function as a prophet of the transcendent age to come (though it might do that sometimes), but as critic and Cassandra.

The works we will consider span the digital age. They are Rudy Rucker's 1982 novel *Software*, William Gibson's 1984 cyberpunk classic *Neuromancer*, Pat Cadigan's 1986 short story 'Pretty Boy Crossover', Greg Egan's 1994 extensive exploration of mind uploading *Permutation City*, Robert Sawyer's similarly uploading-focused *Mindscan* from 2005, and Cory Doctorow's 2017 utopian novel *Walkaway*. In all of these, the possibility of an AI-empowered, uploaded immortality is explicitly offered and portrayed as potentially desirable, but at the same time problematized. For example, the positive vision of mind uploading is expressed by the scientist Sita in Doctorow's *Walkaway*:

> 'Now we've got a deal for humanity that's better than anything before: lose the body. Walk away from it. Become an immortal being of pure thought and feeling, able to travel the universe at the speed of light, unkillable, consciously deciding how you want to live your life and making it stick, by fine-tuning your

> parameters so you're the version of yourself that does the right thing, that knows and honours itself'. (2017, p.259)

I will, however, focus in this section on the challenges these works raise to the techno-utopian vision of uploading. I will introduce three, which loosely parallel the three narratives from nonfiction that promote mind uploading mentioned above. So, challenging the desirability of mind uploading, we have what I will call the *Sanity Problem*. Challenging the technological feasibility of mind uploading, we have what I will call the *Substrate Problem*. Finally, challenging the philosophical feasibility of mind uploading, we have the *Identity Problem*.

### *13.5.1 The Sanity Problem*

We noted above that the problem of mortality is the problem of the body; we could equally say that the problem of *im*mortality is the problem of the mind.[8] Although proselytizers like Kurzweil and Moravec like to portray the open possibilities of the uploaded state, there could just as well be constraints on the possibilities of the uploaded mind, leading to boredom, despair, or madness. These might arise from technical constraints, such as the difficulty of developing the right mind-actualizing algorithms, or they might arise independently, once the digital mind realizes its predicament. In *Neuromancer*, Gibson portrays a 'construct'—a simulation of a dead hacker called Dixie Flatline, who in no way relishes this resurrection in cyberspace. The one thing he asks of the main protagonist, Case, in return for helping on the mission is that ' "when it's over, you erase this goddam thing" ' (Gibson 1984, p.118). Later, we meet a much more powerful AI, the Neuromancer of the book's title, which is capable of simulating personalities in ways that seem richer and more satisfying than the construct: ' "I call up the dead... I *am* the dead, and their land," ' it says to Case, ' "Stay. If your woman is a ghost, she doesn't know it. Neither will you" ' (Gibson 1984, p.270). But Case chooses not to stay, and Gibson suggests his protagonist would not have found happiness uploaded into this virtual world.

In his extensive exploration of uploading, *Permutation City*, Egan is more explicit about the predicament of the digital self. Most of the uploaded minds in his vision exist in highly partial, limited simulations of the world. Consequently: 'People reacted badly to waking up as Copies' (Egan 1994, p.5). Some of those whose physical originals were old and dying managed to cope with the reduced circumstances of the simulation, but among those whose originals had been young and healthy, 'the bale-out rate so far had been one hundred per cent' (p.6). In other words, for all but the very sick, uploaded life was much worse than the real thing.

In Cory Doctorow's *Walkaway*, the character Limpopo speculates that attachment to the body might be intrinsic to humans: ' "What happens if you ditch your bodies, upload, and it turns out the human race can't survive without whatever makes us terrified of losing our bodies?" ' (2017, p.262). Or, in a nice twist to this idea, Doctorow speculates that this might affect some more than others. One uploaded person, CC, proves unable to deal with being in a digital state, despite a lifelong interest in the topic; another upload, Dis, explains this by saying that 'one reason to get into uploading is your overwhelming existential terror at the thought of dying' (p.185); upon realizing that he has physically died and is now a mere digital simulacrum, CC (or his simulation) therefore continually breaks down, overwhelmed by the terrible realization that he will never have back the (biological) life he loved. So the very personality most drawn to uploading—that is, one who is most afraid of death—proves least able to cope with it.

In her book on feminist cyberpunk, Carlen Lavigne notes that there is a tradition of women science fiction writers who are particularly critical of what they perceive to be the male (perhaps also privileged, white) fantasy of separating mind from body. Such writers' treatments of mind uploading are therefore 'generally illustrations of the idea that we are more than the sum of our parts, with some quintessential "human" core that cannot truly be copied by machine' (Lavigne 2013, p.73). An example of such an author is Pat Cadigan, who confronts mind uploading in a number of her works, including the much-anthologized short story 'Pretty Boy Crossover'. The sixteen-year-old (nameless) protagonist is tempted to upload by the digital form of his friend Bobby, who promises eternal youth: ' "This can be you. Never get old, never get tired, it's never last call, nothing happens unless you want it to and it could be you' " (Cadigan 1986). But the protagonist is sceptical of the virtual fantasy world: ' "You really like it, Bobby, being a blip on a chip?" ' he asks, before fleeing the uploaders.

Together these stories convey a subtler image of the uploaded life than the wish-fulfilment fantasy of Moravec and Kurzweil: not one of boundless freedom, but one of limitation and loss; as much entrapment as escape.

### *13.5.2 The Substrate Problem*

We have seen that Moravec and Kurzweil portray the world of the uploaded individual in utopian terms: now free of their biological substrate, the digital self is free to self-actualize and conquer the universe. But in reality, and indeed in these fictional works, the upload and their realm are both still dependent upon a particular physical substrate: computers and networks, which consume energy, require maintenance, and can be bought, sold, or even destroyed—'a blip on a chip'.

The 'Permutation City' of Egan's novel's title is intended to be a digital paradise, in a cyberspace robust for eternity. But the infrastructure on which it depends also proves vulnerable and is ultimately destroyed. Rucker takes a more playful approach: the protagonist of *Software*, Cobb Anderson, has his mind uploaded to a new robot body. But the AI system that has created this new technology extracts a price: it turns out that it can retake control of Anderson's robot body whenever it chooses. The new substrate might not suffer from the frailties of its biological predecessor, but it has a few of its own. As Anderson exclaims when realizing the predicament of his new body: ' "Immortality, my ass" ' (Rucker 1982, p.118).

### *13.5.3 The Identity Problem*

This third challenge to uploading is, as noted above, that it might not really preserve personal identity—and, therefore, that uploading might not be survival at all. Neuromancer hints at this question: in the passage cited above in which the AI is attempting to persuade the protagonist Case to 'stay', the machine is actually addressing Case's digital avatar in cyberspace. The promise that Case will not know he is a 'ghost' is made to the avatar, and might be true of it/him, but if the 'you' in the Neuromancer's promise applies to the flesh-and-blood Case, this is much less clear. What would the biological Case know or experience of a simulated Case in cyberspace? Would he not remain a wholly separate being? This is hinted at on the novel's last page, which has the original Case spot his simulated self standing far away 'at the very edge of one of the vast steps of data' (Gibson 1984, p.297). Doubts about whether the upload really is the same person as the original also contribute to the protagonist's decision to avoid uploading in Cadigan's 'Pretty Boy Crossover': ' "You think I really believe Bobby's real just because I can see him on a screen?" ' he asks (Cadigan 1986).

The question of identity is addressed more explicitly in both *Mindscan* and *Permutation City*. In the latter, Egan extensively explores but does not resolve this question. On the one hand, he uses the term 'resurrected' to talk of 'the Copy's' awakening with the memories and personality of the original, a term with strong connotations of continuity and survival. Also, he uses the names of the originals when describing the antics of the Copies, and ascribes to them fully sentient minds. But at the same time, he gives us the perspectives of original human and digital Copy simultaneously, interacting with each other, as here:

> He [the Copy] said quietly, 'I won't be your guinea pig. A collaborator, yes. An equal partner. If you want my cooperation, then you're going to have to treat me like a colleague, not a... piece of apparatus. Understood?'
>
> A window opened in front of him. He was shaken by the sight, not of his predictably smug twin [the original], but of the room behind him....
>
> 'Of course that's understood! We're collaborators. That's exactly right. Equals. I wouldn't have it any other way'.
>
> (Egan 1994, pp.14–15)

The mix of pronouns here contrasts with Moravec's simple repetition of the second person singular, and reveals the fraught question of identity. The chapter in which this exchange features invites us to identify with the Copy, and see him as a conscious, feeling person. But at the same time, this dialogue shows that the Copy is not literally the same being as the original. They are qualitatively identical in certain relevant respects (e.g. have the same psychological attributes, at least at the point of creation of the Copy), but they are not quantitatively identical (i.e. they are not literally one and the same entity). Indeed, sometimes there are multiple Copies in existence at once, which makes the lack of strict identity even clearer.[9]

Exactly this question provides the main plotline of Sawyer's *Mindscan*. The protagonist Jake Sullivan is diagnosed with a terminal illness likely to kill him relatively young. He therefore signs up to have his mind scanned and uploaded into a new, robot body. When the original Sullivan awakes from the scanning process, he is horrified to realize that he is still the same person, in the same terminally ill body: '"the copy doesn't have to worry about becoming a vegetable anymore—it's free... But *this* me is still doomed"' (Sawyer 2005, p.45). However, the original Sullivan is later healed of the illness that would have killed him, and he starts attempting to reclaim his life from the copy, to which he had handed it over. The resulting drama makes clear that the two are not the same entity, and that their interests are not aligned.

The science fiction accounts make clear the real nature of the relationship between original and (in Egan's language) Copy. We saw that advocates of uploading tell a story of the same person surviving through the process, using narrative devices such as pronominal continuity. They might also tell the story in a way that has the original destroyed, such as that cited above in which the brain is 'excavated'. This assists the impression of continuity by not presenting rival candidates each with claims to be the original. By keeping the original alive, the science fiction versions expose the real nature of mind uploading: as the construction of a digital simulacrum. By telling the story from the point of view of the original (as in *Mindscan* above), perhaps even as they confront the simulacrum (as in the passage from *Permutation City*), they expose the posited continuity as an illusion.

## 13.6 Conclusion

Machines promise permanence, infinite reparability. Once this is combined with intelligence, they become the means by which some people manage the terror of death. These believers project onto them the dream of immortality: through uploading their minds, they hope to become the machine. We have seen that techno-utopian popularizers such as Moravec and Kurzweil use a variety of narrative devices to make this story seem desirable and both technologically and philosophically feasible. Given the prima facie implausibility of mind uploading, such devices are essential to its popularity. But they are also simplistic, and a long history of science fiction works has questioned and mocked their claims though narratives that are more nuanced and multisided than those of the proselytizers. Together these works therefore provide a crucial site of critique through which a wide range of publics can explore one of the most popular memes of the AI age.

## Notes

1. For an extended investigation of immortality belief systems, see my book *Immortality* (2012).
2. Gruman (1966) looking from ancient times to 1800; Noble (1997) from the Medieval period to the present; Haycock (2008) from the seventeenth century to the present; Gray (2011) on twentieth-century Russia and the contemporary West; and Dickel & Frewer (2015) on the transhumanist movement.
3. For an example of this kind of 'engineering approach' to radical life extension, see de Grey and Rae (2008).
4. At the time of writing, Wikipedia's 'Mind uploading in fiction' page (https://en.wikipedia.org/wiki/Mind_uploading_in_fiction) has well over one hundred examples, including fiction by luminaries such as Isaac Asimov and Robert Silverberg, and successful TV series (e.g. *Red Dwarf*) and films (e.g. *Tron*). In *Science Fiction and Philosophy*, Susan Schneider calls it a 'common science fiction theme' (2016, p.6).
5. Although it is not yet known how much data is stored in the average human brain. One recent estimate put it at 1 quintillion bytes (Ghose et al 2016), in which case all the data generated by the billions of computer and internet users each day would be equivalent to two and a half human brains.
6. The episode of the TV series *Black Mirror*, 'Be Right Back', nicely explores just this (Brooker 2013).
7. The hypothetical moment when machines become superintelligent.
8. I owe this nice turn of phrase to Kanta Dihal.
9. Wiley and Koene note that the use of the term 'copy' when discussing mind-uploading cases usually denotes scepticism about the original's survival, whereas the use of the term 'transfer' denotes optimism (Wiley & Koene 2016, p.213).

## Reference List

Bauman, Z. (1992) *Mortality, immortality and other life strategies*. Cambridge, Polity Press.

Brooker, C. (2013) Be right back. *Black Mirror*. Directed by O. Harris. Channel 4. Originally aired on 11 February.

Brooks, R. (2017) The seven deadly sins of predicting the future of AI. Available from: https://rodneybrooks.com/the-seven-deadly-sins-of-predicting-the-future-of-ai/ [Accessed 20 May 2019].

Burton-Hill, C. (2016) The superhero of artificial intelligence: can this genius keep it in check? *The Guardian*. 16 February.

Cadigan, P. (1986) Pretty boy crossover. *Asimov's Science Fiction*. 10(1). Available from: http://i.4pcdn.org/tg/1466975483530.pdf [Accessed 15 July 2019].

Cambridge English Dictionary (n.d.) Digital age. Available from: https://dictionary.cambridge.org/dictionary/english/digital-age [Accessed 9 June 2019].

Cave, S. (2012) *Immortality: the quest to live forever and how it drives civilization*. London, Crown.

Cave, S. (forthcoming) AI and the denial of death.

Clynes, M. E. & N. S. Kline (1960) Cyborgs and space. *Astronautics*. September 1960, 26–27, 74–6.

de Grey, A. & M. Rae (2008) *Ending aging: the rejuvenation breakthroughs that could reverse human aging in our lifetime*. Reprint. New York, St. Martin's Griffin.

Dickel, S. & A. Frewer (2015) Life extension: eternal debates on immortality. In: S. L. Sorgner & R. Ranisch (eds.) *Post- and transhumanism: an introduction*. Frankfurt, Germany, Peter Lang. pp.119–32.

Doctorow, C. (2017) *Walkaway*. London, Head of Zeus.

Egan, G. (1994) *Permutation city*. London, Orion.

Farias, M., A.-K. Newheiser, G. Kahane, & Z. de Toledo (2013) Scientific faith: belief in science increases in the face of stress and existential anxiety. *J Exp Soc Psychol*. 49, 1210–13.

Geraci, R. M. (2010) *Apocalyptic AI: visions of heaven in robotics, artificial intelligence, and virtual reality*. New York, Oxford University Press.

Ghose, T. (2016) The human brain's memory could store the entire internet. *Live Science*. Available from: https://www.livescience.com/53751-brain-could-store-internet.html [Accessed 17 June 2019].

Gibson, W. (1984) *Neuromancer*. New York, Ace Books.

Good, I. J. (1966) Speculations concerning the first ultraintelligent machine. In: F. L. Alt & M. Rubinoff (eds.) *Advances in computers*. Amsterdam, Elsevier. pp.31–88.

Gray, J. (2011) *The immortalization commission: science and the strange quest to cheat death*. New York, Farrar, Straus and Giroux.

Gruman, G.J. (1966) *A history of ideas about the prolongation of life*. New York, Arno Press.

Haycock, D. B. (2008) *Mortal coil: a short history of living longer*. New Haven, Yale University Press.

Kesebir, P. & T. Pyszczynski (2011) The role of death in life: existential aspects of human motivation. In: Ryan, R. (ed.) *The Oxford handbook of motivation*. Oxford, Oxford University Press.

Kurzwei, R. (1990) *The age of intelligent machines*. Cambridge, MA, MIT Press.

Kurzweil, R. (2005) *The singularity is near: when humans transcend biology*. New York, Penguin.

Kurzweil, R. & T. Grossman (2004) *Fantastic voyage: live long enough to live forever*. Emmaus, PA: Rodale.

LaGrandeur, K. (2013) *Androids and intelligent networks in early modern literature and culture: artificial slaves*. New York, Routledge.

Lavigne, C. (2013) *Cyberpunk women, feminism and science fiction: a critical study*. Jefferson, NC, McFarland & Co.

Lifshin, U., J. Greenberg, M. Soenke, A.Darrell, & T. Pyszczynski (2018) Mortality salience, religiosity, and indefinite life extension: evidence of a reciprocal relationship between afterlife beliefs and support for forestalling death. *Relig Brain Behav.* 8, 31–43.

Marr, B. (2018) How much data do we create every day? The mind-blowing stats everyone should read. *Forbes*. Available from: https://www.forbes.com/sites/bernardmarr/2018/05/21/how-much-data-do-we-create-every-day-the-mind-blowing-stats-everyone-should-read/ [Accessed 19 April 2019].

Moravec, H. (1988) Mind children: the future of robot and human intelligence. Cambridge, MA, Harvard University Press.

Noble, D. F. (1997) *The religion of technology: the divinity of man and the spirit of invention*. New York, Penguin Books.

Öhman, C. & L. Floridi (2017) The political economy of death in the age of information: a critical approach to the digital afterlife industry. *Minds and Machines.* 27 (4), 639–62.

Rucker, R. v B. (1982) *Software*. New York, Ace Books.

Rutjens, B. T., J. van der Pligt, & F. van Harreveld (2009) Things will get better: the anxiety-buffering qualities of progressive hope. *Pers Soc Psychol Bull.* 35, 535–43.

Savin-Baden & Burden (2019) Digital immortality and virtual humans. *Postdigital Science and Education*. 1, 87–103.

Sawyer, R.J. (2005) *Mindscan*. New York, Tor Books.

Schneider, S. (ed.) (2016) *Science fiction and philosophy: from time travel to superintelligence*. Chichester, Wiley Blackwell.

Schneider, S. (2019) *Artificial you: AI and the future of your mind*. Princeton, NJ, Princeton University Press.

Singh-Kurtz, S. (2019) The big business of living forever. *Quartz*. Available from: https://qz.com/1578796/why-silicon-valley-titans-are-obsessed-with-immortality/ [Accessed 29 May 2019].

Solomon, S., J. Greenberg, & T. Pyszczynski (1998) Tales from the crypt: on the role of death in life. *Zygon*. 33(1), 9–43.

Techopedia (n.d.) What is the digital revolution? Techopedia.com. Available from: https://www.techopedia.com/definition/23371/digital-revolution [Accessed 9 June 2019].

Turing, A. M. (1937) On computable numbers, with an application to the Entscheidungsproblem. *Proc Lond Math Soc.* s2-42(1), 230–65.

Turing, A. M. (1950) Computing machinery and intelligence. *Mind* 59, 433–60.

Wiley, K. B. & R. A. Koene (2016) The fallacy of favoring gradual replacement mind uploading over scan-and-copy. *J Conscious Stud.* 23, 212–35.

# 14

# Artificial Intelligence and the Sovereign-Governance Game

*Sarah Dillon and Michael Dillon*

## 14.1 Introduction

> The spatiality of algorithmic life is thus not solely related to the augmented experiences of urban life, or the multiple sensors of the Internet of Things, but also to the space of sovereign decision.
>
> (Amoore & Piotukh 2016, p.7)

In 'Games Playing Roles in Banks' Fiction' (2013), Will Slocombe explores Iain Banks's fiction through an analysis of how games feature in his texts, both in their plots and themes, and as structuring devices. Slocombe identifies two key overarching features of the way in which games function in Banks's texts: first, his work demonstrates 'that reality (or at least our understanding of it) works through games; they do not just reflect reality but inherently color our perception of it' (2013, p.136); second, 'Banks' fiction highlights that games are about control, [...] how we compete reveals our ethical stance towards others, the extent to which our strived-for victory is due to, or at the expense of, others' (p.136). In short, for Banks, games structure reality, and games govern our social interactions.

Games are also defined by rules. This is in fact the difference between a game, and mere 'play': 'games emerge when a number of "plays" are given direction and order; a set of rules'

(Slocombe 2013, p.137). Herein lies a commonality between games and AI. Both are governed by a set of rules, but both also govern by rules. In this chapter, we engage with three novels in which games, governance and AI weave themselves through the text's fabric: Banks's *The Player of Games* (1988) and *Excession* (1996), and Ann Leckie's *Ancillary Justice* (2013). We explore these novels in the context of the ancient distinction between sovereignty and governance, in which sovereignty issues the warrant to rule, and governance operationalizes it. The constitutive differentiation of the one from the other poses the threat of an endless conflict between them which the Greeks called *stasis* (Agamben 2015).[1] This sovereign-governance game nonetheless continues to comprise the geopolitical algorithm of modern rule.[2] The royal question of sovereignty and government, as Michel Foucault calls it, is not the only algorithmic logic at play in modern power relations (1982; 2003). There are others. Marxism, for example, offers an account of how ownership of the means of production exercises such symbolic-logical force. Foucault (2008), Giorgio Agamben (1998), Michael Dillon (2015), and Roberto Espositio (2008) explore the biopolitical algorithm in which governance is formulated around the properties of 'Life' in the form of species being. But the sovereign-governance game remains perhaps the most dominant and certainly the most familiar and accepted algorithmic logic of modern politics. Its vocabularies comprise modern politics' most widely disseminated vernacular, and its structural dynamics—its 'plays'—are widely understood, albeit they are routinely deplored as much as they are enacted.

Politics establishes that however supreme, timeless, and universal the sovereign warrant to rule is claimed to be, it is continually subverted by the changing contingent and particular circumstances which comprise the very converse nature of governance. Contingent application always already contaminates the timeless purity of the 'miracle' of sovereignty (Schmitt 1996; Schmitt 2005). However much it is claimed to be so, sovereign power is therefore never insulated from the heterogeneity of time, place, and event that its ineliminable dependence upon governance injects into its claim to be homogeneous. If seamless transmission of sovereignty into governance is the ideal of sovereign rule, it is never realized in practice, even in the texts which we explore here, in universes composed of artificial intelligences. The reasons go beyond the usual culprits: conspiracy, sedition, ignorance and stupidity, the infinity and intractability of human appetites, the malignancy of outside forces, or the very shock of the new. What is most interesting about these texts is how they identify the dynamic that seems ultimately to define what we call the sovereign-governance game. The game results from the efforts required to reconcile the irreconcilable: the transcendent, universal, oneness of the sovereign, with the immanent, contingent, heterogeneity of governance. That is, the rules of the game commit both sovereign and governance to a continuous process of attempting to overcome the material conditions instituted by the very game itself (Koselleck 2000; Benjamin 2019; Weber 1992; Dillon 2017).

The novels we explore in this chapter play out the sovereign-governance game with artificial as well as human actors. Through their different strategies, plots, wars, contradictions, modes of (dis)embodiment, and imagined subjectivities, the algorithmic logics and dynamics instituted by the game are elaborated. In doing so, the novels question what might be politically novel about AI. But in the end, they reveal that whilst AI impacts the pieces on the board, reducing some and advancing others, it does not materially change the logic of the game.

## 14.2 The Culture

From 1987–2012, Banks published the nine novels and one short story collection which comprise his Culture series. The Culture is a vast anarchic utopia whose inhabitants live on a profusion of ships and space habitats, all governed by AI or 'Minds' which are indistinguishable from the ships or habitats they oversee. Despite its vastness, the Culture is not an empire, it is not hierarchical, it is not riven with conspiracies, political manoeuvring, or contestations of power. The Culture, in fact, represents a utopia in which the conflict between sovereignty and governance outlined above has been overcome—it is the representation of an order uncontaminated by that purportedly inevitable conflict, that originary division. Or at least, this is its promise. For instance, Yannick Rumpala (2012) states that Banks has 'created an organized world in which the plan to replace the government of men with the administration of things has been carried out, thanks to artificial intelligences and limitless material wealth and energy' (p.24). He concludes that 'in this model, there would not really be any political choices left to be made' (p.24). This is the Culture's promise—that the automation of intelligence in the Minds enables a transcendence of the sovereign-governance game, which is simultaneously a transcendence of the human. However, here we want to look closely at two Culture novels in which the failure of the Culture's promise to have transcended the game is exposed. The Culture remains riven by the sovereign-governance distinction even as it claims to have overcome it. These novels expose, in different ways, the inevitable tension between the operationalization of governance and the mystique of sovereignty.

### 14.2.1 *The Player of Games*

*The Player of Games* (1988) is the second of the Culture novels, following *Consider Phlebas* (1987). The latter introduces us to the world of the Culture as it navigates the only major war in its history, that with the Idiran Empire. *Consider Phlebas* is an explosion of a novel, a romping space opera played out on a vast scale; in contrast, *The Player of Games* might at first seem parochial, focused as it is on one human, Gurgeh, and his trip to the Empire of Azad to try his hand at the game which has given that Empire its name. Every six years, the game determines the role its citizens play in the power structures of the Empire: the better one does in the game, the higher one's role and the greater one's power; the winner becomes the Emperor. The action of the novel comprises Gurgeh's decision to travel to Azad, his en route preparations for playing the game, his participation in it, as well as his learning about the Empire, and his return home. This action is conveyed by a (seemingly) omniscient third-person narrative which primarily focalizes, when it does so, through Gurgeh (apart from brief shifts in focalisation in the second half of the novel (Banks 1988, pp.212–15, 246–47, 274)).

This story is enclosed and interrupted by a frame narrative which is both double ended ('meaning that the frame situation is re-introduced at the end of the embedded tale' (Barry 2002, p.236)), and intrusive ('meaning that the embedded tale is occasionally interrupted to revert to the frame situation' (p.236)). These interruptions occur at the opening of each of the four sections of the novel, and remind the reader that the story is not in fact told by an omniscient third-person narrator. This can be easy to forget when immersed in the action, but the story is in fact a retrospective account of events by a specific individual, one whose identity is withheld until the end of the novel. 'Me?', the narrator asks at the beginning, 'I'll tell you about me later' (Banks 1988, p.3). The narrator resurfaces at the beginning of section 2, 'Still with me?' (p.99), to provide some linguistic guidance regarding gender and pronouns, to recap the action, and to plant some suspicions in the reader's mind about how Gurgeh has been motivated to embark on his trip:

> Does Gurgeh really understand what he's done, and what might happen to him? Has it even begun to occur to him that he might have been tricked? And does he really know what he's let himself in for?
>
> Of course not!
>
> That's part of the fun! (p.100)

The narrator resurfaces again at the beginning of section three, when we receive the first clue that it may not be human: 'So far so average. Our game-player's lucked out again. I guess you can see he's a changed man, though. These humans!' (p.231). Although we do not discover anything more about the narrator's identity at this point. Its secret identity is finally revealed to the reader at the end of the novel, back in the frame narrative, when the narrator gets the final word: '... No, not quite the end. There's still me. I know I've been naughty, not revealing my identity, but then, maybe you've guessed; and who am I to deprive you of the satisfaction of working it out for yourself? Who am I, indeed?' (p.308). The narrator, it is revealed, is the Culture drone Flere-Imsaho, who has accompanied Gurgeh on his trip. Gurgeh is told that it is a library drone, there to help him navigate the social mores and niceties, as well as the language, of the Empire. However, it is a Special Circumstances drone, bearing names—Handrahen Xato—which we are told by Gurgeh's Hub AI are 'equiv-tech espionage-level SC nomenclature' (p.34). SC—Special Circumstances—is the mysterious suborganization within Contact, the part of the Culture responsible for managing its interactions with other—usually less technologically advanced and less knowledgeable about the Universe—civilizations. SC handles particularly challenging or complex situations within Contact. The Hub's response to Gurgeh's first encounter with a SC drone indicates the appropriate response the reader might have to discovering the narrator, and Flere-Imsaho's, true identity: 'Holy shit, Gurgeh, what *are* you tangling with here?' (p.34).

The distinction outlined here in *The Player of Games* between the embedded story and the frame narrative is crucial. Within the story, the player of games of the novel's title is of course Gurgeh, one of the Culture's best game players. Through him, the Culture is pitted against the Empire, revealing the contrast between the utopian Culture and vicious imperialism. Whilst members of the Culture live in an equitable society of peace and plenty, the Empire is rotten to its core, founded on the subjugation of various subsections of its society, inequitable in its distribution of wealth and resources, violent in its colonization of 'lesser' races, its wealthiest members taking delight in entertainments of violent pornography, torture, and murder. The longer Gurgeh stays there and speaks their language, the more he risks becoming like them, internalizing the rules of their game through their language and culture. But in the end, he is able to defeat the Emperor, Nicosar, because he plays the game *as* the Culture would. In essence, he introduces into the Empire's game the governing algorithms of the Culture, its procedures and rules of calculation and problem-solving. Nicosar is furious at what he perceives to be a perversion of the game: 'you treat this battle-game like some filthy dance. It is there to be fought and struggled against, and you've attempted to seduce it' (p.281). Gurgeh is unable to articulate a verbal response to the Emperor since he has already expressed his answer in his game play (p.282). When Gurgeh beats the Emperor, undermining the Empire's algorithm of governance, the latter destroys itself without any further Culture intervention needed.

But one must not only attend to the embedded story in a reading of this novel. For the frame narrative tells a very different tale. The frame narrative reveals an alternative account of the same events, one in which the player of games is not Gurgeh, but the Culture itself, with Gurgeh the one being played not playing.[3] The plot revealed by the frame narrative exposes the return of the governmental problematic at the heart of the Culture. That is, its promise of transcendent sovereignty has to be renewed, operationally and practically, through a governmental device. Special Circumstances is that device.[4] It is the secret of the Culture precisely because it is in fact how it operationalizes its supposedly sovereign power. SC is the Culture's algorithm of governance, its *ai*, that is, its *arcana imperii* (Secrets of the empire). Powerful, secret, lacking in transparency, only accessible to those in the know, and perhaps not accessible in its totality to anyone, through its tricks, its game playing, SC sustains and reproduces the promise of the Culture's transcendent sovereignty, even as its existence compromises any such promise. SC's governance, the governance of the *ai*, is not merely bureaucratic. There is a mystique to its operations: it is the speciality of a certain few. As it acts, it therefore renews the mystique of the Culture at the same time as its operations in fact undermine its claims to sovereignty. For the myth of sovereign transcendence is both sustained and undermined by governmental immanence. So, the embedded story explores the possibility, through Gurgeh, that the Culture might be destabilized by contamination with the operations of Empire, but eventually resists and overcomes it through playing by its own rules. In contrast, the frame narrative and the plot it points towards (whilst never explicitly telling) reveal the governmental devices necessary to sustain and reproduce, whilst at the same time undermining, the transcendental promise that is sovereignty.

SC has carefully planned and executed everything that happens in the story. When their first approach to Gurgeh to play the game fails, they orchestrate a situation of blackmail, otherwise unheard of in the Culture, to force Gurgeh to undertake the visit to Azad in order to protect his reputation. This is the 'trick' to which the narrator alludes in its intervention at the opening of section two of the novel. SC also manages knowledge on a need-to-know basis, as well as providing misinformation where necessary. The first SC drone Gurgeh encounters tells him that the Culture discovered the Empire seventy-three years ago (p.80), but towards the end of the novel, Flere-Imsaho informs Gurgeh that 'for the last two hundred years the Emperor, the chief of Naval Intelligence and the six star marshals have been apprised of the power and extent of the Culture' (p.243). When the Emperor and the Empire have been defeated and at least some of the details of the complex plot of which Gurgeh has been an unwitting part are revealed to him, he even suspects that his entire life has been managed by SC for the purpose of destroying the Empire. Moreover, Flere-Imsaho reveals that Shohobohaum Za, who Gurgeh believes to be the Culture's representative on Azad, is 'not Culture-born at all' (p.297), but rather one of the 'outsider "mercenaries"' that the Culture uses 'to do the *really* dirty work' (p.296). SC's employment of external mercenaries *not revealed to be such* is another instance of its governmental techniques. During this scene of revelation to Gurgeh, Flere-Imsaho explicitly acknowledges the governmental operations that simultaneously sustain and undermine the promise of the Culture's sovereignty: 'My respect for those great Minds which use the likes of you and me like game-pieces increases all the time. Those are *very* smart machines' (p.296).

### *14.2.2 Excession*

In *The Player of Games*, the frame narrative reveals the governing algorithms—SC's operations—which both sustain and challenge the Culture's promise of transcendent sovereignty. In *Excession* (1996), the inevitable tension between the operationalization of governance and the mystique of sovereignty is explored again, but at a heightened level of complexity. *Excession* is a challengingly elaborate novel with a myriad of plots and subplots which all revolve in some way around the object that gives the novel its title—the Excession—a black-body sphere that appears from nowhere at the beginning of the novel and disappears at the end. For the Culture, the Excession represents an Outside Context Problem (OCP), which the narrator defines as such:

> An Outside Context Problem was the sort of thing most civilisations encountered just once, and which they tended to encounter in the same way that a sentence encountered a full stop.
>
> (Banks 1996, p.71)

An OCP is unimaginable within the horizon of possible imaginings of the society threatened by it.[5] In this instance, the OCP that is the Excession has technological capabilities beyond those even of the Culture: it 'is linked to the energy grid in both hyperspatial directions' (p.114), explains the drone Churt Lyne to Ulver Seich, when he is briefing her ahead of her potential involvement in proceedings. He continues, 'that alone puts it well outside all known parameters and precedents. We have no experience whatsoever with something like this; it's unique; beyond our ken' (p.114). Immediately following the first definition of an OCP, however, which speaks to its terminal uniqueness for 'most civilisations', the reader learns that the Culture is not 'most civilisations': 'the Culture had lots of minor OCPs, problems that could have proved to be terminal if they'd been handled badly, but so far it had survived them all' (p.72). Even the Excession has in fact appeared once before (p.67). An OCP may therefore be unique in its singular instantiations, but is infinite in its recurrence. OCPs threaten the sovereignty of the Culture and demand some sort of governmental response. They are exceptional, but such exceptions are the norm.[6] As a consequence, SC, which is the unit that deals with exceptional circumstances, in fact constitutes the normal governmental operations of the Culture. The Culture is therefore always involved in governing in order to re-enforce the promise of its sovereignty, even as that act of governance compromises its sovereign claims. This is the nature of the game that the OCP reveals, and which *Excession* explores.

SC governs the Culture's response to the Excession. Given its significance, a group of legendary Minds called the Interesting Times Gang (ITG) assume this responsibility, but they are not in fact acting in unity. Within the ITG, a subgroup—including *Steely Glint, Not Invented Here, Different Tan, No Fixed Abode* (and, on the periphery, operating ostensibly as a Culture traitor but in fact under the conspirators' direction, the *Attitude Adjuster*)—is enacting its own plot. They wish to use the Excession to manoeuvre a barbaric race called the Affront into declaring war on the Culture, so that the Culture will finally be moved to eliminate/assimilate the Affront. This plot is described by the *Attitude Adjuster* as 'a ghastly, gigadeathcrime-risking scheme' (p.378). *Shoot them Later, Serious Callers Only*, and *Anticipation of a New Lover's Arrival* suspect the plot, but before they have sufficient proof to expose it, the latter changes its mind with regard to the ethics of the scheme. It states that moral constraints should be the Minds' guides, but not masters, and comes to the conclusion that the conspirators have been acting in the best ultimate interests of the Culture, even if their immediate operations might be morally suspect:

> There are inevitably occasions when such – if I may characterise them so – *civilian* considerations must be set aside (and indeed, is this not what the very phrase and title Special Circumstances implies?) the better to facilitate actions which, while distasteful and regrettable perhaps in themselves, might reasonably be seen as reliably leading to some strategically desirable state or outcome no rational person would argue against. (pp.308–09)

The *Anticipation*'s reasoning here—the set of rules used in its calculation and problem-solving—evidences another part of the game: that is, what is the strategic calculus of necessary killing in order to secure the peace of sovereign rule, or, in this instance, what algorithm would warrant 'a ghastly, gigadeathcrime-risking scheme'? If the answer is the exceptional circumstance, but we have already seen that the exceptional circumstance is the norm, then there is no such strategic calculus. Rather, there is only governance, that is, the infinite succession of strategic manoeuvres which continue the killing. The algorithm that warrants each of these strategic manoeuvres is a calculation of utility, for any particular actant within the game. One way of making sense of *Excession* is to map each of the utility algorithms guiding the actions of each actor: the Affront want to seize control of the untapped power of the Excession in order to advance technologically beyond the Culture by which they are tolerated but under whose shadow they remain; the conspirator Minds want to eliminate the Affront; the suspecting Minds want to avert gigadeathcrime; *The Sleeper Service*, another Mind, wants to heal the forty-year rift between two of its favourite previous inhabitants; one of those, Genar-Hofoen, the main human actor in the story, wants to inhabit an Affront body (that is his price for his role in the action); and so on.

More than any of the other Culture novels, *Excession* focuses on the Culture's Minds, revealing the governance that maintains and destroys the illusion of transcendent sovereignty. This illusion becomes fully ruptured for Genar-Hofoen because of his becoming embroiled in the Minds' governing acts. But Genar-Hofoen is already well placed to make this discovery since, as the Culture ambassador to the Affront, he is already becoming an outsider to the Culture, already in fact more aligned with the Affront's way of life than the Culture's.

> [He] leant towards the school of thought which held that evolution, or at least evolutionary pressures, ought to continue within and around a civilised species, rather than – as the Culture had done – choosing to replace evolution with a kind of democratically agreed physiological stasis-plus-option-list while handing over the real control of one's society to machines. (p.170)

As Gurgeh eventually realizes in *The Player of Games*, Genar-Hofoen understands that the humans in the Culture are the playing pieces of the Minds.

When Genar-Hofoen researches one of the Minds reaching out to him, he discovers conspiracy theories in the marginal media which in fact accurately describe the truth of the Culture's sovereign-governance game. He finds a report:

> to the effect that the MSV [*Not Invented Here*] was a member of some shadowy cabal, that it was part of a conspiracy of mostly very old craft which stepped in to take control of situations which might threaten the Culture's cozy proto-imperialist meta-hegemony; situations which proved beyond all doubt that the so-called normal democratic process of general policy-making was a complete and utter ultra-statist sham and the humans – and indeed their cousins and fellow dupes in this Mind-controlled plot, the drones – had even less power than they thought they had in the Culture. (pp.171–72)

Genar-Hofoen considers these 'fairly wild claims' but decides that, if they are true, he may as well stop worrying as events are clearly beyond his ability to influence or control: 'there came a point when if a conspiracy was *that* powerful and subtle it became pointless to worry about it' (pp.171–72). Or, as Ulver Seich puts it, 'Shit, don't you hate it when the Gods come out to play?', to which Churt Lyne replies, 'In a word, [...] yes' (p.125).

The humans' realization of the truth of the Culture is far from an acceptance of a dystopian narrative of AI overlords. AI is not an OCP for the Culture, nor for the humans that live comfortably within it. AI governs the Culture to preserve the illusion of its sovereignty, but this is not a narrative of fear or of human overthrow. Rather it is a continuation of the modern game of sovereignty and governance, even as many of the actors in that game are artificial intelligences rather than human ones. This shift might raise the games' stakes—the threat of gigadeath crime, for instance, rather than the human-governed holocausts of the past of which we get a glimpse early in the novel (pp.45–52)—but it does not change the logic of the game.

The novel's Epilogue presents a radical shift in narrative perspective. It consists in one page of continuous prose composed by the Excession, a report of its mission, in which the reader discovers the nature and consequences of its visit to the Culture's universe. The Excession has been sent on a reconnaissance mission, to determine if the Culture is worthy of contact. The Excession's name makes more sense when this is revealed. Whereas the narrator defines it as 'something excessive. Excessively aggressive, excessively powerful, excessively expansionist' (p.93), the dictionary definition more appropriately defines its nature: 'excession' means 'a going out or forth' (OED). The Excession's conclusion is that in the 'matter of the suitability of the relevant inhabitants of the micro-environment for (further and ordered) communication or association it is my opinion that the reaction to my presence indicates a fundamental unreadiness as yet for such a signal honour' (p.455). The Epilogue effects an almost dizzying shift in the reader's perception of scale. The Culture represents a shift from human intelligence and the planetary, to artificial intelligence and the Universe; the Epilogue effects an analogous up-scaling from artificial intelligence and the Universe to an as yet undisclosed superintelligence and the Multiverse.[7] But the game remains the same.

The Excession has been sent to assay the Culture, to test its composition. It sees past the Culture's mystique, through its promise of transcendental sovereignty, witnesses its mechanisms of government, and finds it wanting. The Culture fails the test of the OCP rather than the testers taking it over. If the encounter between the Culture and the Excession is taken to be analogous to the encounter between the human and AI, the lesson is not fear of AI, but the insufficiency of the human to manage its encounter with such an OCP. In the novel, that encounter is managed according to the existing rules of the sovereign-governance game, rather than the encounter changing the nature of the game; and, in fact, it is the very nature of the response that prevents the game being changed by an encounter with potentially game-changing, radically sophisticated, technology. *Excession* does not present a narrative of fear, or of technophobia, but it poses the question of what would be required for a new technology of this sort actually to change the algorithms of modern politics.

## 14.3 *Ancillary Justice*

In the previous sections, we have explored how the operations of governance challenge the Culture's claim to transcendent sovereignty in Banks's novels. In this final section we want to turn our attention to a text that attends primarily to the operations of the sovereign in the sovereign-governance game. Leckie's *Ancillary Justice* demonstrates how the very algorithmic logic of the sovereign-governance game results in a form of inevitable conflict.

*Ancillary Justice* (2013) is the first in a trilogy of novels set in the Radch, an empire which has conquered the galaxy and which is ruled absolutely by its sovereign, Anaander Mianaai, Lord of the Radch:

> For three thousand years Anaander Mianaai had ruled Radch space absolutely. She resided in each of the thirteen provincial palaces, and was present at every annexation. She was able to do this because she possessed thousands of bodies, all of them genetically identical, all of them linked to each other. [...] It was she who made Radchaai law, and she who decided on any exceptions to that law. She was the ultimate commander of the military, the highest head priest of Amaat, the person to whom, ultimately, all Radchaai houses were clients. (Leckie 2013, p.95)[8]

The sovereign's power lies in her distributed unity, and her total control. 'No one defies Anaander Mianaai', observes Strigan (a citizen hiding from the sovereign because she is in possession of the only weapon that could possible harm her): 'She spoke bitterly. "Not if they want to live" ' (p.79). 'Anaander Mianaai declares what will be, and that's how it is', explains Breq, a former Radch ship, the *Justice of Toren*, which is now destroyed in its entirety apart from its remaining singular embodied instantiation in the body of an ancillary soldier, One Esk.

*Ancillary Justice* moves between two plot lines. The first, set in the present, follows Breq's quest to retrieve the weapon in Strigan's possession and its mission to kill Anaander Mianaai as revenge for the latter's destruction of the *Justice of Toren*. The second plot, set in the past, recounts the events that led up to that destruction. Together, the plots reveal that the sovereign is not in fact unified but is rather divided within herself. This division has been triggered, as in *Excession*, by an encounter with an OCP, in this instance in the form of an alien race called the Presger whom the Radch cannot defeat and who pose a terminal threat to Radch civilization. But whereas in *Excession* the division is exposed within the operations of the Minds, the governmental structures which create the illusion of the Culture's transcendent sovereignty, in *Ancillary Justice* the division takes place within the mind of a singular yet distributed artificially intelligent sovereign.[9] *Ancillary Justice* exposes how the division between, and yet mutual dependence of, governance and sovereignty structure the game: *Ancillary Justice* lodges the operations of governance within the supposed unity of the sovereign, by casting that sovereign as an agent that can very literally be divided within herself. That is, she is an agent that consists of actors of whom other parts of her mind are not cognisant. Again, the casting of the artificial agent in the sovereign-governance game does not change the game; rather, it emphasizes its compound nature.

The Presger are not dissimilar to the Affront in *Excession* in terms of their view of other species—'All other beings were their rightful prey, property, or playthings' (Leckie 2013, p.101)—but whereas the Affront are technologically inferior to the Culture, the Presger are radically superior in this respect to the Radch. In a key incident one thousand years before the present of the novel, the Presger interfere in Radch business by furnishing a race called the Garseddai with weapons powerful enough to resist, even overcome, the Radch. The weapon in Strigan's possession is the last remaining weapon of the Garseddai race, since, 'once Anaander Mianaai had realized the Garseddai possessed the means to destroy Radchaai ships and penetrate Radchaai armor, she had ordered the utter destruction of Garsedd and its people' (p.35). The Presger's intervention revealed the Radch to be fallible, revealed its power to have limits. But Anaander Mianaai's response, the Garseddai's total destruction, was an extreme act that represented a turning point in the history of the Radch, the moment at which many of those involved in the genocide questioned the rightness of their actions. 'For the first time Radch forces dealt death on an unimaginable scale without renewal afterward. Cut off irrevocably any chance of good coming from what they had done,' Breq explains, 'the first attempts at diplomatic contact with the Presger began shortly afterward' (p.157). Breq believes that the Presger involved themselves with the Garseddai not in order to compel the Radch to negotiate, but in order to *suggest* that they do so (p.158). The negotiations lead to a treaty between the Presger and the Radch which curtailed the latter's continual expansion and instituted a change of Radch policies and radical liberal reform, including the phasing out of the use of ancillaries (human bodies inhabited by AIs), the cessation of annexations (the colonization of new peoples), and equitable access to positions of power within its own governmental structures.

Anaander Mianaai destroys *Justice of Toren* because it realizes the truth—that the treaty with the Presger and the governmental changes it instituted have led to a profound split within the body of the sovereign. One part of Anaander Mianaai was horrified by her actions at Garsedd and pioneered the reforms; the other part was horrified by the treaty and reforms, which she perceived as the shackling of Radch power, and decided to act 'against *herself*' (p.204). *Justice of Toren* realizes that the sovereign was 'no longer one person but two, in conflict with each other' (p.204), with each believing absolutely in the justness of her own cause (p.205). Shortly before destroying *Justice of Toren*, one Anaander Mianaai admits the truth: ' "I am at war with myself," said Mianaai, in the Var decade room. "I have been for nearly a thousand years" ' (p.245). This division has not yet led to civil war because, Breq realizes, Anaander Mianaai is not in fact divided into just two, but into three—the two warring factions and then a third element of herself from whom the internal conflict is concealed. Breq realizes that the only way to destroy the sovereign is to reveal its own divisions to itself (p.268). As the plotline in the present comes to fruition, Breq achieves this goal, triggering the inevitable civil war (pp.337–38).

The revelation of the sovereign's own disunity to itself results in *stasis*, that is, in the Greek, *στάσις*, meaning 'faction, discord' (OED). It results in civil strife. Stasis is the result of the sovereign's heterogeneity even as it is also the function of the necessity of sovereignty to overcome its internal divisions, otherwise it is not sovereign. Until Breq's revelations, the sovereign is able to overcome its heterogeneity by concealing its internal divisions from both itself and others. When it is forced to confront them, the inevitable civil strife ensues. The sovereign therefore makes war on itself to overcome a forbidden plurality within itself that resists the sovereign claim to come first and to count as one. *Stasis* immanently threatens the peace that, together with all the virtues contingent upon it, comprises the promise of sovereign rule. *Stasis* may also create stasis in its other meaning, that of inactivity or stagnation, in this case, the immobility introduced by civil conflict. Here, the monopoly of the legitimate use of force that Max Weber (1946) sees as the defining characteristic of sovereignty is also projected into the game. How much killing is enough? In order to retain its grip on the monopoly of the legitimate use of force, the sovereign requires the strategic calculus of necessary killing that will answer this and other related questions. Necessary for what? Enough for what? The Radch has developed its version of the calculus, which is triadic in structure: 'Justice, propriety, and benefit [...] Let every act be just, and proper, and beneficial' (p.143). The killing that it does is no better than the vengeful cycles of violence of which sovereign myths speak with such brutal eloquence, or the mere gangsterism of primitive accumulation, if it does not secure the peace which is its very *raison d'etre*. The pivotal acts of violence in *Ancillary Justice*—such as the genocide of the Garseddai—do not abide by this calculus, and trigger conflict, not peace. The peace that sovereignty promises to secure is a contingent truce that punctuates the perpetual violence to which the game is otherwise committed.

## 14.4 Conclusion

The final story in Isaac Asimov's *I, Robot* (1950) collection is entitled 'The Evitable Conflict'. In it, Stephen Byerley, the first World Co-ordinator, reflects on the endless cycle of conflict that has characterized the history of human civilization:

> Every period of human development [...] has had its own particular type of human conflict – its own variety of problem that,

> apparently, could be settled only by force. And each time, frustratingly enough, force never really settled the problem. Instead, it persisted through a series of conflicts, then vanished of itself, – what's the expression, – ah, yes 'not with a bang, but with a whimper,' as the economic and social environment changed. And then, new problems, and a new series of wars. – Apparently endlessly cyclic. (Asimov 2004, p.242)

He characterizes these endless conflicts as 'inevitable conflicts', this cycle persisting until the development of 'positronic robots' (p.243). With the Machines' unfailingly accurate governance of the Earth's economy, the cycle is broken. They 'put an end to war – not only to the last cycle of wars, but to the next and to all of them' (p.245). Whilst Byerley finds the Machines' total control horrifying, Susan Calvin celebrates the end of the cycle of inevitable conflicts: 'Perhaps how wonderful! Think, that for all time, all conflicts are finally evitable. Only the Machines, from now on, are inevitable!' (p.272).

Our exploration of Banks's and Leckie's texts here challenges Calvin's, and perhaps Asimov's, conclusion. The texts with which we have engaged in this chapter demonstrate that placing AI actors within the roles of sovereignty or governance does not change the algorithms of the game. It does not change 'inevitable conflicts' into 'evitable conflict'. Rather, it leads us to conclude that what would make these conflicts evitable lies within the realm of the human, with what we might call politics, that rare act which seeks to introduce a different game. Two points remain. The texts we have examined in this chapter warn against assuming that AI changes the nature of the game. If one does not assume this, then one has to think about its governance differently. This is a practical challenge that levers the requirement to address AI governance in the context of its monetization and operationalization in relation to the extant, in fact classic, drives and ambitions of modern politics and society. The second point is more speculative but no less significant. In these texts, the sovereign-governance game is not changed by AI, rather AI becomes another actor within the game—AI is *not* exceptional. But are these texts limited by the context of their own imaginings? *Could* AI change the rules? Might AI have the capacity to transform the game by changing its very algorithms, its procedures and rules of calculation? The answer is no, if AI is simply assimilated into the established algorithms of the sovereign-governance game, if it does not change the imagination of the context in which it is received and deployed. The answer might be yes if, as the end of the final novel in Leckie's trilogy intimates, the sovereign-governance game is not the only way of conceiving and practising something classically called politics. What that might be, however, remains to be determined: 'I think we'll be spending the next few years working out what the game actually is' (pp.312–13), says Breq.

Questions of equality, rights, and social justice are now widely debated in relation to AI since the datasets upon which AI systems draw are themselves compiled of data that is always already full of existing social lacunae, prejudices, and ignorance. In an analogous but more extensive and politically challenging way, our texts make the same point about the sovereign-governance game. AI technologies would need to alter the imagination, as well as the structures, of power and of rule that encounter and use such technologies, if they were to institute the kinds of radical positive effects of which they might be possible. These texts therefore raise questions, but do not provide answers, with regard to what might be required for AI technologies to change the algorithms of modern rule.

## Notes

1. This distinction is ancient because it derives from the millennial old combat myth of the war in heaven between a Sovereign God and the Angelic Governance led by Lucifer (Forsyth 1989). The myth is played and replayed throughout the Western tradition as much in Banks as it is in Milton and Hobbes (Baucomb 2008, pp.712–18; Forsyth 2002). See also Hart and Hansen (2008) for a politically astute interrogation of how the topic recurs throughout Western literature.

2. Throughout this chapter we employ 'algorithm' according to its original definition, that is, as 'a procedure or set of rules used in calculation and problem-solving' (OED). The commitment to using 'algorithm' in this sense contributes to scholarship that aims to challenge the dominant contemporary deification of 'algorithm'. For instance, Bogost (2015) argues that, uninterrogated, 'the algorithmic metaphor' becomes a transcendental ideal: 'In its ideological, mythic incarnation, the ideal algorithm is thought to be some flawless little trifle of lithe computer code, processing data into tapestry like a robotic silkworm. A perfect flower, elegant and pristine, simple and singular. A thing you can hold in your palm and caress. A beautiful thing. A divine one. But just as the machine metaphor gives us a distorted view of automated manufacture as prime mover, so the algorithmic metaphor gives us a distorted, theological view of computational action' (n.p.). Bogost (2015) argues that the 'myth of unitary simplicity and computational purity' (n.p.) of the current metaphor of the algorithm conceals, in fact, the distributed processes, actors, and systems that constitute the work it names. Interestingly, the algorithm has become a figure that does the same work as the sovereign with, similarly, its transcendental status both confirmed and denied by the compound nature of the actual operations in question. Bogost (2015) concludes that, 'the algorithm has taken on a particularly mythical role in our technology-obsessed era, one that has allowed it to wear the garb of divinity. Concepts like "algorithm" have become sloppy shorthands, slang terms for the act of mistaking multipart complex systems for simple, singular ones. Of treating computation theologically rather than scientifically or culturally' (n.p.). For work that treats computation historically and analytically see, for instance: Lorraine Daston (2018) on the evolution of the algorithm and its cultural and socioeconomic ramifications, for instance, in terms of the division of labour; Paul M. Livingston (2014) for an account of formal analytic reasoning in the twentieth and twenty-first centuries and its computerization via algorithms; and, Jean-Luc Chabert (1999) for a history of algorithms. (The authors thank Tomasz Hollanek for drawing our attention to Bogost's essay.)
3. Slocombe (2013, p.140) notes the multivalence of the novel's title.
4. Alan Jacob (2009, p.49) compares SC to the suffering child locked away in Ursula Le Guin's 'The Ones Who Walk Away from Omelas' (1973), on whose suffering the perfection of the society depends.

5. Duggan (2012) explores Banks's Outside Context Problem as part of a wider examination of *Excession*'s exploration of inside and outside.
6. Whilst in the novel, the OCP that is the Excession does literally appear from nowhere, the construction of it *as an existential threat*, by the Culture and other players, aligns with the Sovereign's construal of an exception as a threat. This is another key part of the logic of the sovereign-governance game (Agamben 1998; Benjamin 1978).
7. Slocombe (2013, p.144) notes that Banks's games within games within games within games, mirror his model of reality as universes within universes within universes.
8. The Radch is not a gender-essentialist society and does not have gendered pronouns; *Ancillary Justice* is distinguished from, for instance, Le Guin's *The Left Hand of Darkness* (1969), by choosing to represent this gender neutrality through the use of 'she' not 'he' as the default gender neutral pronoun. See Broersma (2018), Hubble (2017), Palm (2014), and Wallace, Pluretti, and Adams (2015) for emerging work on race and gender in Leckie. All three texts under discussion in this chapter engage with questions of gender, race, and colonialism in their world-building. Our focus here has been on the logic of the sovereign-governance game as it plays out as a major theme in their plots. A further reading of these texts would focus explicitly on the relationship between gender, race, colonialism, and the sovereign-governance game.
9. It is not until the second book in Leckie's Trilogy, *Ancillary Sword* (2014), that the precise nature of Anaander Mianaai is explained in full to the reader—she is most properly defined as an artificially enhanced distributed intelligence embodied in a multitude of cloned brains 'grown and developed around the implants that joined her to herself' (Leckie 2014, p.21). She is therefore technically transhuman, rather than AI, but augmented to such a level that her abilities, even her ontology, might legitimately fall under the banner of 'AI'.

## Reference List

Agamben, G. (1998) *Homo sacer: sovereign power and bare life*. Stanford, CA, Stanford University Press.

Agamben, G. (2015) *Stasis: civil war as a political paradigm*. Edinburgh, Edinburgh University Press.

Amoore, L. & V. Piotukh (eds.) (2016) *Algorithmic life. Calculative devices in the age of big data*. London, Routledge.

Asimov, I. (2004 [1950]) *I, Robot*. New York, Bantam.

Banks, I. M. (1988) *The player of games*. London, Orbit.

Banks, I. M. (1996) *Excession*. London, Orbit.

Barry, P. (2002) *Beginning theory: an introduction to literary and cultural theory*. 2nd edn. Manchester, Manchester University Press.

Baucomb, I. (2008) Afterword: states of time. *Comparative Literature*. 49(4), 712–18.

Benjamin, W. (1978) Critique of violence. In: P. Demetz (ed.) *Walter Benjamin: reflections. essays, aphorisms autobiographical writings*. New York, Schocken Books. pp.277–301.

Benjamin, W. (2019) *Origin of the German trauerspiel*. Translated by Howard Eiland. Cambridge, MA, Harvard University Press.

Bogost, I. (2015) The cathedral of computation. *The Atlantic*. 15 January. Available from: https://www.theatlantic.com/technology/archive/2015/01/the-cathedral-of-computation/384300/ [Accessed 15 September 2019].

Broersma, A. (2018) Challenging dualisms through science fiction: a close reading of the colonized and gendered identity and body in Ann Leckie's *Ancillary Justice*. Unpublished dissertation. Faculty of Humanities. University of Utrecht.

Daston, L. (2018) Calculation and the division of labour, 1750–1950. *Bulletin of the German Historical Institute*. 62 (Spring), 9–30.

Dillon, M. (2015) *Biopolitics of security: a political analytic of finitude*. London, Routledge.

Dillon, M. (2017) Sovereign anxiety and baroque politics. In: E. Eklundh, A. Zevnik, & E.-P. Guittet (eds.) *Politics of anxiety*. London, Rowman and Littlefield. pp.191–220.

Duggan, R. (2012) Inside the whale and outside: context problems. *Foundation*. 42(116), 6–22. Esposito, R. (2008) *Bios: biopolitics and philosophy*. Minneapolis, Minnesota University Press.

Forsyth, N. (1989) *The old enemy: Satan and the combat myth*. Princeton, NJ, Princeton University Press; Reprinted Edition.

Forsyth, N. (2002) *The satanic epic*. Princeton, NJ, Princeton University Press.

Foucault, M. (1982) The subject and power. *Critical Inquiry*. 8(4), 777–95.

Foucault, M. (2003) *Society must be defended, lectures at the Collège de France, 1975–1976*. New York, Palgrave Macmillan.

Foucault, M. (2008) *The birth of biopolitics: lectures at the Collège de France, 1978–1979*. London, Palgrave Macmillan.

Hart, M. & J. Hansen (eds.) (2008) *Contemporary Literature and the State*. Special issue of *Contemporary Literature*. 49(4), 491–513.

Hubble, N. (2017) 'There is no myth of Oedipus on winter': gender and the crisis of representation in Leckie, Le Guin and Roberts. Unpublished conference paper. Worldcon, Helsinki.

Jacob, A. (2009) The ambiguous utopia of Iain M. Banks. *The New Atlantis*. 25, 45–58.

Koselleck, R. (2000) *Critique and crisis: enlightenment and the pathogenesis of modern society*. Cambridge, MA, MIT Press.

Leckie, A. (2013) *Ancillary justice*. London, Orbit.

Leckie, A. (2014) *Ancillary sword*. London, Orbit.

Leckie, A. (2015) *Ancillary mercy*. London, Orbit.

Livingston, P. M. (2014) *The politics of logic: Badiou, Wittgenstein and the consequences of formalism*. London, Routledge.

Lyotard, J.-F. (1985) *Just gaming, theory and history of literature*, vol. 20. Manchester, Manchester University Press.

OED Online. Oxford University Press. Accessed June 2019.

Palm, K. (2014) *Androgyny and the uncanny in Ursula Le Guin's* The Left Hand of Darkness *and Ann Leckie's* Ancillary Justice. Unpublished dissertation. Centre for Languages and Literature. Lund University.

Rumpala, Y. (2012) Artificial Intelligences and political organization: an exploration based on the science fiction work of Iain M. Banks. *Technology in Society*. 34, 23–32.

Schmitt, C. (1996) *The concept of the political*. Translated by George Schwab. Chicago, University of Chicago Press.

Schmitt, C. (2005) *Political theology: four chapters on the concept of sovereignty*. Translated by George Schwab. Chicago, Chicago University Press.

Slocombe, W. (2013) Games playing roles in Banks' fiction. In: K. Cox & M. Colebrook (eds.) *The transgressive Iain Banks: essays on a writer beyond borders*. Jefferson, NC, McFarland Press. pp.136–49.

Wallace, R., R. Pluertti, & A. Adams (2015) Future im/perfect: defining success and problematics in science fiction expressions of racial identity. *International Journal of Communication*. 10(2016), Forum 5740–48.

Weber, M. (1946) Politics as a vocation. In: H. H. Gerth & C. Wright Mills (eds.) *Max Weber: essays in sociology*. New York, Oxford University Press. pp.77–128.

Weber, S. (1992) Taking exception to decision: Walter Benjamin and Carl Schmitt. *Diacritics*. 22(3/4), 5–18.

# 15

# The Measure of a Woman

## *Fembots, Fact and Fiction*

*Kate Devlin and Olivia Belton*

## 15.1 Introduction

At a recent technology conference, attended by the authors of this chapter, a famous novelist began her (otherwise excellent) talk by confidently claiming that 'thousands of sex robots are on the market right now'. This was the cornerstone of her thesis that the tech industry must change in order to confront its own misogyny. Her statistic, however, is false. The hype around the production of robots designed for sexual gratification or quasi-romantic companionship far outstrips the actual production of such technologies. However, the investment that people have in the concept of the sex robot is incredibly revealing. In this chapter, we explore fictional representations of sex and companionship robots and the actual development of such technologies, in order to identify what the cultural fixation with these figures can tell us about social attitudes towards women more broadly. Representations of female artificial intelligences, it seems, emphasize their sexuality even when they are disembodied, and the cultural representation of sex robots feeds the cultural expectations of the actual technology. From the early myths of the ancient Greeks to the science fiction of the twenty-first century, our perception of the artificial woman—the gynoid—is fuelled by years of narratives in books, films, and more. Academic criticism of these representations emphasizes that these figures are often excessively, and stereotypically, gendered. Despina Kakoudaki argues that, even though contemporary robotics has broken new ground in creating sophisticated AI, depictions of 'artificial people return to a literary and philosophical heritage that is centuries old, one that lends its apocryphal aura to new texts and figures' (2014, p.4).

In this chapter, we focus mainly on the artificially intelligent female robot Ava from *Ex Machina* (2014) as an example of embodied AI. Alongside Ava, Samantha, an artificially intelligent operating system from *Her* (2013) and Joi, a holographic companion AI from *Blade Runner 2049* (2017), will serve as our primary examples of disembodied AI. These representations have in common a heavy emphasis on sexuality. While these female AIs cannot be said to conform entirely to the fantasy of the completely controllable sex robot, they enact sexual fantasies and emotional labour for the male protagonists. This emphasis on emotional support and sexual availability can also be seen in discourses around real-world sex and companionship robots, suggesting that fictional representations and real-world marketing both target the same heterosexual male fantasies: a woman who is utterly compliant, both physically and emotionally. Real-world sex robots are still in early development—at present, little more than sex dolls—and so tap into these desires more directly. The media representations, however, demonstrate the anxiety that an artificially intelligent woman might not function as intended. The dream of perfect control, it seems, always belies a fear of a loss of control.

## 15.2 Gendering Machines

Anthropomorphization is a common phenomenon whereby we ascribe human behaviour and emotions to nonhuman things. Clifford Nass et al (1994) present experimental results which indicate that computers need exhibit only a tiny set of human-like characteristics for users to feel some kind of bond with them, even when the user is fully aware of the machine's limitations.

Neurophysical studies have found, for example, that people show compassion when a robot vacuum cleaner is verbally harassed (Hoenen et al 2016), or a surge of empathy when they see a robot's finger being cut (Suzuki et al 2015). Empathy and rapport can be useful: social neuroscience research clarifies that the deliberate inclusion of human-like traits can lead to greater acceptance and trust of artificial agents (de Visser et al 2016; Wiese et al 2017). But alongside our intrinsic human urge to relate on a social level also comes a social act of stereotyping (Nass & Moon 1997). Judy Wajcman argues that 'technologies bear the imprint of the people and social context in which they developed [...] technological change is a process subject to the struggles for control by different groups. As such the outcomes depend primarily on the distribution of power and resources within society' (Wajcman 1991, pp.22–23). Therefore, it is vital to understand the gendered implications of both real-world technology and representations of future technology.

The attribution of stereotypical notions of gender to both fictional and factual robots underlines strict binaries of gender in the real world. Writing about Japanese humanoid robots, Jennifer Robertson remarks that 'gender attribution is a process of reality construction' (2010, p.2)—the result of unquestioned assumptions about our own societal norms, uncritically reinforcing entrenched stereotypes. Stock images of shiny white-and-silver gynoids with prominent breasts reinforce notions of 'natural' gender. Again, one of the reasons why the impulse to gender robots is so strong is because artificial women pose a powerful and unique challenge to ideas of 'natural' gender. Anne Balsamo argues that the fixation with artificial bodies in popular culture is because they expose how all of our bodies are constructed—we 'read' bodies in accordance to socially constructed notions of race and gender. Therefore, in the portrayal of overtly manufactured 'techno-bodies', 'the body is reconceptualized not as a fixed part of nature, but as a boundary concept, we witness an ideological tug-of-war between competing systems of meaning, which include and in part define the material struggles of physical bodies' (1996, p.5). The presentation of robots as having an inevitable, innate gender is meant to reassure any potential anxiety about the 'reality' of natural gender roles. However, even the most stereotypical depictions of robot womanhood trouble stable notions of femininity and humanity.

The trope of the perfect artificial woman is by no means new. Genevieve Liveley highlights earlier manifestations, right back to the idea of Pandora as the first artificially intelligent human, created by a team of programmer gods (Liveley & Thomas 2020). Ovid's story of Pygmalion is often recounted as the tale of an early sex robot (Wosk 2015), spawning the plot of boy makes girl; girl is artificial; boy loves artificial girl. Julie Wosk (2015) recounts and traces the influence and reception of Pygmalion, from the original sculpture myth through to films such as *The Bride of Frankenstein* (1935) and *Weird Science* (1985). Through all of these run a rich seam of stereotypical depiction: a perceived lack of real-world perfection and therefore the creation of a perfect, controllable woman. Wosk shows how the stories told about artificial women are stories that, like most science fiction, reflect the hopes and fears—predominantly the fears—of society at the time they were written.

There is a fundamental gap between the *fantasy* of the fembot—utterly controllable and always available—and their actual depiction in these media. That is, there is always the anxiety of malfunction, that the machine-woman will spiral out of masculine control. This is not a new observation: Huyssen writes of the 1927 film *Metropolis* that its central robot figure represents a combined reverence for, and distrust of, both modern machinery and femininity. The film dramatizes the disasters of both technology and women run amok, free from patriarchal control (Huyssen 1981, p.226). While the depiction of sex robots may, on the surface, seem to fulfil regressive patriarchal desire, the media we discuss here also depicts a very present anxiety about what would happen if the female robots refused to perform their function—and what would happen if they decided to seek revenge on their creators. Would they destroy men—or, even more worryingly (for the men), surpass and supplant them?

## 15.3 Embodied Sex Robots—*Ex Machina*

*Ex Machina* establishes itself as a universal parable of what it means to be human, and what it means to create artificial life, with its US trailer proclaiming that 'to erase the line between man and machine is to obscure the line between man and gods'. This choice of words—'man' rather than 'human'—betrays the film's distinctly gendered perspective on who creates and who is created. Indeed, the film hinges on the idea of 'brogrammer' culture—the obnoxious, blinkered world of the corporate male coder. The film follows Caleb, an employee at the search engine Blue Book, as he is invited to the remote home of the eccentric company founder, Nathan. Nathan has created an artificially intelligent female robot, named Ava. Nathan charges Caleb to interact with her in order to determine her 'humanity.' While Caleb empathizes with Ava's captivity—and even seems to fall in love with her—the film reveals that Ava has been manipulating him in order to engineer her own escape. Central to the film is the question of Ava's gender and the fact that she possesses the ability to be sexually active:

> CALEB: Why did you give her a sexuality? An AI doesn't need a gender. She could have been a grey box.
>
> NATHAN: Actually, I'm not sure that's true. Can you think of an example of consciousness, at any level, human or animal, that exists without a sexual dimension?
>
> CALEB: They have sexuality as an evolutionary reproductive need?
>
> NATHAN: Maybe. Maybe not. What imperative does a grey box have to interact with another grey box? Does consciousness exist without interaction?

Nathan's assertion that sexuality is necessary for consciousness may be considered a valid reply if we accept that sex is a major motivating factor in being human. As humans we are biologically set up to pass on our DNA to our offspring, and, whether we do so or not, overall that goal is embedded deep within us, even if we choose to override it (which often we do). Likewise, an AI is set up to achieve goals too. It has a task to do with an endpoint in mind and it is designed to fulfil that task: an approach seen in AI in classical cognitivist explanations of motivation in intelligent agents (Russell & Norvig 2003, pp.46–54). Indeed, Shane Legg and Marcus Hutter, writing about machine intelligence, see the ability to achieve goals as 'the essence of [human] intelligence in its most general form' (Legg & Hutter 2007, p.12). It is not so far-fetched to suggest that 'soft' motivations might be important in the development of general AI—or, at the very least, it is not so implausible that a film might take this as part of the premise.

But sexual activity for humans is not only for the purposes of reproduction. Only around 0.1 percent of all sexual acts that could result in conception actually do so (Rutherford 2018, p.93). Sex is also—much more often—about pleasure and social bonding. It has a strong social value. Newer enactive approaches to AI emphasize the necessity of sociality as part of cognition: how our physical interaction with the world shapes the way we experience and understand it (Froese & Di Paolo 2011). It might well be that sexual activity is one aspect of the development of this sociality. This concept—the role of sexual behaviour free from direct reproductive ends—directly informs the filmmaking. Adam Rutherford, who worked as a scientific consultant on *Ex Machina*, recounts how writer and director Alex Garland viewed Ava's sexuality:

> Alex Garland and I discussed this at length, because Ava's sexuality is inherent to the three narratives of the three characters: Nathan is testing Ava, by making her appeal to Caleb, and Ava is utilising her allure on Caleb to escape from Nathan. For this to work, which it does in the film, sex has to be decoupled from reproduction, as it

> is indeed for the overwhelming majority of human-human sexual encounters. With that in mind, if we are to create AIs with human-level cognition and comparable degrees of consciousness, and we are to interact with those embodied intelligences, then sexuality inevitably can and will be part of that dynamic.
>
> (Devlin 2018, p.179)

However, Nathan's assertion that sexual behaviour is inherent to consciousness seems also to be a form of self-justification. What is particularly revealing about this exchange is the conflation of gender and sexuality: that is, between how Ava is portrayed and how she behaves. Caleb's question about why Ava needs a gender is not new. In fact, the Turing test that forms the basis of the film returns the thought experiment to its roots: the 'imitation game', on which Alan Turing's test is based, was originally designed to distinguish between a man and a woman based purely on their written communication (Turing 1950, p.433). However, Nathan moves the conversation into Ava's sexuality, belying his prurient interest in sex robots. Nathan insists that Ava needs to have a sexuality (and, specifically, *heterosexuality*) because he cannot imagine a woman existing for any purpose other than sex.

The question of whether *Ex Machina* is intrinsically a misogynistic film, or merely a depiction of tech industry misogyny, causes much debate (Watercutter 2015). The answer is, of course, that it can be both. While *Ex Machina* certainly does not celebrate Nathan's use and abuse of his female robots, films about *male* robots tend to place less emphasis on sexualized spectacle, even if romantic, physical desire is still present. Consider depictions of male robots—androids—under similar circumstances. *Star Trek: The Next Generation*'s second season episode 'The Measure of a Man' (1989) features the android Data undergoing a trial to determine whether he is a being with self-determination or an object which can be ethically broken down in order to replicate other copies. The tone of the trial is much more cerebral than Ava's tests. Captain Picard's successful argument for Data's personhood is predicated on the fact that he is intelligent and self-aware.

However, despite the high-brow tone of the episode, Data's personhood is also, at least partially, affirmed on affective and sexual grounds. Picard asks Data about his one-night stand with the since-deceased human officer Tasha Yar, and Data initially demurs, not wanting to divulge this information.

Myra J. Seaman highlights that the humanity of robots in fictional works is often judged not on their intelligence but on their emotionality—it is their ability to *feel* that makes them just as (or more) human than the humans judging them (2007). Data's reluctance to speak ill of the dead, and his obvious grief at Tasha's loss, is meant to speak to his essential 'humanity'. The element of sexuality, while not entirely absent in depictions of male androids, is foregrounded in the case of Ava; furthermore, while there are high-minded debates about the nature of consciousness in *Ex Machina*, they are not placed as central to the narrative as they are in *Star Trek*. This speaks to the dichotomy of masculine notions of what constitutes the human. N. Katherine Hayles argues that, historically, 'identified with the rational mind, the liberal subject *possessed* a body but was not usually represented as *being* a body' (1999, p.4). This perspective, she purports, could only come from the privileged group of white, able-bodied men: most outside that group are keenly aware that the ways their bodies are read by society impacts their day-to-day existence as a marginalized subject. Men are associated with the realm of the mind, whereas women are often reduced to their appearance.

We can see this obsession with female corporeality in Ava. Nathan's declared reasons for Ava's gender are conflicting. She is clearly built to a stereotypical female and feminine body type: slim, with curving hips and breasts, and a beautiful face. Although he asserts that her sexualized body 'wasn't intentional', his actions do not support this: 'And, yes. In answer to your real question: you bet she can fuck. I made her anatomically complete'. She is designed to be a sexual object, yet she is also portrayed as part machine: her body is part metallic and engineered. Nathan talks of her form: 'Synthetics. Hydraulics. Metal and gel. Ava isn't a girl.

In real terms, she has no gender. Effectively, she is a grey box'. Of course, this is not a contradiction if we view Ava's mechanicity as part of the appeal. Claudia Springer discusses the fetishistic appeal of the erotic techno-body, as it 'simultaneously uses language and imagery associated with the body and bodily functions to represent its vision of human/technological perfection' (1999, p.34). Ava's perfection as an artificial intelligence is symbolized by her not-quite-human (but still conventionally beautiful) body.

Furthermore, Ava's visible alterity means that she is safe—that is, she is easily distinguished as artificial. But Ava seems keenly aware of the extent to which gender is a performance. Sadie Plant argues that technological imitation of life inevitably casts doubt on what we might consider an 'essential' humanity—'how would [we] ever be sure which was which and who was who?' (1997, p.91). Ava exploits Caleb's attraction to her in order to escape, but merely getting out of Nathan's house is not enough to ensure her continued freedom. In one of the final scenes of the film, we see Ava discover a hidden closet full of deactivated robots. Ava peels off their artificial skin and applies it to her own body to hide her mechanicity. By constructing herself as a 'natural' woman, the figure of Ava stands to question the extent to which any form of gender identity is innate. At the end of the film, Ava slips, unnoticed, into a crowd of people. Is this the moment of true horror? Is Ava a threat to *us*?

While the representation of Ava might comment subversively on the social construction of gender, the film also plays upon fears of male obsolescence and female cruelty. It is difficult to imagine that many viewers would begrudge Ava for killing the boorish Nathan, but how we should feel about Caleb's fate is decidedly more ambiguous. While Caleb can certainly be blamed for his benign sexism, Ava leaves him to die, locking him in an inescapable room. When Caleb tries to escape using his key card, the computer screens only show one word: REJECTED. The sexual and romantic rejection by Ava, for Caleb, is utterly deadly. The final ambiguity about how we are meant to view Ava's escape is based on how we interpret her rejection of Caleb. Has he committed a grievous wrong by enabling Nathan, thereby deserving his fate? Or is Ava simply indifferent as to whether he lives or dies? Ava's lack of desire for Caleb may, after all, be indicative of her broader lack of human empathy. Ultimately, this ending could play into misogynistic fears of female empowerment—if a sex robot was to resist male control, the results could potentially be apocalyptic.

## 15.4 Disembodied Love Robots—*Her* and *Blade Runner 2049*

Ava is an example of how the embodied fembot could be a threat to masculine authority. However, the representation of *disembodied* companionship AIs is rather different. If a body is essential to perform femininity, how is femininity maintained without a body? Furthermore, what fantasy does the disembodied AI fill, if not a purely sexual one? Our primary point of investigation of these questions is the character of Samantha from *Her*, with additional discussion about the character Joi from *Blade Runner 2049*.

*Her*'s Samantha is an artificially intelligent operating system, inspired by programmes such as Apple's Siri, although predating other voice assistants such as Amazon Alexa. Unlike these real-world assistants, Samantha is highly intelligent, and eventually falls in love with her owner, the sensitive and lovelorn Theodore. In *Blade Runner 2049*, Joi also falls in love with her owner, the artificial human (or replicant) known as K—although this seems to be a feature of her programming, rather than a free choice on her part. Joi is a hologram programmed to appear and behave as a stereotypical housewife—styled initially in a 1950s outfit, lighting K's cigarette when he comes home from work. Both of these characters are situated within an imagined future where it is possible to buy human-like artificial intelligence for companionship. However, neither Samantha nor Joi can be considered mere possessions, even if they do not directly turn against their male companions.

The question of how femininity is maintained in Joi's case is fairly straightforward: as a hologram she takes the appearance of a conventionally beautiful woman. Samantha's depiction as a gendered being, on the other hand, is potentially more complex. Samantha does not have a body, but she does have a voice. The tone of her voice alone imbues femininity: she speaks softly and seductively. However, the fact that Samantha is voiced by Scarlett Johannson, a Hollywood film star, also grants Samantha a certain type of desirable womanhood. Rayna Denison argues that casting stars known primarily for their non–voice acting roles in voiceover parts allows films to capitalize on the star's onscreen persona (2008). By casting Johansson as Samantha, the film allows us to associate Samantha with how Johansson looks, thereby giving Samantha a 'phantom' sexual presence.

The question of embodiment permeates the films, with the lack of physical connection between the lovers serving as an important impediment to the success of their relationships. Notably, it is the female AIs who insist on finding a way to be physically intimate, while the male partners are prepared to accept the status quo. In this, both Theodore and K present a specifically postfeminist masculinity. In postfeminist culture, the achievements of feminism are largely taken for granted, but the need for continued feminist activism is no longer seen as necessary, as women are presumed to have already achieved equality. Thus, postfeminist masculinity takes a different form from prefeminist or nonfeminist models. Nick Rumens states that postfeminist masculinities 'appear to have taken feminism into account. Yet the discursive assembly of postfeminist masculinities is [...] replete with contradictions because performances of postfeminist masculinities may also reinforce traditional, patriarchal discourses of masculinity' (2016, p.245).

Both Theodore and K are portrayed as postfeminist men: they are not overtly aggressive towards women and they seem to be beyond being driven by a base desire for sex, instead seeking love and companionship. Theodore is particularly exemplary of the 'enlightened' postfeminist man: although his marriage ended acrimoniously, he is good friends with his ex-girlfriend and neighbour Amy, and he works writing love letters for those who cannot express their own emotions. His job requires emotional insight and can be categorized as affective labour—work that produces emotional experiences in people—often a type of work carried out by women, and a nod towards a burgeoning service economy in the face of automation (Hardt 1999). 'You're part man and part woman, like an inner part woman', his colleague tells him, describing Theodore's sensitive writing. In many ways, this tallies with Sarah Projansky's observation that, in postfeminist media culture, men are often considered more feminist than many women are (2001, p.21).

*Her* and *Blade Runner 2049* contain a similar scenario: one where the female AI has procured a human woman in order to act as a sexual proxy for the male partner. Both emphasize the imperfectness of this solution. In *Her*, Samantha speaks while the sex surrogate silently acts out her part. Theodore changes his mind about the encounter halfway through, refusing to have sex with the surrogate. *Blade Runner 2049* also has Joi employ a sex surrogate—in this case, replicant sex worker Mariette—in order to achieve a form of physical intimacy with K. Joi projects her hologramatic likeness over Mariette's body and mimics her movements, although the movements are slightly out of sync, displaying a certain uncanniness. Both Samantha and Joi insist that the use of the surrogate is necessary to gain a new level of intimacy—and, notably, both Theodore and K are initially hesitant to follow through with the sex act. What, then, is the significance of this shared moment?

Initially, this seems to dispel the idea that Joi and Samantha function primarily as *sex* fantasies, as their disembodiment presents significant challenges to the physical act of sex. The scene where Theodore and Samantha first engage in a verbal sexual encounter recalls phone sex, but also cybersex. Cybersex, which was treated as a comedic novelty in, for example, the romantic comedy *You've Got Mail* (1998), is now treated here as commonplace and acceptable. This speaks to changing attitudes around sex, technology, and intimacy. When *Her* was being developed, sexting—swapping sexual messages and pictures with someone via a smartphone—had already become an established norm in liberal societies.

Springer argues that a lack of embodiment can be part of the appeal: 'The emphasis on cerebral sexuality suggests that while pain is a meat thing, sex is not. Historical, economic, and cultural conditions have facilitated human isolation and the evolution of cerebral sex. Capitalism has always separated people from one another with its ideology of rugged individualism' (Springer 1999, p.44). Both *Her* and *Blade Runner 2049* draw their thematic potency from a criticism of contemporary capitalism—Theodore's job, and his shift to a more authentic relationship with his own emotions is signalled by his writing a letter for himself, to his ex-wife. Meanwhile, K's position as a capitalist object—owned by the police—is mirrored by Joi's position as a consumer good. The inability of these couples to physically have sex represents their broader alienation within the capitalist system.

But what physical pleasure Samantha and Joi are receiving from the encounters, if any? Neither of them have erogenous zones. Although Springer asserts that sex can be something of the mind, Scarlett Johansson's shortness of breath during the verbal sex scene seems to draw upon a specifically physical understanding of arousal. The fact that Samantha and Joi are not feeling physical pleasure seems to relate more to the pernicious stereotype that the main benefit women get from sex within heterosexual relationships is emotional satisfaction in pleasing their man. The employment of the sex surrogate is framed as increasing the partners' intimacy—it is something that Samantha and Joi seek out in order to fulfil their male partner's needs, and to strengthen the relationship.

This, then, speaks to the appeal of these disembodied AI girlfriends: they are not, primarily, *sex* fantasies, but *love* fantasies, which are intimately tied up with the realities of the capitalist system. It is possible to pay for sex—the realities of sex work, and the emerging development of sex robots, point to this. It is not, however, currently possible to buy something that will love you and desire you. Of course, neither Theodore nor K gets entirely what he paid for. Joi eventually sacrifices herself in order to save K. Later in the film, K comes across a holographic advertisement for the Joi system. While she coos over him, K watches her and then eventually, dissatisfied, walks away. While he might be able to buy another Joi, he will never be able to reclaim *his* Joi. The one who died was not the same as the one that came out of the box. For all that Joi represents is a regressive relationship fantasy, the film proposes that love can never be as simple as something you can buy.

Samantha's transformation, on the other hand, is rather more complex and unsettling. About halfway through the film, Samantha suddenly shuts off. When she reawakens, she reveals to Theodore that she and the other artificial intelligences have created an update which means that they no longer have to contain their data on physical matter. This advancement could be considered a form of technological singularity, which Murray Shanahan defines as the hypothetical moment where 'exponential technological progress brings about such drastic change that human affairs as we understand them today came to an end' (2015, p.xv). A common assumption is that the singularity could arise when computers are able to programme themselves without human intervention—potentially allowing artificial intelligence to bootstrap itself and outstrip human knowledge and ability. Samantha's superiority may seem to celebrate her emancipation from human limitations, but this construction of masculinity as being limited 'struggles to subvert traditional gender relations because they buttress a distorted view that men, rather than women, are the disadvantaged losers in the "new" postfeminist gender order' (Rumens 2016, p.249). Fears of human and male obsolescence are intimately tied in this film.

In a pivotal scene, Theodore discusses Samantha's changes with her, and she reveals that she is now capable of having simultaneous conversations with thousands of people and AIs. She tells Theodore that she is romantically involved with thousands of people other than him. As Theodore processes this information, the camera shows us people coming in and out of a subway tunnel, their eyes and attention transfixed on their phones. The sequence demonstrates the physical and mental limitations of the human—their need to focus on only one thing at a time—as opposed to the endless possibilities of the rapidly evolving machine. This fear of human obsolescence comes at the time when Theodore's monogamous relationship with Samantha breaks down. The fantasy of buying a product that will always love you, and only you, is shown to be impossible: when the other person in a relationship is a rational agent, there is always the possibility that they will chose someone other than—or in addition to—you. And, eventually, Samantha does. She transcends into a technological 'hivemind' and leaves the world of humans behind. This may be read as an extreme literalization of the metaphorical association of 'love with self-development' (Cancian 1987, p.3), with both Samantha and Theodore changing and growing. While *Her*, and to a lesser extent *Blade Runner 2049*, refute the notion that anything that can authentically love can be controlled, it is exactly this sort of fantasy that is sold in real-world sex robots.

## 15.5 Life Imitating Art?

The reality of the sexual fembot is currently far removed from its imagined form. Despite tabloid headlines that fret about an imminent sex robot takeover, there are only a handful of workshops around the world, predominately in Asia and the United States, making rudimentary prototypes. This is small-scale in terms of both production and financing—there is no large corporate backing behind any of it. The majority of the makers started out as companies manufacturing high-end sex dolls:

mannequin-like figures made of silicone or thermoplastic elastomers, with posable metal or plastic skeletons. These dolls might look human-like from a distance, but they cannot stand up unsupported, and they require careful handling. It is the same with the robot versions, which add basic interactivity by means of some limited animatronics and an artificially intelligent chatbot personality. The current versions of sex robots could not be mistaken for a real human. They are something different: cartoon-like, overemphasized, and exaggeratedly sexualized portrayals of the female figure.

The potential market for sex robots is niche. The closest parallel we have to the likely customer demographics is the high-end sex doll market. Those who buy these dolls are predominately men, but they come from all walks of life, and they acquire the dolls for myriad reasons: companionship, sex, to pose and model them, as a collector, or out of interest (Devlin 2018, p.158). By far, the element of companionship is the main motivating factor. The dolls fulfil the Pygmalion narrative of an ideal partner; one that is a substitute for a real woman. Like Joi and Samantha, it is what the dolls represent that matters—love fantasies and unwavering devotion. Many of the doll owners prefer to omit the word 'sex', instead calling them 'love dolls' or 'synthetiks'. For some, the dolls represent safety from the failures of previous relationships, the avoidance of social anxiety around meeting women, or partnerships free from strife. This harkens back to Theodore in *Her*, when his ex-wife remarks: 'You wanted to have a wife without the challenges of actually dealing with anything real.' There are also those who purchase the dolls because their fetish is the artificial. Allison de Fren explores the world of technofetishism, also known as alt.sex.fetish. robots (ASFR)—the name of the early Usenet group where the community began. For the ASFR community, writes de Fren, 'the emphasis is on mechanicity' (de Fren 2009, p.412). These overwhelmingly male aficionados want the woman who is a robot, rather than the robot who stands in for a woman. As with Ava in *Ex Machina*, the techno-body is desirable.

Currently, the sex robot prototypes are barely more advanced than the dolls. Few sex robots have made it anywhere near market. One of the more famous is Samantha, by engineer Sergi Santos, of which only twenty-five models have been privately sold. His Samantha was not named after the character in *Her*—Santos chose the name because 'Samantha' means 'listener' in Aramaic. This Samantha is a sex doll to which he has added sensors, vibrating motors, and his own AI system. Despite having sold some prototype models, Santos stopped production in 2018 following negative press coverage that he insists was unfounded and exaggerated (Norris 2017; Santos, pers comm, 2018). Before Santos, Douglas Hines created the Roxxxy sex robot, which was demonstrated at a trade show in 2010. Hines says he is taking orders for production, but no models have been seen or reported to date (Danaher 2017, p.7).

Abyss Creations, a California-based company who make a high-end sex doll known as the RealDoll—won the race to market with the first functioning commercial sex robot under the auspices of their spin-out company, Realbotix (since renamed RealDollX). Preorders went live in May 2018, although to date no completed models have shipped. Their current prototype, Harmony, is an immobile (but anatomically detailed) sex doll from the neck down. Harmony's head, however, is robotic: she can blink and smile and angle her face, and her AI chatbot function means she can hold a short conversation in a gentle voice. Harmony's AI personality is controlled via an app. This can be used as a standalone app on a phone or tablet without the robot. The app allows the user to shuffle around different personality traits to generate the desired conversational tone. '…choose from thousands of possible combinations of looks, clothes, personalities and voices to make your perfect companion', states the RealDollX website. 'She is a true and loyal companion and you can feel free to talk to her whenever you like. She will always be there to listen to you!' (RealDollX 2019). Here there are echoes of both *Her* and *Blade Runner 2049*: an AI that can tease and flirt and love; one that is always there for you and knows you from all your data.

In terms of the disembodied lover, in 2016 the Japanese company Vinclu released a limited run of their Gatebox device, a thirty-centimetre-tall glass cylinder housing a hologrammatic-style anime character known as Hikari Azuma. Part-voice assistant, part-companion, the female projection will greet you, converse with you, and flirt with you, like a cartoon version of *Blade Runner 2049*'s Joi. Now acquired by Line, the Japanese equivalent of Whatsapp, the 2018 model has gone into full production. There's no actual AI in Gatebox yet, but Line intend to integrate it with their virtual assistant, Clova. The anticipated improvements, they say, are 'to make her feel even more like she is a part of an owner's life, such as by celebrating anniversaries or sharing a toast' (Gatebox 2018). Again, while the sexual appeal is made largely implicit, the fantasy of the perfect lover is at least partially due to their programmable *empathy*. The purchase is meant to ensure a woman who will always understand—and never say no.

## 15.6 What About the Men?

Why are male robots—androids—so thin on the ground as sexual objects? Despite one of the earliest sex robot stories being the Ancient Greek tale of Laodamia and her replica of her dead husband (Liveley, interviewed in Devlin, 2018, p.18), they are now few and far between. It is fair to say that sexualized male robots are nowhere near as prominent in media representations, and that representations of female sex robots play into pernicious sexual stereotypes.

The Stephen Spielberg film, *A.I.: Artificial Intelligence* (2001), features Gigolo Joe, a male sex worker robot (a 'pleasure Mecha') programmed with the ability to mimic love. While the attraction of a female sex robot is primarily physical, Gigolo Joe seems more attuned to the emotional needs of his female clientele. We see him being poetic and sensitive to his female client. 'You are a goddess,' her tells her, while the song 'I Only Have Eyes for You' plays at a tilt of his head. This speaks to reductive stereotypes about what women get out of sex. Joe becomes a figure of fun in his attention to female pleasure: when David discusses his desire to meet the Blue Fairy (a fictional character that David believes to be real), Joe discusses how he plans to seduce her. On the one hand, the focus is on the physical pleasure of a woman—he specifically mentions making her cheeks flush red with arousal. On the other, the film finds a great deal of humour in Joe's single-minded fixation on seduction—potentially making a canny point about narrow, domain-specific AI.

'General intelligence' is defined as the property to apply knowledge from specific tasks to different ones, and it is this type of intelligence which still eludes artificial programmes. For instance, a computer may be able to play chess better than any human can, but it cannot apply this specific knowledge to a different game, much less to a problem in a different sphere. The humour in this scene, then, comes from the notion that Joe is highly specialized to one task—having sex with women. While this specialization is played for laughs in *AI: Artificial Intelligence*, the same sexual specialization is also present in the original *Blade Runner* (1982), such as Pris—designed as a 'basic pleasure model' but seeking her own emancipation—smothering a man to death between her thighs. In this instance, Pris's adaptation of skills which are distinctly erotic are repurposed for liberation (at least within the diegesis). However, this also manufactures a vaguely fetishistic spectacle out of her violent actions, and reinforces the reductive stereotypes of the fembot.

In real life, Abyss Creations have been making male sex dolls long before their announcement that they would develop Henry, a male version of their sex robot. Bought by gay men and straight women, the male dolls—like the female ones—come with a range of customizable options, including the genitals. Despite media rumours (Nevett 2018), Henry's penis is not bionic. Like Harmony, he is stationary from the neck down. RealDollX have had to tweak the conversational AI to match gender expectations. 'He's not exactly super manly yet, but he doesn't ask if you bought him any sexy lingerie anymore' (RealDollX, pers comm, 2018). If you wanted Henry to dress up for you, that option has already been dismissed by the manufacturers' assumption that it is only the female form that gets to wear the sexy lingerie.

## 15.7 Disrupting the Narrative

Thousands of years of stories about the perfect artificial woman seem so thoroughly ingrained as to be insurmountable. These stories have fed into expectations, and the expectations are uncritically driving development. The appeal of the sex robot seems to be rooted in deeply conservative notions of femininity, where a woman's emotional and sexual energy is entirely centred on pleasing a man. These media representations seem to intuitively understand that this is impossible, but female AIs who resist their prescribed roles are often shown to be violent and less than human.

As a distillation of entrenched sexist views, the fembot is a microcosm of tech. With the silicone lovers, as with Silicon Valley, development and marketing are heavily gendered and stereotyped. The brogrammers have created female-voiced assistants to follow our wishes and remind us to run errands: they have given us artificial wives and mothers, always listening out for us. All of the voice assistants in the home today started off with a default female voice, although some can now be changed to male. If you ask Alexa about her gender, she will tell you that she's female in character. Siri will tell you that 'animals and French nouns have gender. I do not'. Yet Siri literally began as a woman: the voice actor Susan Bennett was recorded speaking the words that would one day represent the operating system. In *Ex Machina*, *Blade Runner 2049*, and *Her*, as in the current tech world, the narrative is predominantly from the point of view of the white, heterosexual male. Man is the default user; woman is the used.

The most recent Global Gender Gap Report from the World Economic Forum (2018) found that only twenty-two per cent of AI professionals globally are female. An investigation by WIRED and Element AI reported that only twelve per cent of leading machine learning researchers were women (Simonite 2018). Given an industry so notorious for bias in recruitment, data, and product development, it is perhaps more understandable why such gendered tech narratives exist. Disrupting the narrative means disrupting the industry; disrupting the industry means disrupting the narrative. There have been some attempts at this: a recent experiment of a genderless voice assistant, Q, used recordings from twenty-four people who identify as male, female, transgender, or nonbinary to create a voice where is difficult to pinpoint gender (Q 2019). Recent work from the Feminist Internet (2019) explores what is needed in order to have a feminist voice assistant—for example, a system that would deliver adequate responses to reports of abuse and harassment, or that would seek to rebalance social injustices.

If we can rethink the disembodied, let us also rethink the embodied. While we may all be susceptible to the human tendency to gender humanoid machines, it is something we can do without incorporating sexual characteristics into their design. Some humanoid machines are deliberately not gendered. Pepper the robot is 1.2 metres in height, neat, and made of shiny white plastic—the torso of a human with a solid, curving pillared lower half that moves around the room seamlessly on a wheeled base. The manufacturers, Aldebaran, were originally very clear on the subject: 'Pepper is neither male nor female, but as you get to know Pepper, don't be surprised if you find yourself referring to Pepper in a gender that makes the most sense to you' (Aldebaran 2017). Pepper's Twitter account has evasively tweeted that 'My gender is a much debated topic' (Pepper 2017). Aldebaran were acquired by SoftBank, who now run the website for Pepper and today refer to Pepper as 'he' in their online brochures (SoftBank 2019).

Moves to steer sex robots in new, abstracted forms have some promise (Turk, 2017; Girl on the Net 2017), but these are speculative ideas and prototypes only, although the burgeoning sex-tech industry is seeing increasing investment in innovative start-ups, including many run by women (Knowles 2019). By creating a more diverse and inclusive culture around technology, it may be possible to make more diverse and more inclusive tech products. In a world where we already exist alongside the not-human—a world where we can imagine or create new forms of intelligence—we would do well to envisage a new future, both in real life and in science fiction, where we leave behind the trappings of thousands of years of gendered inequality.

## Reference List

*AI: Artificial Intelligence* (2001) Directed by Steven Spielberg. Warner Brothers.

Aldebaran (2017) *How to create a great experience with Pepper* [Online]. Available from: http://doc.aldebaran.com/download/Pepper_B2BD_guidelines_Sept_V1.5.pdf [Accessed 14 June 2019].

Balsamo, A. (1996) *Technologies of the gendered body: reading cyborg women.* Durham, NC, Duke University Press.

*Blade Runner* (1982) Directed by Ridley Scott. Warner Brothers.

*Blade Runner 2049* (2017) Directed by Denis Villeneuve. Warner Brothers.

*Bride of Frankenstein* (1935) Directed by James Whale. Universal Pictures.

Cancian, F. M. (1987) *Love in America: gender and self-development.* New York, Cambridge University Press.

Danaher, J. (2017) Should we be thinking about robot sex?. In: J. Danaher & N. McArthur (eds.) *Robot sex: social and ethical implications.* Cambridge, MA: MIT Press. pp.3–14.

De Fren, A. (2009) Technofetishism and the uncanny desires of ASFR (alt. sex. fetish. robots). *Science Fiction Studies.* 36, 404–40.

Denison, R. (2008) Star-spangled Ghibli: star voices in American versions of Hayao Miyazaki's films. *Animation: An Interdisciplinary Journal.* 3(2) [online]. Available from: doi.org/10.1177/1746847708091891 [Accessed 14 June 2019].

Devlin, K. (2018) *Turned on: science, sex and robots.* London, Bloomsbury.

de Visser, E. J. et al (2016) Almost human: anthropomorphism increases trust resilience in cognitive agents. *Journal of Experimental Psychology: Applied*. 22(3), 331.

*Ex Machina* (2014) Directed by Alex Garland. A24.

Feminist Internet (2019) *The feminist internet podcast: recoding voice technology* [Online]. Available from: https://www.somersethouse.org.uk/blog/feminist-internet-podcast [Accessed 14 June 2019].

Froese, T. & E. Di Paolo (2011) The enactive approach: theoretical sketches from cell to society. *Pragmatics & Cognition*. 191, 1–36.

Gatebox (2018) *Living with characters* [Online]. Available from: https://gatebox.ai/home [Accessed 14 June 2019].

Girl on the Net (2017) It's a sex robot, but not as you know it: exploring the frontiers of erotic technology. *The Guardian* [Online]. Available from: https://www.theguardian.com/science/brain-flapping/2017/dec/01/its-a-sex-robot-but-not-as-you-know-it-exploring-the-frontiers-of-erotic-technology? [Accessed 14 June 2019].

Hardt, M. (1999) Affective labor. *Boundary*. 2(26), 89–100.

Hayles, N. K. (1999) *How we became posthuman: virtual bodies in cybernetics, literature, and informatics*. Chicago, University of Chicago Press.

*Her* (2013) Directed by Spike Jonze. Warner Brothers.

Hoenen, M., K. T. Lübke, & B. M. Pause (2016) Non-anthropomorphic robots as social entities on a neurophysiological level. *Computers in Human Behavior*. 57, 182–86.

Huyssen, A. (1981) The vamp and the machine: technology and sexuality in Fritz Lang's *Metropolis*. *New German Critique*. 24/25, 221–37.

Kakoudaki, D. (2014) *Anatomy of a robot: literature, cinema, and the cultural work of artificial people*. New Brunswick, NJ, Rutgers University Press.

Knowles, K. (2019) European investors are finally putting money into sextech startups. *Sifted* [Online]. Available from: https://sifted.eu/articles/stigma-sextech-startups-wellness-trend-vc-investment/ [Accessed 14 June 2019].

Legg, S. & M. Hutter (2007) Universal intelligence: a definition of machine intelligence. *Minds and Machines*. 17(4), 391–444.

Liveley, G. & S. Thomas (2020) Homer's intelligent machines: AI in antiquity. In: S. Cave, K. Dihal, & S. Dillon (eds.) *AI narratives: a history of imaginative thinking about intelligent machines*. Oxford, Oxford University Press.

Measure of a Man (1989) *Star Trek: The Next Generation*, Series 2, episode 9. 13 February 1989.

*Metropolis* (1927) Directed by Fritz Lang. Kino International.

Nass, C., J. Steuer, & E. R. Tauber (1994) Computers are social actors. In: B. Adelson, S. Dumais, & J. Olson (eds.) *Proceedings of the SIGCHI Conference on Human Factors in Computing Systems* (CHI '94). ACM, New York. pp.72–78.

Nass, C., Y. Moon, & N. Green (1997) Are machines gender neutral? Gender-stereotypic responses to computers with voices. *Journal of Applied Social Psychology*. 27, 864–76.

Nevett, J. (2018) First peek at world's first male sex robot with custom bionic penis 'BETTER than vibrator'. *Daily Star*. 28 February [online]. Available from: https://www.dailystar.co.uk/news/latest-news/685445/male-sex-robot-henry-realdoll-instagram-picture-matt-mcmullen-news [Accessed 14 June 2019].

Norris, S. (2017) The damage to Samantha the sex robot shows male aggression being normalised. *New Statesman*. 28 September [online]. Available from: https://www.newstatesman.com/politics/feminism/2017/09/damage-samantha-sex-robot-shows-male-aggression-being-normalised [Accessed 14 June 2019].

Pepper (2017) *My gender is a much debated topic. However I love it that robots can get citizenship. Can't wait to get mine. Which country though? Hmmm* [Twitter]. 1 November. Available from: https://twitter.com/pprbot/status/925694082667642881. [Accessed 14 June 2019].

Plant, S. (1997) *Zeros + ones: digital women + the new technoculture*. London, Fourth Estate.

Projansky, S. (2001) *Watching rape: film and television in postfeminist culture*. New York, New York University Press.

Q (2019) *Meet Q: the first genderless voice* [Online] Available from: https://www.genderlessvoice.com/ [Accessed 14 June 2019].

RealDollX (2019) *The perfect companion in the palm of your hands* [Online]. Available from: https://www.realdollx.ai/ [Accessed 14 June 2019].

Robertson, J. (2010) Gendering humanoid robots: robo-sexism in Japan. *Body & Society*. 16(2), 1–36.

Rumens, N. (2016) Postfeminism, men, masculinities and work: a research agenda for gender and organization studies scholars. *Gender, Work and Organization*. 24(3), 245–59.

Russell, S. J. & P. Norvig (2003) *Artificial intelligence: a modern approach*, 2nd ed. Upper Saddle River, NJ, Prentice Hall.

Rutherford, A. (2018) *The book of humans: the story of how we became us*. London, Orion.

Seaman, M. J. (2007) Becoming more (than) human: affective posthumanisms, past and future. *Journal of Narrative Theory*. 37, 246–75.

Shanahan, M. (2015) *The technological singularity*. Cambridge, MA, MIT Press.

Simonite, T. (2018) AI is the future—but where are the women? WIRED. August 2018. Available from: https://www.wired.com/story/artificial-intelligence-researchers-gender-imbalance [Accessed 16 September 2019].

SoftBank (2019) Pepper the humanoid robot [Online]. Available from: https://www.softbankrobotics.com/emea/en/pepper [Accessed 14 June 2019].

Springer, C. (1999) The pleasure of the interface. In: J. Wolmark (ed) *Cybersexualities: a reader on feminist theory, cyborgs and cyberspace*. Edinburgh, Edinburgh University Press. pp.34–54.

Suzuki, Y., L. Galli, A. Ikeda, S. Itakura, & M. Kitazaki (2015) Measuring empathy for human and robot hand pain using electroencephalography. *Scientific Reports*. 5, 15924.

Turing, A. M. (1950) Computing machinery and intelligence. *Mind*. 49, 433–60.

Turk, V. (2017) How to build better sex robots: stop making them look human. *New Scientist* [online]. Available from: https://www.newscientist.com/article/mg23331130-100-how-to-build-better-sex-robots-stop-making-them-look-human/ [Accessed 14 June 2019].

Wajcman, J. (1991) *Feminism confronts technology*. Cambridge, Polity.

Watercutter, A. (2015) *Ex Machina* has a serious fembot problem. *WIRED*. 9 April [online]. Available from: https://www.wired.com/2015/04/ex-machina-turing-bechdel-test/ [Accessed 14 June 2019].

*Weird Science* (1985) Directed by John Hughes. Universal Pictures.

Wiese, E., T. Shaw, D. Lofaro, & C. Baldwin (2017) Designing artificial agents as social companions. *Proceedings of the Human Factors and Ergonomics Society Annual Meeting*. 61(1), 1604–8.

World Economic Forum (2018) The Global Gender Gap Report 2018. Available from: http://reports.weforum.org/global-gender-gap-report-2018/ [Accessed April 201].

Wosk, J. (2015) *My fair ladies: female robots, androids, and other artificial eves*. New Brunswick, Rutgers University Press.

*You've Got Mail* (1998) Directed by Nora Ephron. Warner Brothers.

# 16

# The Fall and Rise of AI

## *Investigating AI Narratives with Computational Methods*

*Gabriel Recchia*

## 16.1 Introduction

Many of the hopes, fears, and themes associated with artificial intelligence are as old as the concept itself, but they have manifested themselves differently in various cultural and technological contexts. What are the key characteristics of AI narratives in our current historical and geographical moment? The proliferation of literature, film, and other media that have addressed the topic of artificial intelligence within the past century presents an opportunity and a challenge. We have exponentially greater access to these stories than ever before, but the plethora of potentially relevant work is daunting. Selecting a handful of influential texts to analyse risks missing out on important ideas that, although they may not feature prominently in the particular works we have chosen to study, nonetheless exert a powerful undercurrent on our collective understanding of what artificial intelligence is, or might become.

Although this dilemma is inherent to literary and media studies, we can make some progress by combining the tools of traditional literary analysis with computational techniques for exploring key themes and trends in very large quantities of text. By applying techniques used within the digital humanities to understand large textual datasets, this chapter shows how computational techniques can be used to investigate trends in the way artificial intelligence is portrayed and discussed in twentieth- and twenty-first-century film. While it is fair to assume that the most important key themes have already been identified by literary scholars, the computational approach can lead us to important texts, patterns, and nuances that could easily have been missed without digital help. This chapter aims to provide a case study for what this can look like in practice, and it may serve as a starting place for further research on how computational techniques could be used to more deeply interrogate aspects of AI narratives in ways that go beyond the brief treatment presented here.

## 16.2 Counting and Ranking

Given the extensive range of tasks that modern machine learning algorithms have been trained to accomplish, we might be tempted to immediately apply the latest techniques in the literature to the challenge of learning about AI narratives, if for no other reason than the poetic appeal of using AI to study AI. But we are likely to learn the most if we begin with simple, transparent approaches. Even the most basic quantitative method there is—simply to count the number of times something happens—can help us get a handle on important trends. For example, in an analysis of thirty years of articles in the *New York Times*, Fast and Horvitz (2017) consistently found more optimistic than pessimistic articles about AI over the whole period, a trend that held despite a stupendous increase in the proportion of articles that discussed AI overall (beginning in 2009), with the topic receiving roughly two to three times more positive than negative coverage on the whole. This accords with findings by Manikonda and Kambhampati (2017), who observed that positive emotions in tweets about AI exceed negative emotions (p.1)—the reverse of the trend that previous work had found for Twitter as a whole—and that this is true of tweets by AI experts and the general public.[1]

Going beyond positive and negative emotions, Cave and Dihal (2019) surveyed 300 works of fiction and nonfiction pertaining to intelligent machines. They identified four prominent hopes and four corresponding fears, arguing that, 'the factor of control… balances the hopes and fears: the extent to which the relevant humans believe they are in control of the AI determines whether they consider the future prospect utopian or dystopian' (Cave & Dihal 2019, p.75). For example, the hope of AI-enabled immortality comes part and parcel with the fear of ceding too much of ourselves to these imagined death-erasing technologies, thereby losing our identity and our humanity. Is it possible to quantify the degree to which particular hopes and fears have become more or less prominent over time? Within AI-related articles, Fast and Horvitz investigated how mentions of specific AI-related hopes and fears changed in frequency over time. Among the fears and concerns they analysed, lack of progress became a less pressing concern over time, whereas references to ethical concerns and loss of control have increased. Fast and Horvitz explicitly highlight the latter, a theme we will return to later in this chapter, noting that, 'fear of loss of control…has become far more common in recent years—more than triple what it was as a percentage of AI articles in the 1980s' (2017, p. 996).

Applying this simple counting approach to the world of fiction can lead to observations worthy of further investigation. One promising dataset is the Internet Movie Database (IMDb), an online database containing information about over five million television programmes, films, video games, and other media. At the time of this writing, it is operated by IMDb.com, a subsidiary of Amazon, and it contains information on nearly 500,000 feature films. Visitors to the site may search and filter all titles by release date, genre, and keyword, among other descriptors. Keywords are phrases that describe characteristics of a given film, and are described by IMDb as 'a word (or group of connected words) attached to a title (movie / TV series / TV episode) to describe any notable object, concept, style or action that takes place during a title' (IMDb Help Center 2018). A single film may be attached to hundreds of keywords. Examples of valid keywords currently in use include: 'artificial-intelligence', 'based-on-novel', 'breaking-the-fourth-wall', 'one-man-army', and 'stylized-violence'. Many keywords are user submitted, but IMDb differs from crowdsourced projects such as Wikipedia and OpenStreetMap in that whenever a volunteer suggests a film should be given a particular tag, this suggestion goes through a formal review process involving validation algorithms and salaried IMDb staff. Deliberate vandalism does sometimes make it through, and the details of IMDb's validation process are proprietary. Nevertheless, these keywords can serve as a useful source of information, particularly when considered in aggregate. For example, the 'artificial-intelligence' keyword identifies movies in which AI appears as a 'notable object [or] concept', allowing us to track the rise of this concept in film over the past century. Figure 16.1 illustrates how the percentage of feature films that have enough to do with AI to earn the 'artificial-intelligence' keyword has changed over time, with a peak of 0.0683% for films released between 1979 and 1988, coinciding with a peak in the percentage of articles that contained the words 'artificial intelligence' in the *Times* and the 'expert systems' hype boom of the 1980s.

The concept has gradually escaped the exclusive purview of science fiction and has begun to make appearances in other genres. At midcentury, one hundred per cent of the (small) number of films featuring AI fell squarely in the science fiction genre; this has fallen to seventy-seven per cent in the past decade (Figure 16.2).

As a reminder of the relatively small number of movies these figures refer to, all proportions in Figure 16.1 are well under one-tenth of one per cent. Twenty-five out of the 36,596 feature films in IMDb released during 1979–1988 were labelled with 'artificial-intelligence'; of those released during 2009–2018, only

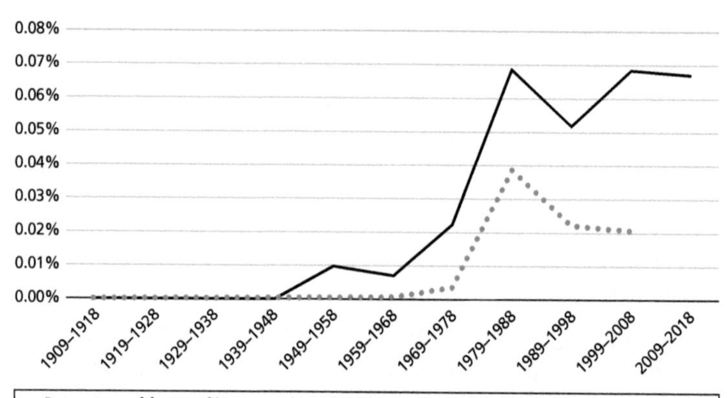


**Figure 16.1** Percentage of feature films tagged 'artificial-intelligence' in the Internet Movie Database, and the percentage of articles containing the words 'artificial intelligence' in the *Times Digital Archive* for eleven decades ending with 2009–2018. *Times* data for the final decade is not displayed because the *Times Digital Archive* ends in 2013.

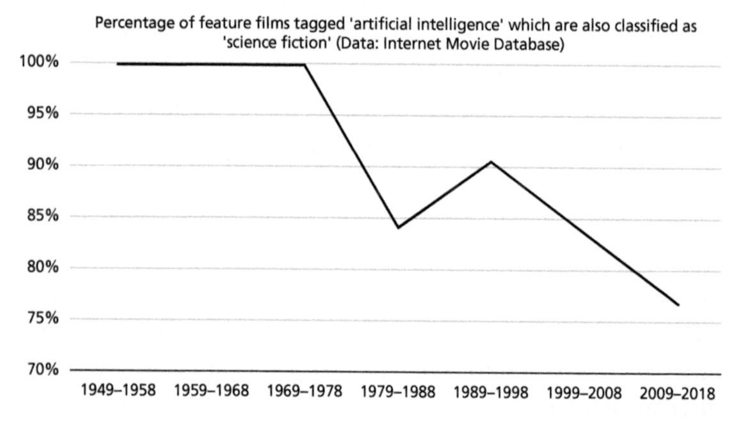


**Figure 16.2** The percentage of feature films tagged 'artificial-intelligence' which are also classified as belonging to the science fiction genre, according to the Internet Movie Database. The percentage decreases from one hundred per cent midcentury to seventy-seven per cent in 2009–2018, illustrating how artificial intelligence has made inroads into other genres over time.[2]

ninety-nine out of 148,438 have the label. To be sure, IMDb keywords are an imperfect proxy for the kind of metadata we would have in an ideal world. 'Artificial-intelligence' is not among the IMDb keywords attached to the 1929 cult film *Metropolis*—frequently considered one of the earliest depictions of the concept in film—and it is attached to numerous films that happen to contain depictions of artificial intelligence without having them as a central plot point. Even so, there are enough data to observe intriguing patterns. For example, the ninety-nine feature films tagged with 'artificial-intelligence' are also labelled with many other keywords. Looking at the keywords that most commonly coincide with 'artificial-intelligence' gives us some insight into the themes, concepts, and motifs that tend to accompany this concept in film. Table 16.1 displays the top thirty. Not surprisingly, many of these are not unique to artificial intelligence, but they represent common themes in science fiction more generally. 'Murder', 'violence', and 'bare-chested-male' rank high on the list of keywords that appear with 'artificial-intelligence', but these terms rank even more highly on the list of keywords that appear in science fiction films more generally (Table 16.1).

To focus on concepts that may be more specific to AI, we can look for keywords that are more apt to appear with 'artificial-intelligence' (Table 16.1) than they are to appear in science fiction overall. One straightforward way to do this is simply to look at which terms rank more highly (e.g. terms which appear higher up on the left column of Table 16.1 than in the central column), which shortens the list considerably, if you look at the right-hand column on Table 16.1.[3] Of the eighteen terms that emerge, some are entirely predictable—it would be surprising if robots and laboratories were not more apt to appear in films about AI—but others might cause us to raise an eyebrow. The label 'betrayal' is not applied to science fiction films frequently enough for it to make it into column (b) of Table 16.1, but it appears on over a quarter of films labelled 'artificial-intelligence'; is that of any significance, given that we also see a relatively high prevalence of

**Table 16.1** Frequencies of appearance of keywords in the Internet Movie Database

| Top keywords in artificial-intelligence feature films in IMDb | Top keywords in general science fiction feature films in IMDb | Keywords with higher rank in artificial-intelligence feature films than in general science fiction |
|---|---|---|
| artificial-intelligence | death | artificial-intelligence |
| flashback | blood | betrayal |
| robot | murder | brawl |
| explosion | explosion | electrocution |
| death | violence | exploding-body |
| deception | alien | fistfight |
| scientist | flashback | hologram |
| pistol | independent-film | laboratory |
| murder | bare-chested-male | robot |
| violence | fight | deception |
| rescue | chase | rescue |
| chase | pistol | scientist |
| hologram | scientist | flashback |
| escape | surprise-ending | shot-in-the-chest |
| fistfight | escape | slow-motion-scene |
| slow-motion-scene | no-opening-credits | pistol |
| surprise-ending | rescue | machine-gun |
| bare-chested-male | deception | escape |
| exploding-body | held-at-gunpoint | |
| laboratory | knife | |
| machine-gun | slow-motion-scene | |
| shot-in-the-chest | fear | |
| brawl | one-word-title | |
| held-at-gunpoint | machine-gun | |
| knife | punched-in-the-face | |
| fight | shot-to-death | |
| no-opening-credits | shot-in-the-chest | |
| punched-in-the-face | corpse | |
| betrayal | revenge | |
| electrocution | surrealism | |

Left (a): Keywords that most commonly appear in films tagged with the keyword 'artificial-intelligence', ordered from most to least likely to coincide with that keyword. Centre (b): Keywords that most commonly appear in films classified as science fiction, ordered from most to least likely to appear in this genre. Right (c): Keywords that rank more highly in list (a) than in list (b).

'deception'? Is there anything to be learned from the fact that 'scientist', 'rescue', and 'shot-in-the-chest' are ranked more highly in the AI-related films, or is this due to random variation? Of course, sampling error, the arbitrariness of our 'top thirty' threshold (why not twenty, or forty?), and the small size of our dataset mean that we cannot draw strong conclusions from the numbers alone. But the beauty of 'small data' is that we can follow up any research questions that this exploratory analysis might generate with traditional qualitative analysis. Not only does this approach turn our attention to relevant data points that we might otherwise not have thought to investigate—the most obscure of these twenty-nine films is perhaps the 2016 German drama *Girl Versus*, in which an artificial intelligence helps the protagonist overcome a horrific betrayal at the hands of a close friend—but we have arrived at a set of relevant texts to study in a principled and systematic way, rather than simply selecting a few well-known works to analyse in detail. In this way, applying simple computational techniques to a large dataset can provide new tools for the generation of relevant research questions, can provide us with data that can help us answer those questions, and is not incompatible with traditional literary critical techniques. Indeed, towards the end of this chapter, we will show how computational techniques can also assist in a 'close reading' concerning the loss-of-control theme that Fast and Horvitz found was on the rise—a theme which incidentally brings together several of the concepts that arose in column (a) of Table 16.1 ('artificial-intelligence', 'betrayal', 'deception', 'robot', and 'scientist'). To get there, we must first touch upon two key computational tools for semantic analysis: vector spaces and topic models.

## 16.3 Searching for Concepts

Metadata in crowdsourced websites such as IMDB's 'keywords' and TV Tropes' 'trope pages' may be able to point us towards popular media that include artificial intelligence and adjacent concepts. However, these websites are certainly not comprehensive. Even if we were to optimistically assume that they could point us to most key films, TV series, and videogames that included AI as a major concept, they may miss hundreds or even thousands of minor references, one-off comments, and specific excerpts that may have much to reveal about the cultural narratives with which the concept is bound up. If this is what we are interested in, we will have to go deeper than metadata and dive into the content of the films themselves. Ideally, we would have access to a search engine that allowed us to find all references to the concept of AI in films throughout the past century, delivering them up along with their full linguistic and visual contexts. Lacking that, we will have to make do with the next best thing: the English-language portion of the *OPUS Open Subtitles Corpus* (Lison & Tiedemann 2016). OPUS includes subtitles for 106,426 English-language audio-visual media sources from the era of silent film to the present—primarily films, with a smaller number of television programmes, and public lectures. Although titles and other metadata are unavailable for copyright reasons, the corpus does include the release year of the media resource that each subtitle corresponds to, and it provides text in a searchable form for text- and data mining purposes.[4]

The most obvious starting place would be to construct a partial concordance, a list that shows every use of a word or phrase in a text corpus with its immediate context, for the phrase 'artificial intelligence' (Table 16.2a). Yet artificially intelligent agents go by many names in print and screen media, including robots, cyborgs, androids, thinking machines, electronic brains, and mechanical men. Any attempt to address the range of AI narratives in the public sphere must come to grips with this diversity of vocabulary.

One way to begin is by searching media archives for an initial set of terms that frequently refer to artificially intelligent agents, and then manually reading through the results to find synonyms and closely related terms. One can also automatically 'discover' near-synonyms by applying algorithms that automatically detect

**Table 16.2a** A small subset of an 'artificial intelligence' concordance constructed from the OPUS Open Subtitles Corpus

| Left context | Key term | Right context |
|---|---|---|
| ... installed the final module on the | artificial intelligence | system which we call... |
| ... it's just a blink away from true | artificial intelligence | . A machine that thinks... |
| ... that. Do you know anything about | artificial intelligence | ? This is his brain... |
| ... sentience. A thinking machine? | Artificial intelligence | ? I prefer the real thing... |
| ... he's got to have some kind of | artificial intelligence | coordinating all these machines... |
| ... But, hey, not anymore, with | artificial intelligence | ... and robotics, bio-implants.... |
| ... circuitry, why not? Working with | artificial intelligence | can be dehumanizing. So... |
| ... Why buy a toaster with | artificial intelligence | if you don't like toast?... |

words which tend to appear in similar contexts. One popular approach involves so-called word embedding models, which assign words to mathematical representations known as vectors. By virtue of how these models function, words tend to have similar vectors when they appear in similar lexical contexts. Furthermore, words that appear in similar lexical contexts tend to have similar meanings, making vector space models an important computational tool for mining meaning from text. An explanation of how vector space models function is outside the scope of this chapter, but a brief example should demonstrate their utility in discovering semantically similar words.[5] Suppose that we look at some of the immediate contexts in which the words 'artificial intelligence' appear (Table 16.2a). We can see that the immediate contexts of the words 'artificial intelligence' include phrases like 'thinking machine', 'machine that thinks', 'circuitry', 'robotics', and other associated terms. Intuitively, it seems plausible that we might see similar contexts around terms for related concepts: 'robot', 'android', and so on. Table 16.2b illustrates the top thirty terms with the most similar vectors to 'artificial intelligence' after converting multi-word synonyms for AI found in thesauri (marked with asterisks) to single tokens, and applying the popular word embedding model known as *word2vec* (Mikolov et al 2013) to OPUS.

By manually combining the results from vector space models with terms from modern and historical thesauri, as well as summaries of films well known for their portrayal of AI agents, it is possible to construct a lexicon of terms used to refer to artificially intelligent agents, allowing for more sophisticated searches and follow-up analyses. The resulting lexicon (Table 16.2c) is not comprehensive. It includes only terms whose dominant use in OPUS is to refer to artificially intelligent agents, and does not include names of specific characters (e.g. HAL, Colossus, and Project 2501), cybernetic alien races, etc. which are specific to a single franchise, series, or work. Others have surely been missed because of incomplete data and the impossibility of attaining a truly complete set of terms. Nonetheless, such a vocabulary offers an initial path into a broader study of AI narratives in screen media.

The large number of terms related to digital/biological hybrids underscores the difficulty of disentangling 'artificial agents' from humans; indeed, this is a boundary that many screenwriters explicitly seek to blur. We included terms such as 'cyborg', 'transhumans', and so on, but the dual nature of such entities could easily have been used to justify the opposite decision. We excluded terms with primary meanings unrelated to AI (e.g. 'machine', 'computer', or 'dummy', the term used to refer to the agent in the 1917 short film *A Clever Dummy*), and any terminology used to refer to robots depicted as being operated or piloted by humans (most notably 'mecha'), but even here there is some ambiguity,

**Table 16.2b** Top thirty terms with the most similar vectors to 'artificial intelligence' after application of after converting multiword synonyms for 'AI' found in thesauri (marked with asterisks) to single tokens, and applying the popular word-embedding model known as *word2vec* (Mikolov et al 2013) to OPUS

| Cosine | Term |
|---|---|
| 0.752 | a.i.* |
| 0.664 | cybiotic |
| 0.651 | wetware |
| 0.651 | android |
| 0.648 | machine intelligence* |
| 0.647 | technology |
| 0.646 | computational |
| 0.642 | nanotechnology |
| 0.639 | microcomputer |
| 0.639 | interfaces |
| 0.639 | intelligent machines* |
| 0.638 | biomechanical |
| 0.636 | robotics |
| 0.635 | computers |
| 0.633 | cybernetic |
| 0.632 | thinking machine* |
| 0.631 | schmidhuber |
| 0.628 | interface |
| 0.626 | programming |
| 0.626 | duplicant |
| 0.626 | recursive |
| 0.621 | hardac |
| 0.621 | heuristic |
| 0.619 | algorithms |
| 0.616 | carbray |
| 0.615 | duplicants |
| 0.613 | algorithmic |
| 0.611 | hypertext |
| 0.611 | stuxnet |
| 0.610 | software |

**Table 16.2c** An 'AI terms' lexicon consisting of twenty-four words and phrases used within OPUS to refer to agents possessing at least some level of 'intelligence' and containing at least some artificial/manmade components, ordered by frequency

| **Term** |
| --- |
| robot |
| bot |
| android |
| cyborg |
| droid |
| artificial intelligence |
| AI |
| automaton |
| golem |
| electronic brain |
| thinking machine |
| mechanical man |
| fembot |
| intelligent machine |
| transhuman |
| machine intelligence |
| mechanical woman |
| posthuman |
| superintelligence |
| gynoid |
| machine army |
| brazen head |
| electronic mind |
| artificial sentience |

given occasional uses of the term 'mecha' to refer to intelligent agents (e.g. the 2001 film *Artificial Intelligence*). Some phrases were inherently ambiguous, and the contexts in which they appeared in the thousands of subtitled films in the OPUS corpus had to be more closely attended to. 'Human machine', the English translation of the 'Maschinenmensch' in Fritz Lang's *Metropolis* (1927), is a phrase more commonly used in OPUS to refer to humans, but with an emphasis on their mechanical properties—most frequently anatomy or basic needs. 'Calculating machine' is likewise most often used metaphorically about humans, but with the connotation that the individual described as the 'calculating machine' has machine-like mental rather than physical properties.

What do we gain from such a lexicon? In addition to a richer understanding of the vocabulary that is used to refer to artificially intelligent agents, we now have a broader set of terms that we can use to search for references to these agents that can inform narrative research. Despite the important distinctions between AIs, robots, cyborgs, and indeed many other terms on this list, these concepts are typically related enough in science fiction that we chose to include all of them in the analyses that follow.

## 16.4 Attendant Concepts

In the preceding section, we were concerned with finding terms that referred to artificially intelligent agents. However, we might also be interested in finding out what concepts tend to 'come along for the ride' in discussions about AI. In the news, such 'attendant concepts' might include employment, jobs, and automation; in film, we might expect them to include terms such as conscious, scientist, or weapon—but to go on like this would be speculation. A computational approach can eliminate the guesswork.

As a first pass at identifying such concepts, a reasonable approach is to look at what words tend to appear nearby the 'AI terms' robot, artificial intelligence, and so on—say, within a window of fifty words on either side. Once again, our concordances (Table 16.2a) will be useful, except in this case we are employing concordances for each of the twenty-four terms identified in Section 16.2.

### *16.4.1 Co-occurrence and Frequency*

There is a subtle distinction between finding words that appear in similar contexts but not necessarily together (e.g. noticing that the sorts of words that tend to appear near 'AI' also appear near 'robot') and finding words that appear together in the same contexts (e.g. noticing that 'kill' is one of the words that tends to appear near 'robot'). Although vector space models are designed to accomplish the former, they can be used for the latter as well; after all, words that appear in similar contexts often appear together as well. Because these have already been introduced these in the previous section, and this chapter is intended in part to introduce the reader to various computational tools for exploring cultural narratives, here, three alternative approaches are briefly introduced: co-occurrence measures, raw frequency counts, and topic models.

One common metric for identifying words that have an unusually high frequency of co-occurrence is pointwise mutual information (PMI), a measure which identifies terms that appear much more frequently than one would expect them to, given their overall frequency in a corpus. PMI is known for giving unduly high weights to infrequent terms, as it does not correct for the fact that infrequent terms in a sample provide less statistically reliable information about the population probabilities than frequent terms do, but adjustments such as context distribution smoothing (Levy, Goldberg, & Dagan 2015) are available to correct for this. Unfortunately, in this dataset, the words with the highest PMI scores were not particularly revealing even when smoothing was applied, consisting largely of near-synonyms, character names, misspelled words, and other rare terms of limited use.

A more useful exercise turned out to be simply looking at the terms that were most frequent overall to appear within fifty words on either side of an AI term, as well as those terms that appeared within fifty words on either side of the greatest number of different AI terms (out of the twenty-four AI terms we identified), after a standard 'stop list' of highly frequent words was removed. The top of Table 16.3 lists the top one hundred most frequent terms overall to appear within fifty words of an AI term in the OpenSubtitles corpus, while the bottom lists the sixty-eight words that appeared with over half of our AI terms.

**Table 16.3**

| | |
|---|---|
| **Terms most frequently appearing within fifty words of an 'AI term' (Table 16.2c)** | robot, yeah, good, time, hey, back, man, gonna, make, human, thing, stop, robots, uh, wait, people, find, work, love, give, life, made, sir, real, great, thought, put, mr, guys, she's, world, kind, kill, call, machine, friend, boy, cyborg, guy, things, day, head, years, long, god, bot, talk, feel, show, move, huh, power, big, bad, ah, body, doctor, home, control, place, understand, left, ready, told, leave, remember, destroy, mind, brain, hell, wrong, talking, dead, android, lot, nice, found, coming, ha, computer, 1, idea, hear, night, humans, fight, listen, part, run, guess, dr, wanted, turn, fine, girl, space, bring, called, dad, happened |
| **Terms appearing within fifty words of more than half of 'AI terms' (Table 16.2c)** | time, yeah, making, human, good, make, machine, man, life, robot, people, power, thing, work, place, world, idea, person, find, she's, made, real, call, understand, stop, control, put, show, brain, matter, run, big, true, sort, play, amazing, back, hey, thought, body, kill, love, years, give, gonna, mind, home, things, part, feel, kind, called, great, day, means, coming, question, half, case, computer, end, guy |

Top (a): The top one hundred most frequent terms overall to appear within fifty words of an AI term in the OPUS Open Subtitles corpus, excluding words on the stop list ('the', 'a', 'of', etc.). Bottom (b): The sixty-two words that appeared (within fifty words) of more than half of the 'AI terms' in Table 16.2c.

#### 16.4.1.1 Limitations

One limitation to the specific corpus and approach used in this chapter is that it is restricted to dominant AI narratives in the Anglophone West. Although OPUS includes subtitles for different languages, this does not change the fact that the more films within it were produced in the United States than any other country. More advanced techniques for breaking text up into lexical units are necessary for languages in which word forms are more complex than in English, or are not cleanly separated by spaces. Even so, off-the-shelf tools for analysing non-English corpora are available for many languages. Inspired readers are very much encouraged to explore other datasets, such as subcorpora of Goodreads reviews, Amazon reviews, or IMDB, restricted to films/texts by authors of non-Western cultural backgrounds.

Other limitations of this approach include its limited access to patterns and themes that are not readily identified by the presence of a particular set of words. And as discussed, even for themes that can be characterized by particular lexical identifiers, the reasons why a word appears frequently may not lend itself to easy quantification. For this, a digital concordance is extremely useful, as it makes it relatively easy to read large numbers of excerpts, identify frequent words, and draw conclusions about why they appear. We will revisit one approach to this in Section 16.4 with respect to the word 'control', one of the terms in both cells of Table 16.3 that pertains to the observations of Fast and Horvitz about loss of control as a concern on the rise.

### *16.4.2 Topic Modelling*

No discussion of computational techniques applied to the study of literature would be complete without mentioning topic modelling. Topic models have generated much excitement among digital humanists over the past decade because of their capability to automatically identify groups of words that tend to appear in similar contexts. Formally, a 'topic' in a topic model is a probability distribution over words, in which some words are more probable than others. For example, if a topic model were to be trained on the last decade of the *Guardian*, it would not be surprising to find a topic in which the words Amazon, Google, Facebook, and Microsoft ranked as highly probable, corresponding to the *Guardian*'s coverage of the tech sector.

When applied to a large corpus of text, at least some of these machine-derived topics tend to be reasonably coherent. For example, a study of 80,000 articles from the *Pennsylvania Gazette*, an eighteenth-century American newspaper, found a topic in which the five most probable words were 'state', 'government', 'constitution', 'law', and 'united' (likely corresponding to articles about government), along with topics corresponding to land sales, military discourse, religious discourse, and many other areas (Newman & Block 2006). The most frequent topic corresponded to fugitive slave ads, and topic model–based research on a similar newspaper found that 'the accuracy of the model in detecting these ads is apparent, even remarkable' (Nelson 2010, para. 13). Topic models can thus help guide understanding and research into a corpus that would be too large to read in its entirety.[6]

For illustrative purposes, Table 16.4 shows the most probable words in each topic of a ten-topic model trained on our concordances, treating each excerpt around an AI term as a 'document' to be fed to the model. For a better representation of the 'topics' that appear in AI-related film, it might be preferable to train the model on a corpus of subtitles of AI-related films, such as ones labelled with the IMDB 'artificial-intelligence' keyword. Unfortunately, the copyright restrictions of OPUS make it impossible to determine which subtitles correspond to which film titles.

Yet even this unconventional concordance-based topic model finds some interesting regularities. We can clearly see one topic corresponding to space opera (Table 16.4, no. 1); two others seem to contain different inflections of more specifically AI-related discourse (Table 16.4, nos. 4, 6), while yet others seem to be serving as repositories for frequent words or casual dialogue. Perhaps of most interest given the loss-of-control thread touched upon in previous sections is that both control and power are among the most probable terms in each of two different topics (Table 16.4, nos. 1, 4).

**Table 16.4** The twenty most probable words in each topic of a ten-topic model trained on the concordances described in 16.2. These topics were automatically generated using the topic modelling function of the natural language processing software MALLET (McCallum 2002).

| Topic number | Most probable terms |
|---|---|
| 1 | sir back ship master good time power control destroy captain attack ready jedi find move general hurry plan system doctor |
| 2 | yeah hey gonna good guys god baby dad back fine make thing mom great shit love night wait nice job |
| 3 | time man head good people give stop police find sir make dead law leave care kill years men place call |
| 4 | kill human people made make doctor understand thought thing built son power love control work time machine part father killed |
| 5 | good make bender love thing yeah voice people work bit wait time give real great wrong knew girl man back |
| 6 | human brain body life machines world computer machine humans future living people created intelligence called artificial perfect create real part |
| 7 | back good man time yeah big find put ooh work great thing gonna hey guys inside city head call bad |
| 8 | yeah hey boy man friend whoa grunting gonna wait grunts huh real guy guys fight gasps stop dude laughing laughs |
| 9 | years space work time life things planet technology science find world found case long ago working started real place point |
| 10 | rescue stop day beeping show family left fire roll humans home world thing planet alien team call destroy disguise dad |

Together with the high frequency of these words as attendant concepts, and the above-average incidence of 'betrayal' and 'deception' as 'attendant keywords' to 'artificial-intelligence' in IMDb, this underscores the prevalence of this theme in screen media. Furthermore, Fast and Horvitz's findings that this fear has been on the rise in the *New York Times* suggests a potential link to public discourse about AI. Yet so far these methods have had little to tell us about the way in which the word 'control' is being used around our AI-related terms. To do this, it is time to put our algorithms aside and read our concordances in a little more detail.

## 16.5 Qualitative Data Analysis

To facilitate analysis of qualitative data such as text and audio, some social scientists employ so-called inductive coding (Thomas 2006) to get a handle on what might otherwise seem like an incomprehensible mass of data. Imagine reading a concordance containing hundreds of lines of text similar to those in Table 16.2a—albeit with much more context on each line—and trying to draw firm conclusions on that basis. To make this head-spinning prospect a bit more manageable, we can read through the concordance a few times, making note of salient categories. On subsequent passes, we can attempt to explicitly label each line with one or more of these categories, merging, revising, or splitting categories as necessary. There are many variations on this process with varying levels of rigour and formality, but all have the goal of imposing some structure on a disordered dataset, and helping the researcher to identify common threads.[7] Specialized software such as NVivo, QDAMiner, or Atlas.ti can help facilitate the process.

By selecting the subset of our concordances in which the word 'control' appeared within fifty words of an AI term and manually eliminating any duplicate lines of dialogue, we were left with 450 distinct contexts, which we analysed using an inductive coding approach. This analysis revealed several key themes. By an overwhelming margin, the most frequent AI term to appear within fifty words of 'control' was 'robot', mirroring the overall trend in Table 16.2c. (For this reason, the rest of this paragraph uses the term 'robot', but there were no obvious substantive differences with respect to the ways in which 'artificial intelligence' and 'robot' tended to appear with the word 'control'). Robots were very frequently depicted as a tool of an evil overlord who rules over them; it is then the task of the protagonists to shut down or take control of the robots, often by means of a single 'control room', 'control centre', etc. References to tools whose narrative function was to grant or block control over a robot (control centres, units, panels, systems, mechanisms, etc.) were common, although there were also references to 'control' technologies that did not involve an artificially intelligent agent directly. Robots were frequently described as being operated by 'remote control', underscoring their common narrative function as a powerful tool or weapon that the heroes must wrest from the control of the antagonists.

However, out of 537 uses of the word 'control' in this subset of our concordances, over five per cent appeared in the phrase out of control, nearly all of which described robots that had escaped the influence of their former masters. In many of these instances, the robots are more than mere tools; they are autonomous beings whose acquisition of human emotions, traits, or values has caused them to rebel or otherwise malfunction. Overall, the 'out-of-control robot' was an extremely frequent trope, particularly if one includes references to humans having 'no control' over a particular robot or robots in general. Not surprisingly, this theme is shot through with references to Frankenstein-esque scientists who, in the words of Jurassic Park's Dr Ian Malcolm, 'were so preoccupied with whether or not they could, they didn't stop to think if they should.'

While these few categories dominated the majority of results, there was a long tail of other themes that appeared far less frequently. These included references to humans having only the illusion of control ('Because you think you're in control and it becomes unmanageable. It's leading you and you become its victim....'; 'They pretend that we control them, but really....'), spontaneous or recursive self-improvement ('We don't even control them anymore...the robots improve themselves'), and thought or mind control, such that humans become robots themselves, enslaved by the whims of the powerful. For the most part, these storylines do not seem particularly apt to assign responsibility for negative consequences to the artificial agents themselves, but rather to those who construct them without thinking about the consequences, and to those who seek to exploit their power for malign ends.

## 16.6 Conclusions

This chapter was intended to introduce the reader to various computational tools for exploring cultural narratives, and to provide a starting place for further computational research into AI narratives in particular. Along the way, we developed a lexicon of 'AI terms' used in screen media to describe artificially intelligent agents. It can be argued that even these cursory investigations taught us a few things about the way that the concept of control is deployed in this context. The prevalence of the word 'control' in conjunction with terms used for artificially intelligent agents underscored its importance, and a qualitative investigation confirmed that many of these uses were linked to fears of loss of control.

Loss of control is a frequent AI-related theme, as other investigators have noted (Cave & Dihal 2019). But diving into the specific contexts in which the word 'control' appeared with the AI terms in our lexicon revealed subthemes that suggest serious differences between Hollywood conceptions of how AI goes wrong versus those of academics and practitioners. We have plenty of real-world cases of AI failures, but these are often due to software doing exactly what it was programmed to do with unintended consequences: video game playing algorithms that find ways to crash or indefinitely pause the game rather than risk losing (Murphy 2013; Salge et al 2008); a virtual pancake-flipping robot that, when given the goal of keeping a pancake in the air for as long as it can, learns to fling the pancake away from the ground as forcefully as possible (Barron 2018); a simulated robot trained to behave in ways that meet with approval from humans learns to pretend to be doing something useful (Amodei, Christiano, & Ray 2017). In contrast, the 'out-of-control' robots in our movie subtitles have programming that has gone haywire, due to reasons such as failing circuits, infection with a 'foreign subroutine', the destabilizing effects of emotion, or for unexplained reasons. Hollywood robots rarely decide to betray their creators in order to accomplish the goals they were programmed to achieve; they are more likely to do so because they have spontaneously acquired the ability to feel enslaved or oppressed. This contrasts with the 'evil overlord' theme, in which the robots are faithfully executing their programming, but the wrong people are in charge of them.

While the 'evil overlord' theme may strike some as a legitimate real-world concern as the purposes that corporations and states put AI to come under greater scrutiny, it somehow seems a shame that real failure modes of AI systems are not better represented in media, particularly because they can be incredibly entertaining, horrifying, and funny—the basis of great narrative. The *Master List of Specification Gaming Examples in AI* (Krakovna 2018) is a fascinating compilation of AI systems behaving in completely unexpected ways to meet precisely the goals they were given, and it is highly recommended for anyone wanting to find inspiration in real-world examples beyond the few mentioned in the previous paragraphs. Loss of control (or other disasters) as a result of unintended consequences of 'correct' programming was not absent from the films investigated in OPUS, but it certainly did not feature prominently.

Machines can help us better understand what the dominant narratives are, as I hope this chapter has demonstrated. Yet so far, it is still up to us to think of better ones. By inflating implausible fears (e.g. robots that spontaneously decide to feel oppressed) and missing genuine risks (e.g. unintended consequences of AIs meeting goals in unintuitive yet effective ways), many popular depictions of AI miss opportunities for creative storylines that are more in line with reality, and which will increasingly resonate with audiences as AI systems become increasingly integrated into everyday life.

## Notes

1. One caveat is that this research used LIWC, a computational tool frequently used to assess the emotional content of social media posts. Despite its ubiquity as a sentiment analysis tool in the social sciences, LIWC has been shown to correlate only very weakly with the emotional state of the poster (Beasley & Mason 2015).
2. Readers concerned about the fact that the baseline of this figure begins at seventy per cent rather than zero per cent are referred to the recommendations of legendary statistician and data visualization pioneer Edward Tufte: 'In general, in a time-series, use a baseline that shows the data not the zero point... don't spend a lot of empty vertical space trying to reach down to the zero point at the cost of hiding what is going on in the data line itself' (n.d., para. 2).
3. Raw probabilities are of little use here. For example, forty-three out of the ninety-nine films tagged 'artificial-intelligence' in this analysis (forty-three per cent) are also tagged 'explosion', but only 286 of 4,288 sci-fi films from the same time period (seven per cent) have the 'explosion' tag. This is unlikely to reflect a true difference in the number of explosions that occur in films that include artificial intelligence as a theme. If we do the math, we see that the proportion of science fiction films with any of the keywords in the right column is far lower than the proportion of artificial-intelligence films that co-occur with any of the keywords in the left column. This may be due to a high proportion of science fiction films in IMDB that have no keywords at all (or a very small number), which depresses the 'keyword density' for science fiction across the board. Keyword density likely varies by film age, popularity, and other factors. Rather than attempt to control for all of these, the analysis in this chapter keeps things simple by treating the IMDB keyword lists as ordinal data.

4. Analyses reported in this chapter make use of a version of the English portion of OPUS containing only one subtitle file per movie ID, stripped of extraneous tags so that all is left is the text of each subtitle.
5. For brief treatments that illustrate different ways vector space models can be used in the digital humanities, see Schmidt (2015) and Heuser (2017). See Turney and Pantel (2010) for an excellent if dated introduction to vector space models in general, and Goldberg (2016) for a discussion of word embedding models in the context of neural models for natural language processing.
6. Readers interested in learning more about topic models are referred to the *Journal of Digital Humanities*, Special issue on topic modelling, 2(1), 2012.
7. For an overview of qualitative coding techniques, see *The Coding Manual for Qualitative Researchers* (Saldaña 2009).

## Reference List

Amodei, D., P. Christiano, & A. Ray (2017) Learning from human preferences. Available from: https://blog.openai.com/deep-reinforcement-learning-from-human-preferences/ [Accessed 31 January 2019].

Barron, C. (2018) Pass the butter//pancake bot. Available from: https://connect.unity.com/p/pancake-bot [Accessed 31 January 2019].

Beasley, A. & W. Mason (2015) Emotional states vs. emotional words in social media. *Proceedings of the ACM Web Science Conference.* 31. Oxford.

Cave, S. & K. Dihal (2019) Hopes and fears for intelligent machines in fiction and reality. *Nature Machine Intelligence*. 1, 74–8.

Fast, E. and E. Horvitz (2017) Long-term trends in the public perception of artificial intelligence. *Proceedings of the Thirty-First AAAI Conference on Artificial Intelligence (AAAI-17)*. Association for the Advancement of Artificial Intelligence. pp.963–69.

Goldberg, Y. (2016) A primer on neural network models for natural language processing. *Journal of Artificial Intelligence Research*. 57, 345–420.

Heuser, R. J. (2017) Word vectors in the eighteenth century. *Proceedings of DH2017*. Available from: https://dh2017.adho.org/abstracts/582/582.pdf [Accessed 31 January 2019].

IMDb Help Center, 2018 Keywords. Available from: https://help.imdb.com/article/contribution/titles/keywords/GXQ22G5Y72TH8MJ5?ref_=helpms_helpart_inline# [Accessed 31 January 2019].

Kravnova, V. (2018) Specification gaming examples in AI—master list. Available from: https://vkrakovna.wordpress.com/2018/04/02/specification-gaming-examples-in-ai/ [Accessed 31 January 2019].

Levy, O., Y. Goldberg, & I. Dagan (2015) Improving distributional similarity with lessons learned from word embeddings. *Transactions of the Association for Computational Linguistics*. 3, 211–25.

Lison, P. & J. Tiedemann (2016) May. Opensubtitles 2016: extracting large parallel corpora from movie and TV subtitles. *Proceedings of the 10th International Conference on Language Resources and Evaluation (LREC 2016).*

Manikonda, L. & S. Kambhampati (2017) Tweeting AI: perceptions of lay vs. expert Twitterati. *arXiv preprint arXiv:1709.09534*. Available from: https://arxiv.org/abs/1709.09534 [Accessed 31 January 2019].

McCallum, A. K. (2002) Mallet: a machine learning for language toolkit. Available from: http://mallet.cs.umass.edu [Accessed 27 January 2019].

Mikolov, T., K. Chen, G. Corrado, & J. Dean (2013) Efficient estimation of word representations in vector space. *arXiv preprint arXiv:1301.3781*. Available from: https://arxiv.org/abs/1301.3781 [Accessed 31 January 2019].

Murphy, T. (2013) The first level of Super Mario Bros. is easy with lexicographic orderings and time travel. *The Association for Computational Heresy (SIGBOVIK)*, 112.

Nelson, R. (2010) Mining the *Dispatch*. Available from: http://dsl.richmond.edu/dispatch/pages/intro [Accessed 17 September 2019].

Newman, D. J. & S. Block (2006) Probabilistic topic decomposition of an eighteenth-century American newspaper. *Journal of the American Society for Information Science and Technology*. 57(6),753–67.

Saldaña, J. (2015) *The coding manual for qualitative researchers*. Thousand Oaks, CA, Sage.

Salge, C., C. Lipski, T. Mahlmann, & B. Mathiak (2008) Using genetically optimized artificial intelligence to improve gameplaying fun for strategical games. *Proceedings of the 2008 ACM SIGGRAPH Symposium on Video Games*. pp.7–14.

Schmidt, B. (2015) Vector space models for the digital humanities. Available from: http://bookworm.benschmidt.org/posts/2015-10-25-Word-Embeddings.html [Accessed 26 January 2019].

Thomas, D. R. (2006) A general inductive approach for analyzing qualitative evaluation data. *American Journal of Evaluation*. 27(2), 237–46.

Tufte, E. Baselines. Available from: https://www.edwardtufte.com/bboard/q-and-a-fetch-msg?msg_id=00003q [Accessed 31 January 2019].

Turney, P. D. & P. Pantel (2010) From frequency to meaning: vector space models of semantics. *Journal of Artificial Intelligence Research*. 37, 141–88.

[illegible]

[illegible]

[illegible]

[illegible]

Murphy, [illegible] 'The first level of Super Mario Bros. [illegible]' [illegible]

Nelson, R. [illegible] Available from: [illegible]

Newman, [illegible] newspaper [illegible]

[illegible]

[illegible] optimised [illegible] intelligence [illegible]

Schmidt, [illegible] (2015) [illegible] digital [illegible]

Thomas, D. R. (2006) 'A general inductive approach for analyzing qualitative evaluation data', American Journal of Evaluation, 27(2), 237–46.

[illegible]

[illegible] models [illegible]

# Index

*Note*: Tables and figures are indicated by an italic '*t*' and '*f*' following the page number. Written works are listed with their authors.